网络运维从入门到精通
——29个实践项目详解

樊胜民 编

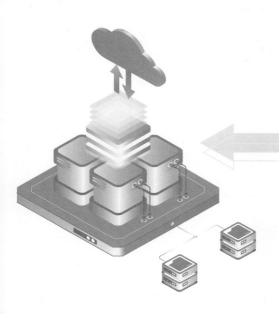

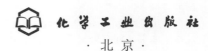

·北京·

内容简介

本书从网络基础知识讲起，介绍了网络传输介质、网卡、模拟器等网络设备，接着通过项目实战案例，详细介绍了网络运维中遇到的问题和解决方案，将用到的知识点随时在项目中进行讲解，做到突出实战，理论够用为度。

本书深入浅出地介绍了计算机网络的多方面知识，注重应用实践，可作为网络从业人员的专业学习和参考用书，也可供高校计算机、通信、网络等专业的师生阅读参考。

图书在版编目（CIP）数据

网络运维从入门到精通：29个实践项目详解/樊胜民编．—北京：化学工业出版社，2022.6（2025.4重印）
ISBN 978-7-122-41064-1

Ⅰ.①网⋯ Ⅱ.①樊⋯ Ⅲ.①计算机网络-基本知识 Ⅳ.①TP393

中国版本图书馆CIP数据核字（2022）第049535号

责任编辑：宋　辉　　　　　　　　　　装帧设计：王晓宇
责任校对：刘曦阳

出版发行：化学工业出版社
　　　　（北京市东城区青年湖南街13号　邮政编码100011）
印　　装：北京缤索印刷有限公司
787mm×1092mm　1/16　印张 $21\frac{3}{4}$　字数547千字
2025年4月北京第1版第7次印刷

购书咨询：010-64518888　　　　售后服务：010-64518899
网　　址：http://www.cip.com.cn
凡购买本书，如有缺损质量问题，本社销售中心负责调换。

定　　价：99.00元　　　　　　　　　　　　版权所有　违者必究

前言 PREFACE

目前华为、新华三❶的网络设备在企业中使用非常普遍,熟练掌握常见命令以及网络协议,才能成为一名合格的网络工程师。本书从最基础的理论开始,循序渐进地介绍企业中的网络运维项目。通过学习本书,读者能熟练掌握堆叠协议 irf、聚合技术 Aggregation、动态路由协议 ospf、冗余网关协议 vrrp、生成树协议 stp 等。同时本书也介绍时间服务器搭建,以及网络设备配置文件上传服务器备份方法。

本书分为两章:

第一章 网络基础知识

从网络传输介质开始讲解,同时介绍华为与华三网络设备模拟器 eNSP\HCL 的安装步骤与使用方法,也介绍了 VMware Workstation 虚拟机网卡的三种设置方法。

第二章 网络项目规划详解

本章基于网络常见协议进行案例讲解,每个项目的介绍都包含拓扑图、项目需求以及项目实施步骤,同时通过"知识加油站"对相关网络知识拓展介绍。

本章共计 29 个项目(包括 2 个电子版呈现的项目),从最基础的远程访问(ssh/telnet)逐渐深入讲解静态与动态路由协议、端口镜像配置方法、ACL 控制列表、NAT 转换、网络设备升级办法、无线局域网搭建

注:❶ 新华三是华三经重组合并后的新企业,为了沿用之前用法,书中有新华三和华三两种称谓。

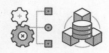

以及文件系统常见命令、irf 堆叠技术、聚合技术等。最后给出一个完整的综合项目规划，全面贯穿所学网络基础知识，达到融会贯通的目的。

本书由樊胜民编写，张淑慧、樊攀、张玄烨、张崇、樊茵、李帅等为本书的编写提供了帮助，在此表示感谢。

本书能顺利完成，离不开妻子的默默支持，离不开单位领导高度信任，将多个重大网络规划实施项目交于笔者负责，其中包括某事业部园区网络规划并实施，以及采用新华三 SDN（ADCampus）网络框架的园区网络升级改造项目，实现了整个园区网范围内"网随人动"。笔者从中积累了大量网络规划建设经验。

本书适合网络运维技术人员以及网络爱好者学习。

由于图书容量有限，为了给读者提供更多应用案例，特将 2 个案例以电子版形式提供给读者，扫描二维码即可下载。

由于水平有限，书中难免有疏漏之处，敬请广大读者朋友批评指正。

编者

项目28

项目29

目录 CONTENTS

第一章 网络基础知识 —— 1

第一节 网络传输介质 / 1
第二节 VMware Workstation 虚拟机几种网卡详解 / 8
第三节 HCL华三模拟器 / 14
第四节 华为eNSP模拟器 / 24
第五节 网络运维人员要具备的技能 / 32

第二章 网络项目规划详解 —— 37

项目1 华三设备telnet/ssh远程访问配置 / 37
项目2 华为设备telnet/ssh远程访问配置 / 45
项目3 华三设备Console认证配置 / 50
项目4 直连路由 / 58
项目5 静态路由 / 67
项目6 交换机MAC地址表 / 77
项目7 vlan与交换机端口模式 Access、Trunk / 83
项目8 H3C链路聚合 / 93
项目9 华三irf堆叠技术 / 98
项目10 H3C端口镜像 / 112
项目11 华为ACL访问控制列表 / 117
项目12 环回接口与H3C 模拟器网络设备通信 / 134
项目13 DHCP动态获取IP地址 / 144
项目14 telnet远程访问的安全性分析 / 155
项目15 NAT网络地址转换 / 161
项目16 虚拟路由冗余协议（vrrp）/ 176
项目17 动态路由ospf / 190
项目18 生成树stp配置 / 210
项目19 使用ftp /tftp升级系统 / 233

项目20　交换机校时/时间服务器
　　　　搭建 / 244

项目21　实战VPN两种配置（L2TP/
　　　　IPsec VPN）办法 / 250

项目22　构建无线局域网 / 257

项目23　交换机配置文件自动
　　　　备份 / 284

项目24　文件管理系统 / 289

项目25　中大型企业网络规划 / 295

项目26　防火墙基础配置 / 314

项目27　防火墙策略路由配置 / 328

项目28　GRE隧道（电子版）/ 342

项目29　防火墙IPsec VPN
　　　　配置（电子版）/ 342

参考文献　342

第一章 网络基础知识

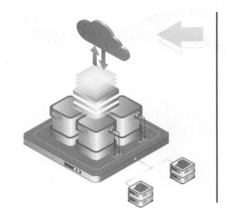

本章主要介绍网络基础知识,包括常见网络传输介质以及工器具,由于项目中部分规划是在模拟器上完成的,因此也介绍了华为 eNSP、华三 HCL 模拟器(绘制网络拓扑图的软件)的安装步骤与使用方法。熟练掌握相关基础内容,便于今后完成网络项目规划建设。

第一节 网络传输介质

网络规划中需要使用的传输介质主要包括网线、光纤、光纤收发器、光模块、光纤配线架(ODF)、光纤终端盒、光纤跳线等,下面让我们逐一认识它们。

一、网线

1. 双绞线

常见的网线为双绞线,家家户户应该都有,如图 1-1-1 所示,这种网线价格低廉,在网络中大量使用,但在传输距离与速度上都有一定的限制(理论上网线传输距离是 100m,长距离使用光纤传输)。

2. 网络跳线

网络跳线如图 1-1-2 所示,主要用于连接电脑、交换机(暂且理解为能提供多个网络接口的设备)等设备,目前使用的大部分网线跳线是超五类线与六类网线,超五类

图1-1-1 网线

图1-1-2 网络跳线

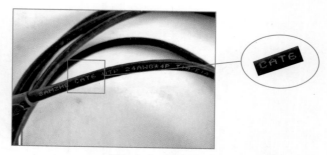
图1-1-3 标注CAT6

网线一般使用CAT5e标注，六类网线用CAT 6标注，如图1-1-3所示。

二、水晶头

水晶头如图1-1-4所示，用于连接网卡或者交换机、路由器等网络设备，采用RJ-45国际标准接口，电话线水晶头采用的是RJ-11标准。

认识了网线与水晶头后，制作"网头"（将网线压制到水晶头）是运维人员基础技能之一，制作步骤如图1-1-5～图1-1-8所示。"网头"线序分为T568A、T568B两种方式，一般情况下都按照T568B标准制作，参照表1-1-1。

图1-1-4 水晶头

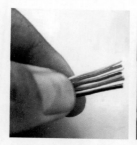
图1-1-5 整理线序

图1-1-6 插入水晶头

图1-1-7 压紧网线

图1-1-8 制作完毕

表1-1-1 水晶头网线线序

类型	1	2	3	4	5	6	7	8
T568A	绿白	绿	橙白	蓝	蓝白	橙	棕白	棕
T568B	橙白	橙	绿白	蓝	蓝白	绿	棕白	棕

三、光纤收发器

网络传输分为光信号与电信号，光纤收发器是将电信号和光信号进行数据互换的一种网络设备。按照传输路径分为多模（支持多种模式数据传输）、单模（支持单一模式数据传输），速率上分为千兆、百兆，在使用中根据情况选择。千兆单模双纤收发器如图1-1-9所示。单模单芯的光纤收发器在使用中需要区分A、B两种型号（A、B之间通过光纤远距离传输信号），配对使用，见图1-1-10。

图1-1-9 千兆单模双纤收发器

图1-1-10 单模单纤收发器TL-FC311A-3 与 TL-FC311B-3

光纤收发器外观有指示灯，了解光纤收发器指示灯状态含义，便于在项目中使用以及故障排查，图1-1-11是一款收发器指示灯。指示灯状态含义见表1-1-2。

图1-1-11 收发器指示灯

表1-1-2 光纤收发器指示灯状态含义

指示灯	亮	灭	闪烁
FX	光纤连接正常	光纤连接不通	—
TX 1000	电口工作在 1000M 速率	电口工作异常	—
FX Link Act	光口链路正常	光口链路异常	光口链路数据传输中
TX Link Act	电口链路正常	电口链路异常	电口链路数据传输中
FDX	全双工模式传输数据	半双工模式	—
PWR	供电正常	供电异常	—

前面我们介绍了光纤收发器，只有一个网口，如果有多台电脑，就需要多个网口，不然，其他的电脑只能排队等候连接网络，因为只有一个网口，不排队是不行的。有解决办法吗？当然有，可以采用1光2电、1光4电、1光8电形式的光纤收发器，或者直接购买交换机。图1-1-12是一款1光8电的光纤收发器。

目前光纤收发器使用比较普遍，尤其是在安防系统中，图1-1-13所示是单模双芯光模块（后续介绍）与单模双芯光纤收发器配合使用，图1-1-14是单模单芯光模块与单模单芯光纤收发器（需要配对使用）的典型应用。

图1-1-12 1光8电收发器（具备8个网口）

图1-1-13 光纤收发器应用

图1-1-14 光纤收发器应用

3

四、光模块

在介绍光模块之前，先认识交换机（后面会有详细介绍）的外观，如图1-1-15所示。交换机有两种接口，蓝色方框内的接口与水晶头连接，红色方框内的接口就需要采用光模块连接，那么什么是光模块呢？光模块是一种电子器件，作用是进行光电和电光转换（与光纤收发器作用类似），光模块的发送端（TX）把电信号转换为光信号，接收端（RX）把光信号转换为电信号。

图1-1-15　交换机外观

光模块按照封装形式分类，常见的有SFP、SFP+等。SFP代表的是10G以下光模块的封装类型，SFP+代表的是10G光模块的封装类型，QSFP+代表的是40G光模块的封装类型，QSFP28代表的是100G光模块的封装类型。

图1-1-16、图1-1-17所示分别是千兆与万兆光模块。

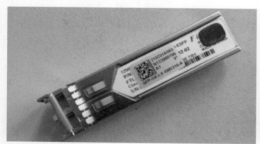

图1-1-16　千兆光模块：SFP-GE-LX-SM1310

图1-1-17　万兆光模块：SFP-XG-LX-SM1310

光模块用英文字母以及数字表示型号，都是什么含义呢？MM指的是多模，多模光模块的中心波长主要为850nm；SM指的是单模，单模光模块的中心波长通常为1310nm、1330nm、1550nm。表1-1-3是一款光模块型号字母以及数字含义。光模块的接口类型一般是LC，如图1-1-18所示，与之配套的光纤跳线LC接头如图1-1-19所示；光模块与尾纤连接如图1-1-20所示。

图1-1-18　光模块LC接口

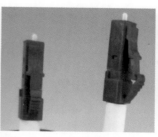

图1-1-19　光纤跳线LC接头

图1-1-20　LC连接示意图

表 1-1-3　SFP-GE-LX-SM 1310 示例型号字母以及数字含义

SFP	GE	LX	SM	1310
封装形式	速率千兆	传输距离 10km	单模	波长

如果交换机是全光口的，当我们需要连接 RJ-45 标准的网络设备，比如一台电脑连接时，那就尴尬了！这时可以采用光转电模块，图 1-1-21 是光转电模块现场使用实例。

图1-1-21　光转电模块工作应用

在实际工作中，还有一种是单模单纤光模块，需要成对使用（安防系统中使用较多），如图 1-1-22 所示。

图1-1-22　单模单纤光模块（成对使用）

五、光纤

如图 1-1-23 所示，光纤在使用中分为单模与多模两种，支持多种传播路径的称为多模光纤，而支持单一模式的被称为单模光纤，单模光纤用于远距离信号传输。多模光纤用于短距离的光纤通信。多模光纤使用 LED 光源，而单模光纤使用激光光源。

六、光纤跳线

光纤连接器俗称光纤跳线，它的两端都有连接器接头，主要作用是连接设备之间的光纤回路，以及设备与光纤布线链路的转接跳线。

从颜色上区分，单模光纤采用黄色外护套，多模光纤采用橙色或水绿色外护套。常用光纤跳线的接口类型有 LC、SC、FC、ST 等。

图1-1-23　光纤

LC-LC 光纤跳线主要用于连接光模块，图 1-1-24 所示是 LC 单模光纤跳线，图 1-1-25 是 LC 多模光纤跳线。

在数据中心用于服务器与存储连接的一般是水蓝色的万兆多模光纤，如图 1-1-26 所示。

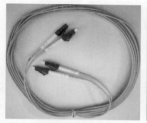 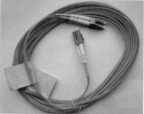 

图1-1-24　LC单模光纤跳线　　图1-1-25　LC多模光纤跳线　　图1-1-26　数据中心水蓝色万兆多模跳线

SC-LC 光纤跳线多用于光纤收发器与光模块连接，如图 1-1-27 所示。

FC-LC 光纤跳线一般在 ODF（光路转换的一种设备）侧采用，与光模块连接，如图 1-1-28 所示。

ST-ST 光纤跳线多用于光纤配线架，目前使用很少，如图 1-1-29 所示。

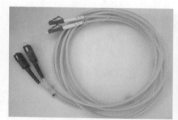  

图1-1-27　SC-LC光纤跳线　　图1-1-28　FC-LC光纤跳线　　图1-1-29　ST-ST光纤跳线

七、配线架

分为双绞线配线架与光纤配线架（ODF）。双绞线配线架（图 1-1-30）一般安装在机柜上，如图 1-1-31 所示。光纤配线架（ODF）一般用在机房，固定在机柜上，具有固定光缆和

图1-1-30　双绞线配线架正面

图1-1-31　双绞线配线架背面

保护功能，如图 1-1-32 所示。图 1-1-33 是光纤配线架（ODF）内部结构。

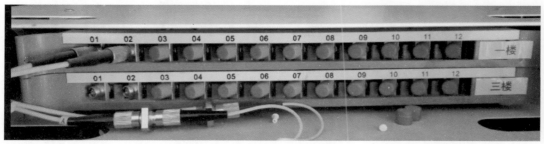

图1-1-32　FC光纤配线架

八、收发器机架

当机柜中有多个光纤收发器，线路会凌乱不堪，如图 1-1-33 所示，使用收发器机架（图 1-1-34）是不错的选择，可以将多个光纤收发器进行收纳管理，电源集中供电，机柜内部看起来美观整齐。

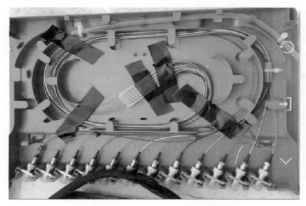

图1-1-33　ODF内部结构

图1-1-34　多个光纤收发器凌乱不堪

九、光纤终端盒

主要作用于光纤与尾纤的熔接，并对光纤及其元件提供保护。图 1-1-36 是一款光纤终端盒，一般挂墙或者桌面。

图1-1-35　收发器机架

图1-1-36　光纤终端盒

十、理线架

一般安装在机柜上，用于整理线缆，与配线架一起使用，如图 1-1-37 所示。现场应用效果如图 1-1-38 所示。

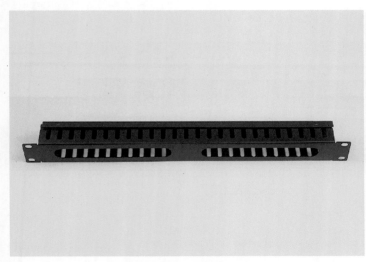
图1-1-37 理线架

图1-1-38 理线架应用

第二节 VMware Workstation 虚拟机几种网卡详解

VMware Workstation 是一款通过软件模拟的具有完整硬件系统功能、独立运行操作的系统，通过 VMware Workstation 在一台物理计算机上模拟出一台或多台虚拟的计算机，这些虚拟机完全就像真正的计算机那样进行工作，可以安装操作系统、程序、访问网络资源等。VMware Workstation 中虚拟机网络连接常见模式分为桥接模式、NAT 模式、仅主机模式，如图 1-2-1 所示。

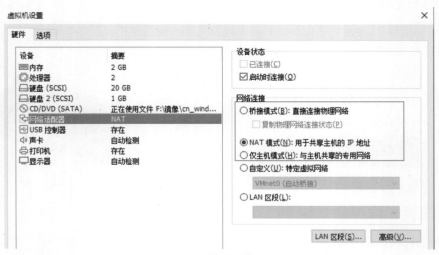
图1-2-1 虚拟机中常见的3种模式

一、桥接模式

虚拟机有独立的 IP 地址，虚拟机和主机之间互相访问资源，相当于接入到一个局域网中，虚拟机与主机 IP 地址在一个网段（例如，主机 IP：10.1.20.66，虚拟机 IP:10.1.20.3），虚拟机的数据通过主机网卡转发。

> IP 地址是 IP 协议提供的一种统一的地址格式，它为互联网上的每一个网络和每一台主机分配一个逻辑地址，以此来屏蔽物理地址的差异。IP 地址就像是我们的家庭住址一样，如果你要写信给一个人，你就要知道他（她）的地址，这样邮递员才能把信送到。计算机发送信息就好比是邮递员，它必须知道唯一的"家庭地址"才能不至于把信送错人家。
>
> IP 地址是一个 32 位的二进制数，通常被分割为 4 个"8 位二进制数"，用"点分十进制"表示成（a.b.c.d）的形式，其中，a、b、c、d 都是 0～255 之间的十进制整数。例如，点分十进 IP 地址（100.4.5.6），实际上是 32 位二进制数（01100100.00000100.00000101.00000110）。

使用场合：IP 资源充足；如果准备利用 VMware Workstation 在局域网内新建一个虚拟服务器，为局域网其他用户提供网络资源服务（ftp 等），则选择桥接模式。新建虚拟机网卡配置见图 1-2-2。虚拟机桥接模式与主机示意见图 1-2-3。虚拟机安装步骤在此忽略，安装完毕，网卡参数配置见图 1-2-4。

图1-2-2　网卡桥接模式配置

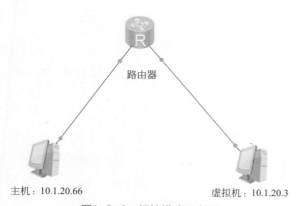

图1-2-3 桥接模式示意图

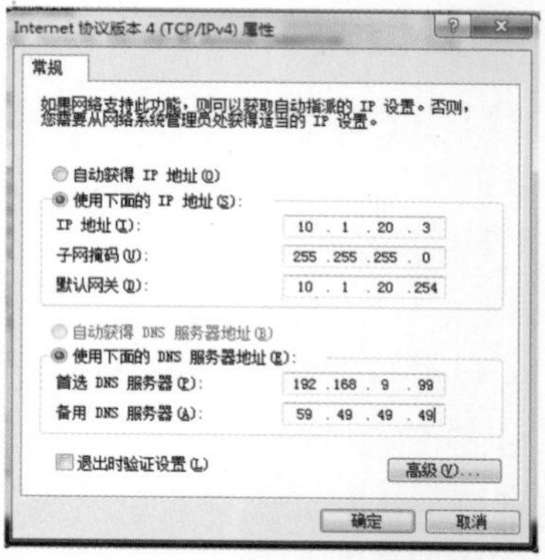

图1-2-4 虚拟机网络配置

若新建的虚拟机网卡采用桥接模式，能否与主机之间通信呢？测试结果如图1-2-5所示，可以通信。当主机申请了访问互联网权限（本例使用访问百度）时，可以通信，参照图1-2-6。若虚拟机网卡采用桥接模式，虚拟机如与互联网通信，同样虚拟机也必须具有互联网权限，才能与互联网通信，见图1-2-7，否则无法与互联网通信。

测试　虚拟机（IP：10.1.20.3）-ping-（10.1.20.66）- 通。

主机（IP：10.1.20.66）-ping- 百度 www.baidu.com（10.1.20.66 前提有外网权限）- 通。

虚拟机（IP：10.1.20.3）ping- 百度 www.baidu.com（10.1.20.3 前提有外网权限）- 通。

图1-2-5 虚拟机与主机ping实验

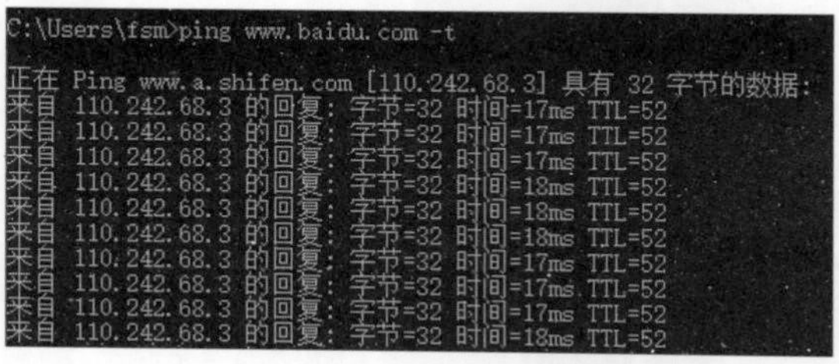

图1-2-6 主机与互联网通信

图1-2-7 虚拟机与互联网通信

二、NAT 模式

最简单的虚拟机上网（互联网）模式，与主机共享 IP 地址。

VMware Workstation 安装完毕后，主机的网络连接出现两张虚拟机网卡，如图1-2-8所示。

图1-2-8 虚拟机网卡

见图1-2-9，将虚拟机网卡设置为 NAT 模式，当采用 NAT 模式的时候，主机、虚拟机之间网络示意参照图1-2-10。VMnet8 虚拟交换机内包含 DHCP 以及 NAT 服务，虚拟机网卡设置自动获取 IP 地址，参见图1-2-11，在虚拟机中通过 win+R，调出"运行"对话框，输入"cmd"，在弹出提示符中输入 ipconfig 命令，查看虚拟机自动获取的 IP 地址，如图1-2-12 所示。NAT 模式中，虚拟机通过物理主机网卡向外转发数据。主机与虚拟主机 IP 地址不在一个网段（本例主机 IP：10.1.20.66，虚拟机 IP：192.168.160.128）。虚拟机与主机之间可以互相通信，如图 1-2-13、图 1-2-14 所示。如主机能访问互联网，NAT 模式下虚拟机就可以访问互联网。当 IP 资源比较紧缺时，NAT 上网的方式是不错的选择。

图1-2-9　虚拟机设置为NAT模式

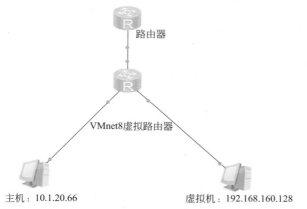

图1-2-10　NAT 模式主机与虚拟机之间网络示意

图1-2-11　虚拟机自动获取IP地址

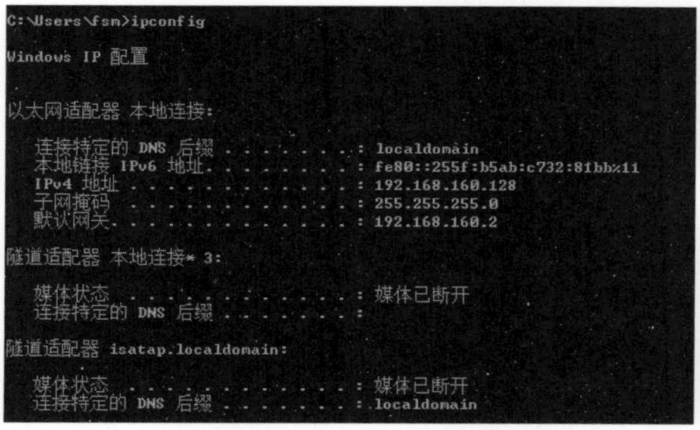

图1-2-12　虚拟机IP地址

图1-2-13　虚拟机与主机ping实验

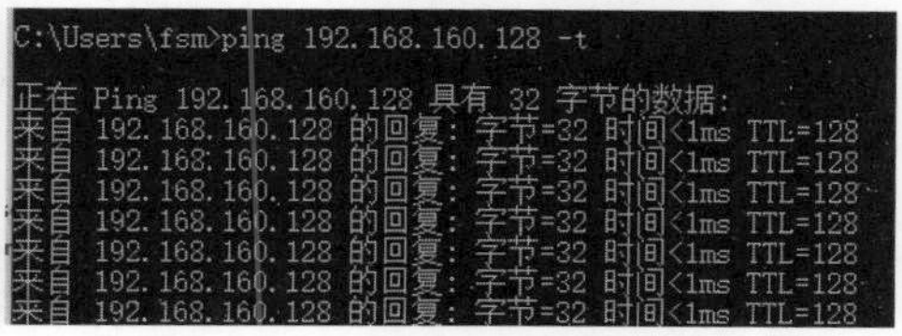

图1-2-14　主机与虚拟机ping实验

11

测试 虚拟机（IP：192.168.160.128）-ping- IP（10.1.20.66）- 通。

主机（IP：10.1.20.66）-ping- 虚拟机（IP：192.168.160.128）- 通。

图1-2-15 虚拟机访问互联网

虚拟机上网测试（只要主机 IP 地址：10.1.20.66 能上外网，NAT 模式虚拟机就可以连接外网，以访问百度为例），见图 1-2-15。

三、仅主机模式

通过主机 VMware net1 网卡进行虚拟服务器之间的通信，VMware net1 虚拟机网卡具有 DHCP 服务器的功能，无 NAT 模式，仅主机模式下各个虚拟机之间互通，但是虚拟机与主机不能互相访问，也就是虚拟机是一个独立的系统。

新建虚拟机 1：设置仅主机模式

新建虚拟机网络连接设置如图 1-2-16 所示，在虚拟机中通过 win+R，调出"运行"对话框，输入"cmd"，在弹出提示符中输入 ipconfig 命令，查看虚拟机自动获取的 IP 地址，如图 1-2-17 所示。

图1-2-16 虚拟机1网卡设置仅主机模式　　图1-2-17 虚拟机1自动获取的地址

新建虚拟机 2：仅主机模式

新建虚拟机网络连接设置如图 1-2-18 所示，在虚拟机中通过 win+R 调出"运行"对话框，输入"cmd"，在弹出提示符中输入 ipconfig 命令，查看虚拟机自动获取的 IP 地址，如图 1-2-19 所示。

仅主机模式，主机（IP：10.1.20.66）与两个虚拟机之间网络示意如图 1-2-20 所示。

虚拟机 1（IP：192.168.211.129）-ping- 虚拟机 2（IP：192.168.211.130）- 通，见图 1-2-21。

虚拟机 1（IP：192.168.211.129）-ping- 主机（IP:10.1.20.66）- 不通，见图 1-2-22。

图1-2-18 虚拟机2仅主机模式

图1-2-19 虚拟机2自动获取的地址

主机（IP：10.1.20.66）-ping-虚拟机1（IP：192.168.211.129）-通，虚拟机1与主机无法通信，但是主机与虚拟机1之间可以通信，见图1-2-23。

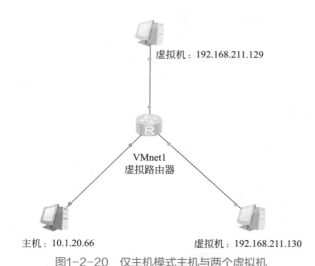

图1-2-20 仅主机模式主机与两个虚拟机之间网络示意图

图1-2-21 两个虚拟机之间通信

图1-2-22 虚拟机1与主机无法通信

图1-2-23 主机与虚拟机1之间通信

图1-2-24 虚拟机2与主机无法通信

图1-2-25 主机与虚拟机2之间通信

虚拟机2（IP：192.168.211.130）-ping- 主机（IP：10.1.20.66）- 不通，见图1-2-24。

主机（IP：10.1.20.66）-ping- 虚拟机2（IP：192.168.211.130）- 通，虚拟机2与主机无法通信，但是主机与虚拟机2之间可以通信，见图1-2-25。

第三节 HCL 华三模拟器

HCL 华三模拟器就是华三云实验室（新华三集团）推出的功能强大的图形化网络设备模拟软件，可以模拟路由器、交换机、防火墙、PC 等。以下是安装步骤以及功能介绍。

一、安装步骤

演示安装环境：操作系统为 Windows 10，内存 16G。

步骤 1　双击运行 HCL_V2.1.2_Setup.exe 安装程序，进入语言选择，选择语言"简体中文"，点击"OK"，见图 1-3-1。

步骤 2　进入欢迎界面，点击"下一步"，见图 1-3-2。

步骤 3　选择接受许可，点击"下一步"，见图 1-3-3。

图 1-3-1　语言选择

图 1-3-2　欢迎向导

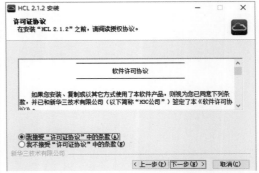

图 1-3-3　接收许可协议

步骤 4　选择安装位置，点击"下一步"，见图 1-3-4。

步骤 5　选择安装组件，默认选择，点击"安装"，见图 1-3-5。

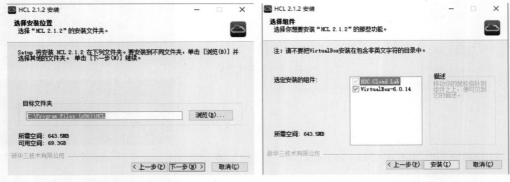

图 1-3-4　安装位置　　　　　　　　图 1-3-5　默认选择

步骤 6　进度条快完时，弹出安装 Oracle VM VirtualBox 对话框，点击"下一步"，见图 1-3-6。

步骤 7　选择默认安装功能，点击"下一步"，见图 1-3-7。

图1-3-6 安装Oracle VM VirtualBox

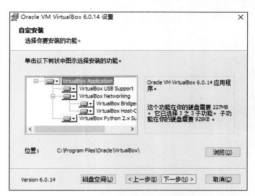
图1-3-7 安装继续，默认选择

步骤8 默认选择，点击"下一步"，见图1-3-8。

步骤9 弹出警告信息，网络临时中断，点击"是"，见图1-3-9。

图1-3-8 默认选择，点击"下一步"

图1-3-9 弹出警告，点击"是"

步骤10 弹出对话框，选择"安装"，见图1-3-10。

步骤11 弹出对话框，"始终信任"前打上对勾，安装插件，选择"安装"，见图1-3-11。

图1-3-10 点击"安装"

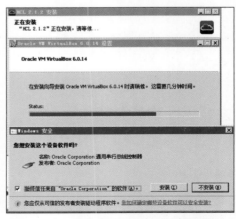
图1-3-11 安装虚拟机插件，选择始终信任

15

步骤 12　去掉对勾，暂时不运行虚拟机，点击"完成"，见图 1-3-12。

步骤 13　点击"完成"，完成 HCL 全部安装，见图 1-3-13。

步骤 14　当安装完毕运行 HCL 软件的时候，有时提示如图 1-3-14 所示的错误，解决办法，选择桌面 HCL 快捷方式，右键选择属性，兼容模式选择 Windows 7，见图 1-3-15。

图1-3-12　去掉对勾，完成安装

图1-3-13　安装完成

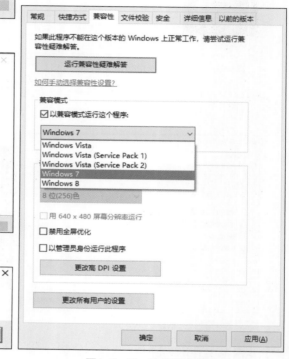

图1-3-14　运行软件报错

图1-3-15　选择兼容模式

注意：如需使用抓包工具，可自行下载安装 Wireshark，HCL 安装包没有集成抓包工具。

二、主要界面功能介绍

1. 打开软件界面

如图 1-3-16 所示。

2. 网络设备

从上到下依次是 DIY 设备▦、路由器▦、交换机▦、防火墙▦、终端▦、连线▦。

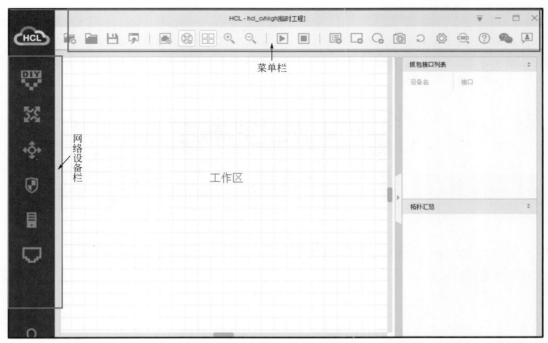

图1-3-16 软件界面

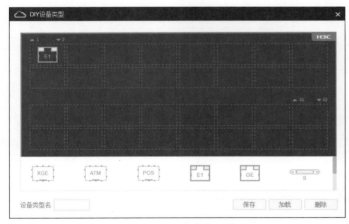

图1-3-17 自定义设备

1) DIY 设备 点击 "DIY Device"，启动创建自定义设备类型弹出框，添加相应的接口，见图1-3-17。一般不使用。

2) 路由器 模拟器提供一个路由器，包含 7 个 GigabitEthernet 接口，4 个 Serial 接口，见图1-3-18，路由器正反面见图 1-3-19。

图1-3-18 路由器图标

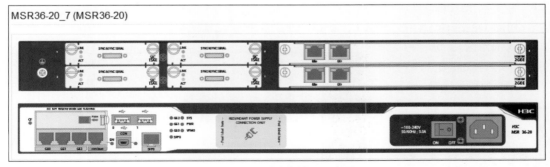

图1-3-19 路由器接口示意图

17

3）交换机　模拟器提供一个型号交换机，包含 48 个 GigabitEthernet（千兆）接口，4 个 TenGigabitEthernet（万兆）接口，2 个 FortyGigE 接口，见图 1-3-20，交换机正反面见图 1-3-21。

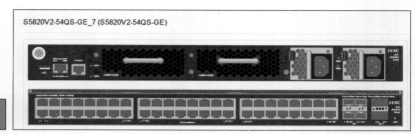

图1-3-20　交换机图标　　　　　　图1-3-21　交换机接口示意图

4）防火墙　模拟器提供一个型号防火墙，包含 24 个 GigabitEthernet（千兆）接口，防火墙图标及接口见图 1-3-22。

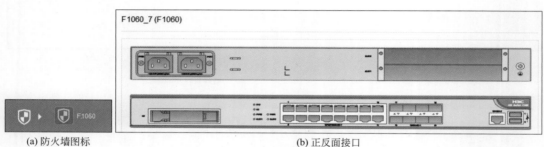

(a) 防火墙图标　　　　　　　　　　(b) 正反面接口

图 1-3-22　防火墙图标及接口示意图

3. 设备启动

将网络设备（比如交换机）拖拽到软件工作区，通过鼠标右键点击"启动"，见图 1-3-23，或者鼠标选择设备 ，点击菜单栏绿色按钮 。

图1-3-23　设备启动

三、如何使用 SecureCRT 软件连接 HCL

使用 SecureCRT 连接华三云实验室（H3C Cloud Lab，简称 HCL）之前，需要安装相应软件：SecureCRT、Named Pipe TCP Proxy。

步骤1　拓扑图

如图 1-3-24 所示，打开 HCL，新建一个交换机（设备区拖拽交换机到工作区），进行相关配置工作，类似这样的界面，称之为拓扑图。

步骤2　运行 Oracle VM VirtualBox 管理器

见图 1-3-25，通过关闭与打开拓扑图中的交换

图1-3-24　拓扑图

机，确认刚新建的交换机在 Oracle VM VirtualBox 管理器中是哪个虚拟机，如图 1-3-26 和图 1-3-27 所示。

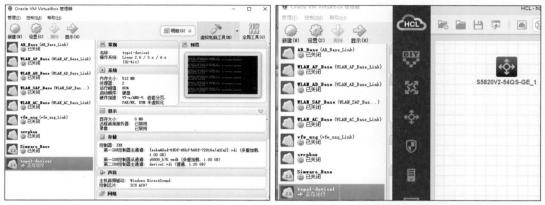

图1-3-25　Oracle VM VirtualBox管理器界面　　图1-3-26　打开交换机观察虚拟对应关系

经过对比确认新建的虚拟交换机后，进行端口设置。

鼠标右键点击刚确认的虚拟机，依次选择设置 - 串口 - 端口 2，将路径/地址框中的内容复制，见图 1-3-28。

图1-3-27　关闭交换机观察虚拟对应关系　　　　图1-3-28　复制路径

启动拓扑图中的交换机，见图 1-3-29。

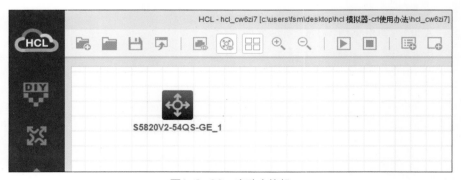

图1-3-29　启动交换机

19

步骤3 打开 Named Pipe TCP Proxy 软件

运行界面见图 1-3-30。

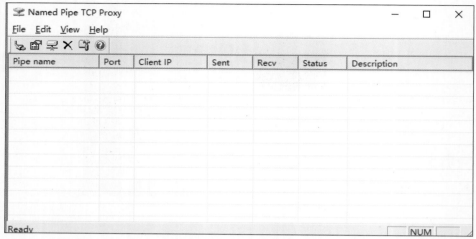

图1-3-30 Named Pipe TCP Proxy运行界面

在图 1-3-30 空白处右键点击，弹出"Add"，见图 1-3-31，路径内容粘贴到 Pipe 的框内，同时在 Port 处填上数字，在 Description 的框内，填写拓扑图中交换机的名字（可以忽略）。

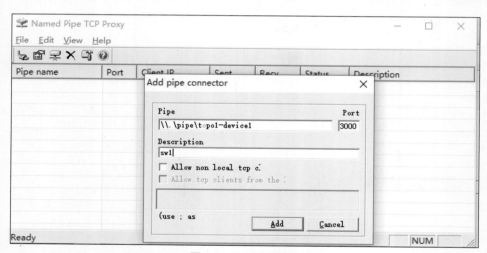

图1-3-31 设置端口

点击"Add"，弹出见图 1-3-32 所示界面。

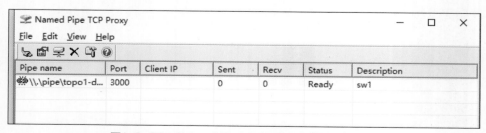

图1-3-32 Named Pipe TCP Proxy软件显示状态

步骤4 打开 SecureCRT 软件（自行安装）

点击"Quick Connect"（快速连接），见图 1-3-33。

进行关联参数设置，设置完毕点击"Connect"，见图 1-3-34。

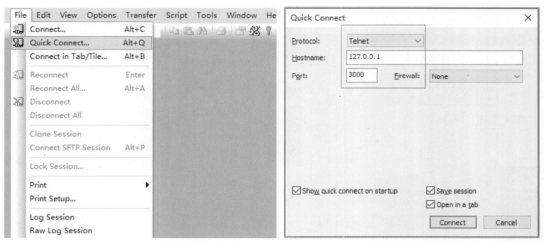

图1-3-33 选择快速连接　　　　　　　　　图1-3-34 相关参数设置

SecureCRT 软件访问 HCL 界面如图 1-3-35 所示，设置完毕。

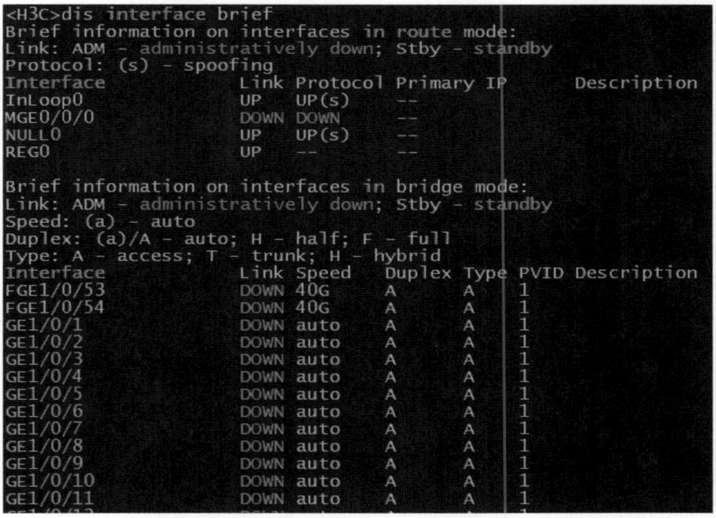

图1-3-35 SecureCRT软件访问HCL

查看 Named Pipe TCP Proxy 软件，见图 1-3-36 查看状态 / 发送 / 接收等数据。

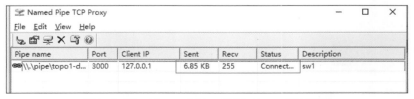

图1-3-36 Named Pipe TCP Proxy显示截图

四、演示如何使用 HCL 软件绘制拓扑图

案例

两台电脑通过交换机网络互联，设备包括两台 PC、一台交换机。实现两台 PC 之间互相通信。

步骤 1　打开软件

选择交换机见图 1-3-37，拖拽至工作区见图 1-3-38。

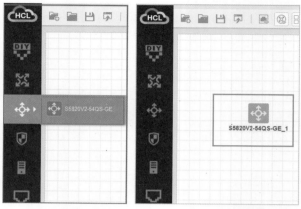

图1-3-37　选择交换机　　图1-3-38　交换机拖拽至工作区

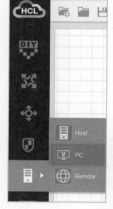

图1-3-39　选择 PC

步骤 2　选择 PC

如图 1-3-39 所示，拖拽至工作区（共计两台 PC），见图 1-3-40。

图1-3-40　工作区放置两台PC

步骤 3　启动所有设备

设备在没有启动时，显示是黑底白字，见图 1-3-41。鼠标左键选中所有设备，见图 1-3-42。右键点击任何一台设备，选择"启动"，见图 1-3-43。设备启动完毕后，软件界面右侧就可以看到设备启动状态（在设备名字前边的圆点由红色变为绿色），见图 1-3-44。

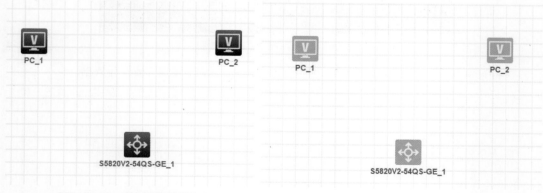

图1-3-41　设备未启动状态　　图1-3-42　选中所有设备

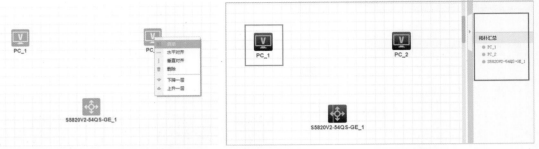

图1-3-43　启动设备　　　　　　　　　　图1-3-44　设备启动完毕

步骤4　连接网络，实现网络通信

按照图1-3-45，选择连接网络，将所有设备按照图1-3-46连接在一起。如需要显示设备连接的端口，在图1-3-47中点击红色框内的图标，即可显示。

步骤5　配置设备

交换机保持默认配置，主要是PC配置IP地址，鼠标右键点击PC1，选择"配置"，见图1-3-48。弹出图1-3-49，接口管理选择"启用"，IPv4配置选择静态，IPv4地址设为192.168.1.1，掩码地址设为255.255.255.0，点击"启用"即可完成PC1的配置工作。同样方法，完成PC2的配置，见图1-3-50。

图1-3-45　选择网络连接线

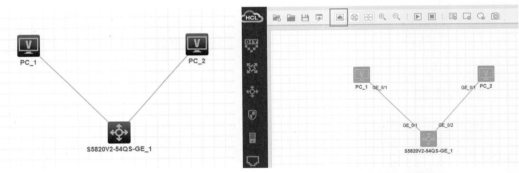

图1-3-46　所有设备连接网络　　　　　图1-3-47　显示设备端口

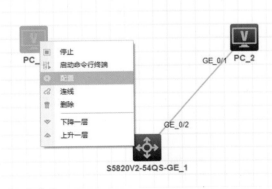

图1-3-48　选择PC配置　　　　　　　　图1-3-49　PC1配置

23

步骤6 网络连通测试

双击PC1图标，进入命令行，输入ping 192.168.1.2，弹出结果如图1-3-51所示，说明网络正常，拓扑图无误。

图1-3-50　PC2配置　　　　　　　　图1-3-51　网络连通测试

第四节　华为 eNSP 模拟器

eNSP（Enterprise Network Simulation Platform，企业网络仿真平台）是华为模拟器仿真平台，图形化操作界面，主要对路由器、交换机、防火墙进行软件仿真实验，方便初学者完成实验操作。

一、安装步骤

演示教程安装环境：Windows 10、内存16G。

步骤1　双击安装包，右键选择"以管理员身份运行"，见图1-4-1，选择中文（简体），见图1-4-2，点击"确定"。

 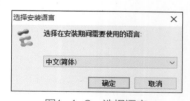

图1-4-1　以管理员身份运行　　　　　图1-4-2　选择语言

步骤2　进入欢迎界面，点击"下一步"；见图1-4-3。
步骤3　选择接受许可，然后点击"下一步"；见图1-4-4。
步骤4　选择安装位置，安装路径最好不要更换，使用英文目录；见图1-4-5。
步骤5　安装eNSP开始菜单选项，默认选择，点击"下一步"；见图1-4-6。

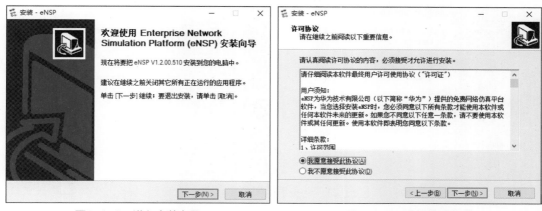

图1-4-3 进入安装向导　　　　　　　　图1-4-4 选择接受协议

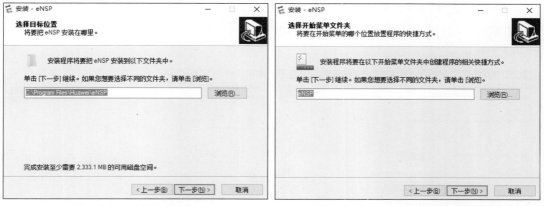

图1-4-5 选择安装路径　　　　　　　　图1-4-6 安装eNSP开始菜单选项

步骤6　安装附加任务，默认选择，点击"下一步"，见图1-4-7。
步骤7　分别安装WinPcap、抓包工具、虚拟系统，全部打上对勾，点击"下一步"，见图1-4-8。
步骤8　安装WinPcap 4.1.3软件，点击"Next"，见图1-4-9。
步骤9　安装抓包分析工具Wireshark，点击"Next"，见图1-4-10。

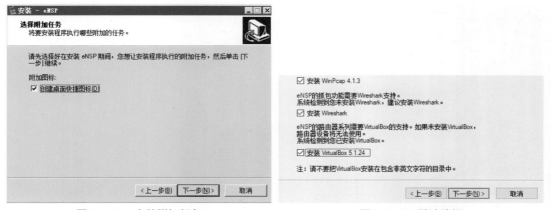

图1-4-7 安装附加任务　　　　　　　　图1-4-8 默认选择

图1-4-9 安装WinPcap

图1-4-10 安装Wireshark抓包工具软件

步骤10 选中Wireshark软件中的选项，默认选择，点击"Next"，见图1-4-11。

步骤11 方框内不要打对勾（因为已经安装WinPcap 4.1.3），点击"Install"，继续安装，见图1-4-12。

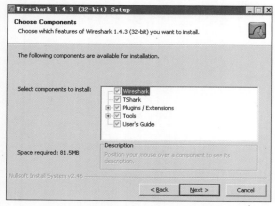
图1-4-11 选中Wireshark软件中的选项

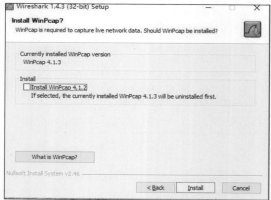
图1-4-12 选择"Install"

步骤12 完成Wireshark软件安装，见图1-4-13。

步骤13 安装虚拟机软件Oracle VM VirtualBox，选择"Next"，见图1-4-14。

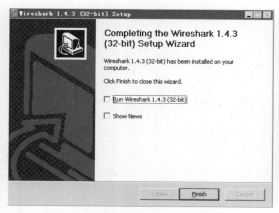
图1-4-13 完成Wireshark软件安装

图1-4-14 安装Oracle VM VirtualBox

步骤14　选择虚拟机 Oracle VM VirtualBox 安装目录，默认选择，点击"Next"，见图1-4-15。
步骤15　在安装虚拟机的过程中可能会出现临时断网警告信息，点击"Yes"，见图1-4-16。

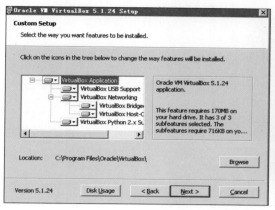

图1-4-15　虚拟机Orale VM VirtualBox安装目录

图1-4-16　临时断网警告信息

步骤16　安装虚拟机插件，"始终信任"前打上对勾，点击"安装"，见图1-4-17。

步骤17　安装完毕，两个对勾去掉，先不要运行 eNSP 软件，见图1-4-18。

图1-4-17　信任安装插件

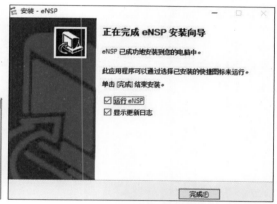

图1-4-18　安装完毕

步骤18　以管理员身份打开 eNSP，依次点击菜单-工具-注册设备，勾选所有项-点击"注册"，见图1-4-19。然后关闭 eNSP。

步骤19　当打开 eNSP 软件，遇到如图1-4-20所示的提示时，在防火墙软件中放行即可，见图1-4-21。

步骤20　检测 eNSP 软件是否安装成功，最可靠的办法是启动一台 AR2220，看能否进入到输入命令的位置，见图1-4-22。

图1-4-19　注册设备

图1-4-20　将eNSP在防火墙允许的提示信息

图1-4-21 防火墙放行eNSP软件　　　　图1-4-22 运行路由器

二、主要功能介绍

启动 eNSP，界面分为 4 大部分，分别是设备区、工作区、常见工具按钮、菜单栏。见图 1-4-23。下面通过实验介绍。

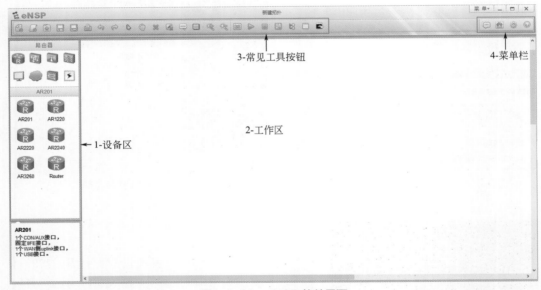

图1-4-23 eNSP 软件界面

 两台电脑经过交换机互联通信。

步骤1　点击"交换机"-选取相应型号的交换机-移动到"工作区"-点击鼠标左键-放置设备。见图 1-4-24。

同样办法放置两台 PC，点击 选取 （自动）或者 （手动）连接网线。

步骤2　启动设备，设置显示接口信息，添加注释描述说明。

在启动设备的时候，可以单个设备右键启动，见图 1-4-25。也可以全部选中启动。启动

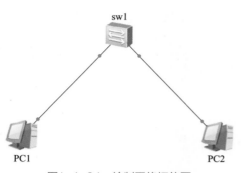

图1-4-24　绘制网络拓扑图

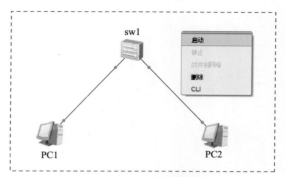

图1-4-25　启动设备

设备也可以点击工具栏▶，或者"菜单"-"工具"-"启动设备"。启动完毕，连接线的红色圆点变为绿色，见图1-4-26。

　　点击常见工具中的进行注释说明（例如 PC 的 IP 地址），点击显示接口信息，见图 1-4-27。

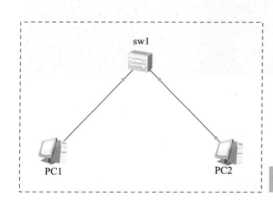

图1-4-26　启动完毕

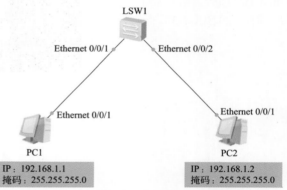

图1-4-27　网络拓扑图标识信息（IP地址以及掩码）

步骤3　配置工作，关闭设备之前一定要保存配置（save）。
① 双击拓扑图中的交换机，即可进入配置界面。

```
<Huawei>system-view
Enter system view, return user view with Ctrl+Z.
[Huawei]sysname sw1
[sw1]quit
<sw1>save
The current configuration will be written to the device.
Are you sure to continue?[Y/N]y
```

② PC 配置

　　PC1 配置：IP 地址配置 192.168.1.1，子网掩码配置 255.255.255.0，参数填写完毕，点击"应用"。见图 1-4-28。

　　PC2 配置：IP 地址配置 192.168.1.2，子网掩码配置 255.255.255.0，参数填写完毕，点击"应用"。见图 1-4-29。

图1-4-28 PC1参数配置

图1-4-29 PC2参数配置

步骤4 测试网络情况。

双击PC2，命令行输入ping 192.168.1.1 -t，测试网络通畅。见图1-4-30。

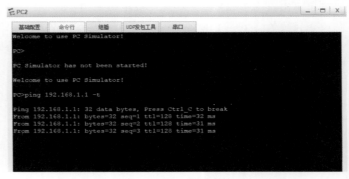
图1-4-30 测试网络

① 快捷设置设备的接口信息

打开软件界面"右工具栏"，见图1-4-31，设置拓扑图中设备的接口信息。在工作中比较实用。

② 如何修改CLI字体大小以及背景颜色

点击 弹出选项卡 – 选择字体设置（图1-4-32）。拓扑图界面文字颜色以及背景也在这儿设置。

图1-4-31 进入软件工具栏

图1-4-32 字体设置界面

三、如何使用 SecureCRT 软件连接 eNSP

在 eNSP 模拟器中，双击设备可以进行配置工作，但是存在快捷键不能友好支持。建议使用 SecureCRT 连接设备进行配置工作，具体步骤如下：

步骤1 在拓扑图中，右键选择如图 1-4-33 所示的路由器 - 设置 - 配置，显示设备串口号。

图1-4-33 查看设备串口号

步骤2 打开 SecureCRT 软件。点击"Quick Connect"（快速连接），见图 1-4-34。
进行参数设置，协议选择"Telnet"，设置完毕点击"Connect"，见图 1-4-35。

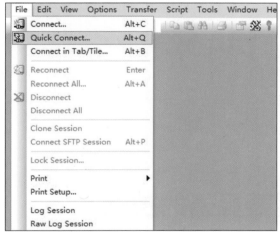

图1-4-34 选择快速连接

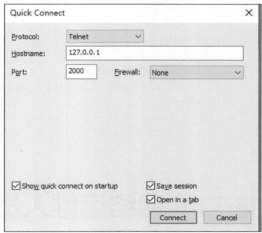

图1-4-35 配置参数

步骤3 启动路由器在 SecureCRT 软件进行相关配置工作。见图 1-4-36。

图1-4-36 SecureCRT软件访问eNSP

技巧　在交换机输入命令时可以简写，比如从用户视图进入系统视图，完整命令为"system-view"，可以简写成"sys"；在输入命令的时候，可以只输入前几个字母，然后按"Tab"补全完整命令。在输入命令时使用"Tab"以及"？"，可以达到事半功倍的效果。

当输入一个命令，不清楚后续如何输入的时候，可以使用"？"查看。比如我们查看接口命令的时候，输入"display interface？"：

```
[sw1]display interface ?
  Ethernet            Ethernet interface
```

```
  GigabitEthernet    GigabitEthernet interface
  MEth               MEth interface
  NULL               NULL interface
  Vlanif             Vlan interface
  brief              Summary information about the interface status and
  configuration
  description        Interface description
  slot               Slot number
  |                  Matching output
  <cr>
[sw1]display interface brief
PHY: Physical
*down: administratively down
(l): loopback
(s): spoofing
(b): BFD down
(e): ETHOAM down
(dl): DLDP down
(d): Dampening Suppressed
InUti/OutUti: input utility/output utility
Interface                PHY     Protocol   InUti   OutUti    inErrors   outErrors
GigabitEthernet0/0/1     down    down       0%      0%        0          0
GigabitEthernet0/0/2     down    down       0%      0%        0          0
```

第五节 网络运维人员要具备的技能

如今是网络飞速发展时代，办公以及各个信息化系统都建立在网络技术基础之上，人们对网络在速度、稳定性以及可靠性上都提出了更高的要求。作为一名合格的网络工程师，需要全面掌握网络数通交换方面的技能，同时还需精通网络安全，及时加强企业出口防火墙安全策略配置等工作，保障企业内网安全高效运行。企业发展对网络依赖越来越高，这就要求运维人员知识储备丰富，在故障出现时，能根据故障现象，快速定位及时解决。

一、熟练掌握数通技能以及常见网络协议与网络架构

网络运维最基本的要求是保障网络畅通以及网络安全。实现网络互通，尤其是跨设备之间网络互通，就需要配置静态路由（Static）或者动态路由协议（比较经典的就是 ospf）。为了实现网络的"健壮性"或者说"可靠性"，一般采用设备堆叠（比如华三的 irf 堆叠协议）。为了避免出口网关单设备运行风险，采用两台以上设备配置 vrrp 虚拟网关冗余协议。考虑单链路在网络规划中的风险性，建议链路冗余，一般采用聚合技术，或者交换机开启生成树协议（防止二层网络环路重要协议 -stp）。企业内网中成百上千的 IP 地址，如何访问互联网，就需要在边界路由器或者防火墙上配置 NAT（网络地址转换），实现公司内网访问互联网。在中大型企业中，一般配置有 DHCP 服务器，也可以在网络设备中配置 DHCP，实现公司内网 IP 地址自动获取，避免人为分配 IP 地址，可能疏忽导致 IP 地址冲突而带来的烦劳。近

几年无线终端普及速度迅猛,无线覆盖的应用越来越多,如何规划无线网络也是运维人员必须掌握的技能。网络设备远程管理配置 telnet/ssh 协议是网络运维基础,在日常网络运维中,检查交换机以及其他网络设备都是远程登录到设备进行,而不是直接到现场处理。

想要快速掌握设备配置命令,最简单的办法就是在 eNSP、HCL 等模拟器中搭建实验环境,绘制各种网络规划布局图,通过实验了解各种网络协议与常见命令。

图 1-5-1 是典型的三层网络架构(前面提到的各种网络协议要

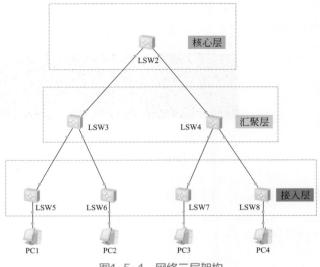

图1-5-1 网络三层架构

在这里大显身手):接入层—汇聚层—核心层。网络中直接与终端连接(比如 PC/摄像头等)的交换机称为接入层交换机,接入层就是允许终端设备连接到网络,接入层交换机一般端口比较多,这是它的特点。位于接入层和核心层之间的部分称为汇聚层,汇聚层是多台接入层交换机的网络汇聚点,它必须能够及时处理来自接入层设备的转发流量,并提供到核心层的上行链路。在企业网络中,网关一般部署在汇聚层,也可以部署在核心层,根据实际情况而定。核心层的主要目的是高速转发数据流量,具备高的可靠性能和数据吞吐量,是企业网络中的"首脑",一般是万兆交换机。

案例:办公楼 6 楼办公室网络不稳定分析的排查过程

步骤 1 检查办公楼 leaf(汇聚)交换机下行接口光强度情况。发现 leaf-port 2/0/9,RX 接收光强度偏离正常值,为 RX-26.20 db,见图 1-5-2,正常值范围应为 −3.00 ～ −22.08db。

```
    28         3.42         4.63         -6.19         -5.24
Alarm thresholds:
      Temp.(°C) voltage(v)  Bias(mA)  RX power(dBm)  TX power(dBm)
High    73        3.80        31.31      -0.00          6.00
Low     -3        2.80        1.00       -18.01        -12.50
[DYJT-BQ-Leaf_BG.4.10]dis transceiver diagnosis interface GigabitEthernet 2/0/7
GigabitEthernet2/0/7 transceiver diagnostic information:
Current diagnostic parameters:
      Temp.(°C) voltage(v)  Bias(mA)  RX power(dBm)  TX power(dBm)
    35       3.42         8.43         -12.00         -6.08
Alarm thresholds:
      Temp.(°C) voltage(v)  Bias(mA)  RX power(dBm)  TX power(dBm)
High    73        3.60        38.58      -3.00          6.00
Low     -3        3.00        1.00       -22.08        -12.50
[DYJT-BQ-Leaf_BG.4.10]dis transceiver diagnosis interface GigabitEthernet 2/0/8
GigabitEthernet2/0/8 transceiver diagnostic information:
Current diagnostic parameters:
      Temp.(°C) voltage(v)  Bias(mA)  RX power(dBm)  TX power(dBm)
    34       3.43         8.39         -13.35         -6.05
Alarm thresholds:
      Temp.(°C) voltage(v)  Bias(mA)  RX power(dBm)  TX power(dBm)
High    73        3.60        38.93      -3.00          6.00
Low     -3        3.00        1.00       -22.08        -12.50
[DYJT-BQ-Leaf_BG.4.10]dis transceiver diagnosis interface GigabitEthernet 2/0/9
GigabitEthernet2/0/9 transceiver diagnostic information:
Current diagnostic parameters:
      Temp.(°C) voltage(v)  Bias(mA)  RX power(dBm)  TX power(dBm)
    36       3.43         8.66         -26.20         -6.00
Alarm thresholds:
      Temp.(°C) voltage(v)  Bias(mA)  RX power(dBm)  TX power(dBm)
High    73        3.60        39.32      -3.00          6.00
Low     -3        3.00        1.00       -22.08        -12.50
[DYJT-BQ-Leaf_BG.4.10]dis transceiver diagnosis interface GigabitEthernet 2/0/10
GigabitEthernet2/0/10 transceiver diagnostic information:
Current diagnostic parameters:
```

图1-5-2 光强度检查

步骤 2 根据经验，光衰过大，一般情况是交换机连接的尾纤故障导致。更换尾纤后，RX 接收光强度变为 −6.82db，符合正常值，见图 1-5-3。但是故障依旧，还是报修不稳定。

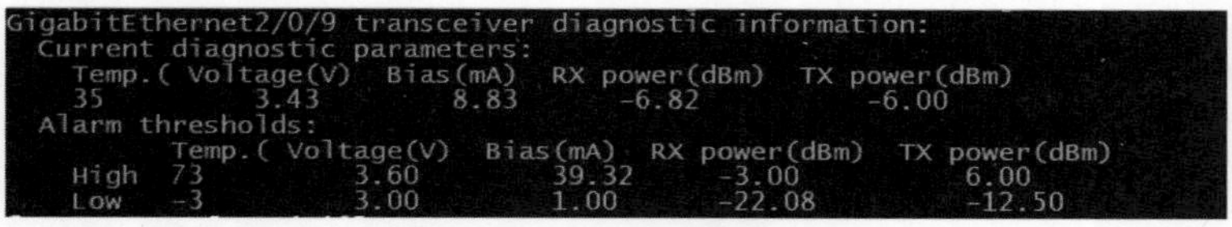

图1-5-3　光强度检查

步骤 3 继续检查 leaf（汇聚交换机）连接的所有接入层交换机。

通过命令"dis stp brief"发现其中一台交换机 port 1/0/11 显示 DISCARDING。DISCARDING 属于端口阻塞，说明这台交换机连接的网络存在环路。继续观察，该接口状态不定期在 DISCARDING（阻塞）与 FORWARDING（转发）交替变化，这就是故障原因，当接口堵塞的时候，port 1/0/11 连接的下行网络中断，见图 1-5-4。

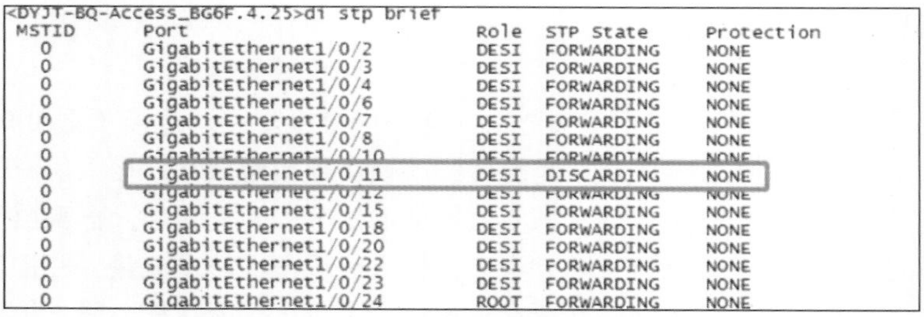

图1-5-4　port 1/0/11显示 DISCARDING

步骤 4 继续排查，发现其中一个办公室网络链路连接异常，见图 1-5-5。

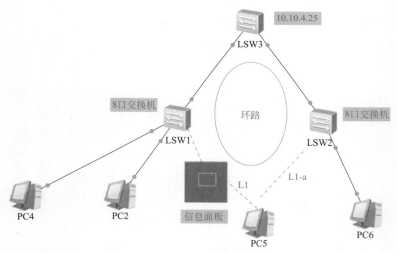

图1-5-5　网络环路

L1 蓝色网线正常情况下与信息面板连接，另一端与 PC5 连接。由于 PC5 电脑移走后，有人不小心将 L1 这条网线一端连接信息面板，另一端连接到 sw2 的交换机，这样，就构成了环路，拆掉 L1 网线，故障排除。注意：sw1\sw2 都是办公室自行配置的 8 口非管控交换机。

原因找到这儿，那又是什么原因导致的不稳定？

步骤5　继续检查。

用测试仪测试 L1 的这条网线，并不断晃动网线，发现网线线序 3 时断时续。到此故障原因彻底明白了，当网线线序 3 正常连接的时候，构成环路，port 1/0/11 就处于阻塞状态，sw2 交换机连接的终端无网络；当网线线序 3 断开的时候，环路破除，port 1/0/11 就处于转发状态，sw2 连接的终端网络正常。

RJ-45 各脚功能，见表 1-5-1。

表 1-5-1　RJ-45 各脚功能

1	传输数据正极 Tx+	5	备用
2	传输数据负极 Tx-	6	接收数据负极 Rx-
3	接收数据正极 Rx+	7	备用
4	备用	8	备用

二、善于使用 ping、tracert、ipconfig、arp、nslookup 命令，以及 Wireshark 抓包等软件排除故障

在日常运维中，ping 以及 tracert 两个命令使用非常多，ping 命令用来检查网络是否通畅，tracert 命令用于跟踪路由信息，可以查出数据从本地到目标主机经过的路径。Wireshark 抓包软件通过抓取不同位置报文来分析故障，在工作中经常使用。

案例：某网段不通的故障排除

前几天，同事联系我说，公司与华为云之间有个网段不通，其他网段数据正常。公司与华为云之间通过 IPsec-VPN 实现数据互通，有正常网段的，说明公司与华为之间隧道正常。通过"tracert"追踪路由信息。图 1-5-6 是公司与华为云之间建立隧道上业务正常网段路由追踪信息。图 1-5-7 是公司与华为云之间建立隧道上业务不正常网段路由追踪信息，发现路由追踪一直在核心交换机（IP：172.99.0.14）与防火墙（IP：172.99.0.13）之间来回切换，无法追踪到 VPN 设备，初步判断应该是路由问题。在防火墙中检查路由信息，仔细排查发现问题所在：缺少路由信息。新增如下路由信息，故障排除。

```
ip route-static 10.10.1.0 24 172.99.0.94
```

路由追踪信息正常见图 1-5-8。

图 1-5-6　"tracert"追踪路由信息（一）

图 1-5-7　"tracert"追踪路由信息（二）

网络丢包是运维中比较棘手的问题，硬件故障、配置问题都有可能引发该故障，比如网络环路、IP 地址冲突等。

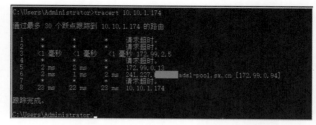

图1-5-8 "tracert"追踪路由信息（三）

案例：一个丢包故障的排除

安防系统在视频回放的时候卡顿严重，通过 ping 视频存储磁盘阵列（共计 4 台），IP 地址有丢包现象，与在磁盘阵列同一个交换机上通过网线连接上 PC，配置与磁盘阵列同网段地址，PC-ping-磁盘阵列不丢包。初步判断故障在上联链路，检查汇聚交换机，发现与安防交换机连接的接口数据流量非常大，于是，通过在交换机上配置聚合技术，解决由于数据流量大导致处理不及时而丢包的故障，见图 1-5-9，这里采用 4 条链路聚合。

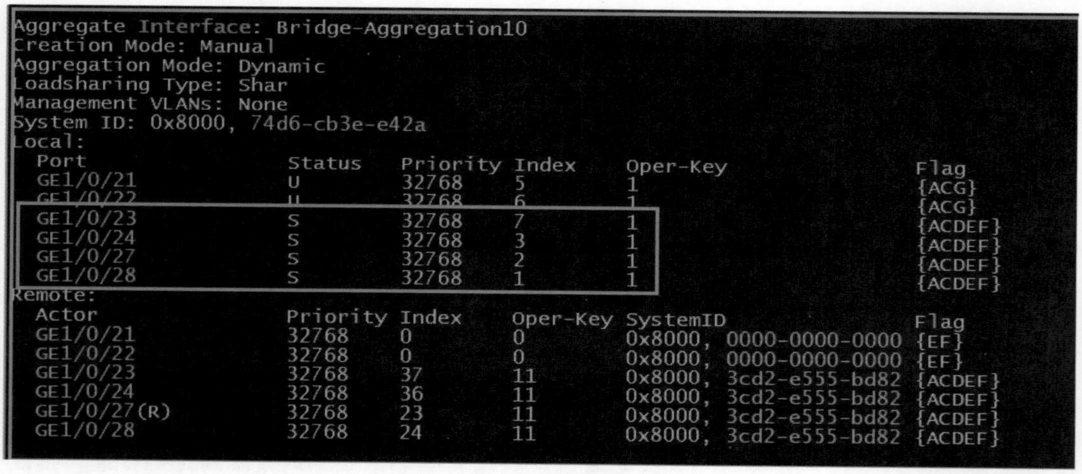

图1-5-9 链路聚合

网络运维人员应该善于总结经验，熟练掌握处理问题的方法与思路，不断学习新的网络架构，比如 SDN 等新技术，同时要具备编写网络规划方案的能力。

第二章 网络项目规划详解

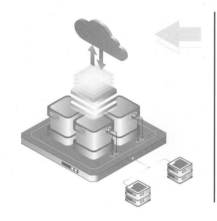

本章通过案例分析讲解网络常见协议，每个案例分析中包含项目介绍、拓扑图、项目需求，以及项目实施步骤，同时通过"知识加油站"对相关网络知识拓展介绍。

项目1 华三设备 telnet/ssh 远程访问配置

新购买的网络设备，一般经过 console 口进行本地配置后，在网络可达的环境下，就可以通过远程协议 telnet/ssh 登录设备（交换机、路由器等）进行调试维护工作，而不需要进入网络机房进行面对面调试。telnet 是一种明文传输协议，用在一般要求的网络中，而 ssh 协议是一种安全协议，采用公钥与私钥加密，它传输的数据经过加密后进行，可以使用户的ID、密码等信息在传输中不泄露。telnet 端口号是 23，ssh 端口号是 22。

 项目简介

本节分别介绍 telnet/ssh 的基本配置步骤。在模拟器以及真机上完成双重配置工作，并介绍交换机如何恢复默认设置，关于 telnet/ssh 远程登录并管理的配置是运维需要掌握的最基础的配置。

一、配置 telnet 协议远程访问（华三网络设备）

1. 通过 telnet 协议远程控制拓扑图

实现功能：如图 2-1-1 所示，AR1 路由器模拟访问的主机（电脑），AR2 路由器通过配置 telnet 远程控制协议，实现远程控制。

网络设备接口以及 IP 地址规划参照表 2-1-1。

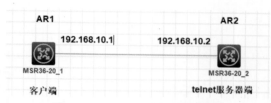

图2-1-1　telnet远程访问拓扑图

表 2-1-1　网络设备接口以及 IP 地址规划

设备	接口	IP
AR1	GigabitEthernet 0/0	192.168.10.1
AR2	GigabitEthernet 0/0	192.168.10.2

2. 项目配置步骤

步骤 1　AR1 路由器基本配置

```
# 进入系统模式
<H3C>system-view
# 修改设备名字
[H3C]sysname AR1
# 进入接口
[AR1]interface GigabitEthernet 0/0
# 配置 IP 地址（或者 ip address 192.168.10.1 24）
[AR1-GigabitEthernet0/0]ip address 192.168.10.1 255.255.255.0
# 保存配置
<AR1>save
```

步骤 2　AR2 路由器配置 telnet 远程访问协议

```
# 进入系统模式
<H3C>system-view
# 修改设备名字
[H3C]sysname AR2
# 启用 telnet 服务（记住，否则是无法使用的）
[AR2]telnet server enable
# 进入接口
[AR2]interface GigabitEthernet 0/0
# 配置 IP 地址
[AR2-GigabitEthernet0/0]ip address 192.168.10.2 255.255.255.0
# 配置本地登录用户名字为 fsm
[AR2]local-user fsm
# 一个用户添加完毕
New local user added.
# 设置远程访问密码（密码为 a123456）
[AR2-luser-manage-fsm]password simple a123456
# 配置服务类型为 telnet
[AR2-luser-manage-fsm]service-type telnet
```

配置用户具有最高权限
[AR2-luser-manage-fsm]authorization-attribute user-role network-admin

默认角色的权限级别如下：

network-admin：具有最高权限，可操作系统所有功能和资源。

Level-n(0-15)：数值越大，权限越大。level-15 相当于 network-admin 权限。
退出
[AR2-luser-manage-fsm]quit
允许用户界面可以同时打开 5 个会话
[AR2]user-interface vty 0 4
设置验证模式为 scheme（可以理解为使用用户名与密码认证）
[AR2-line-vty0-4]authentication-mode scheme
退出系统视图
[AR2-line-vty0-4]quit
退出到用户视图
[AR2]quit
保存命令（华三设备在系统视图下也可以进行命令保存）
<AR2>save

在 AR2 配置完 IP 地址，首先检查 AR2 与 AR1 是否能 ping 通，经检查通信正常。

```
[AR2]ping 192.168.10.1
Ping 192.168.10.1 (192.168.10.1): 56 data bytes, press CTRL_C to break
56 bytes from 192.168.10.1: icmp_seq=0 ttl=255 time=1.000 ms
……省略……
--- Ping statistics for 192.168.10.1 ---
5 packet(s) transmitted, 5 packet(s) received, 0.0% packet loss
round-trip min/avg/max/std-dev = 0.000/0.600/1.000/0.490 ms
```

步骤 3 telnet 远程访问登录测试

在 AR1 路由器的用户视图下，输入 telnet 192.168.10.2

输入用户名：fsm

密码：a123456

如图 2-1-2 中红色方框显示，远程访问到 AR2 设备界面。

图2-1-2 AR1远程通过Telnet登录AR2

二、配置 ssh 协议远程访问华三网络设备

1. 通过 ssh 协议远程控制的拓扑图

实现功能：如图 2-1-3 所示，AR1 路由器模拟访问的主机（电脑），AR2 路由器通过配置 ssh 远程控制协议，实现远程控制。

图2-1-3 ssh远程控制拓扑图

网络设备接口以及 IP 地址规划参照表 2-1-2。

表 2-1-2　网络设备接口以及 IP 地址规划

设备	接口	IP
AR1	GigabitEthernet 0/0	192.168.20.1
AR2	GigabitEthernet 0/0	192.168.20.2

2. 项目配置步骤

步骤 1　AR1 基本配置

```
<H3C>sys
<H3C>system-view
[H3C]sysname AR1
[AR1]interface GigabitEthernet 0/0
[AR1-GigabitEthernet0/0]ip address 192.168.20.1 255.255.255.0
<AR1>save
```

步骤 2　AR2 的 ssh 远程配置

```
<H3C>system-view
[H3C]sysname AR2
# 启用 ssh 服务
[AR2]ssh server enable
[AR2]interface GigabitEthernet 0/0
[AR2-GigabitEthernet0/0]ip address 192.168.10.2 255.255.255.0
[AR2]local-user fsm
[AR2-luser-manage-fsm]password simple a123456
# 配置远程访问服务类型为 ssh
[AR2-luser-manage-fsm]service-type ssh
[AR2-luser-manage-fsm]authorization-attribute user-role network-admin
[AR2-luser-manage-fsm]quit
[AR2]user-interface vty 0 4
[AR2-line-vty0-4]authentication-mode scheme
# 配置 VTY 用户界面支持的协议是 ssh，有些设备可以不配置 protocol inbound ssh。
[AR2-line-vty0-4]protocol inbound ssh
[AR2-line-vty0-4]quit
[AR2]quit
<AR2>save
```

步骤 3　ssh 远程访问登录测试

在 AR1 设备的用户视图下输入 ssh 192.168.20.2

输入用户名：fsm

密码：a123456

```
<AR1>ssh 192.168.20.2
Username: fsm
Press CTRL+C to abort.
Connecting to 192.168.20.2 port 22.
# 提示没有认证，是否继续？输入 y
The server is not authenticated. Continue? [Y/N]:y
# 是否保存公钥，输入 y
Do you want to save the server public key? [Y/N]:y
fsm@192.168.20.2's password:输入密码
```

```
Enter a character ~ and a dot to abort.
************************************************************************
* Copyright (c) 2004-2017 New H3C Technologies Co., Ltd. All rights reserved. *
* Without the owner's prior written consent,                                  *
* no decompiling or reverse-engineering shall be allowed.                     *
------------------------------------------------------------------------
```

通过 ssh 协议远程访问登录 AR2
```
<AR2>
```
在配置 ssh 协议时，关于密钥对配置，默认是 1024 长度，如需更长密钥对，配置时修改。
```
[sw1]public-key local create rsa
The range of public key modulus is (512 ~ 2048).
If the key modulus is greater than 512, it will take a few minutes.
Press CTRL+C to abort.
Input the modulus length [default = 1024]:
Generating Keys...
.
Create the key pair successfully.
```

三、知识加油站

1. 交换机恢复出厂设置的方法

品牌：华三 (H3C)　设备型号：S5130S-28P-EI

用户视图下输入重置交换机配置命令
```
<Access>reset saved-configuration
```
提示将删除当前配置文件，是否继续。输入 Y
```
The saved configuration file will be erased. Are you sure? [Y/N]:Y
```
以下提示配置文件已经清除
```
Configuration file in flash: is being cleared.
Please wait ...
MainBoard:
Configuration file is cleared.
```
用户视图下配置重启命令
```
<Access>reboot
```
输入 N，回车继续
```
Start to check configuration with next startup configuration file, please wait.........DONE!
Current configuration may be lost after the reboot, save current configuration? [Y/N]:N
```
继续重启设备　输入 Y
```
This command will reboot the device. Continue? [Y/N]:Y
```
重启完毕，恢复出厂配置
```
<H3C>
```

2. 交换机 telnet 配置完毕后查看本机是否配置成功的方法（同样适合 ssh）

① 自检办法 1：参照图 2-1-1。

telnet- 127.0.0.1,输入用户名以及密码。
```
<AR2>telnet 127.0.0.1
Trying 127.0.0.1 ...
Press CTRL+K to abort
Connected to 127.0.0.1 ...
```

```
**************************************************************
* Copyright (c) 2004-2017 New H3C Technologies Co., Ltd. All rights re-
served.*
* Without the owner's prior written consent,                 *
* no decompiling or reverse-engineering shall be allowed.    *
**************************************************************
login: fsm
Password:
<AR2>
```

② 自检办法 2：参照图 2-1-1，输入接口 IP 地址

```
<AR2>telnet 192.168.10.2
Trying 192.168.10.2 ...
Press CTRL+K to abort
Connected to 192.168.10.2 ...
**************************************************************
* Copyright (c) 2004-2017 New H3C Technologies Co., Ltd. All rights re-
served.*
* Without the owner's prior written consent,                 *
* no decompiling or reverse-engineering shall be allowed.    *
**************************************************************
login: fsm
Password:
<AR2>
```

3. H3C 交换机 –telnet 协议真机配置方法

图 2-1-1、图 2-1-2 是在模拟器上完成配置工作，真机配置类似。品牌：华三（H3C），型号：S5130S-28P-EI 交换机。

交换机在使用中要与其他网络贯通，还需要配置默认路由信息（路由后续讲解，如果不清楚路由的概念，可以学习完路由项目，再来学习这块内容）。

1）基础配置

```
# 配置路由信息
[sw1]ip route-static 0.0.0.0 0 192.168.10.254（根据实际情况配置）
<H3C>system-view
System View: return to User View with Ctrl+Z.
[H3C]sysname sw1
[sw1]interface vlan 1
# 在 vlan1 接口下配置 IP 地址，不能直接在接口下配置 IP 地址
[sw1-Vlan-interface1]ip address 192.168.10.1 24
[sw1-Vlan-interface1]quit
[sw1]telnet server enable
[sw1]local-user fsm
New local user added.
[sw1-luser-manage-fsm]password simple a123456
The new password is too short. It must contain at least 10 characters.
[sw1-luser-manage-fsm]password simple a123456123456
[sw1-luser-manage-fsm]service-type telnet
[sw1-luser-manage-fsm]authorization-attribute user-role network-admin
[sw1-luser-manage-fsm]quit
[sw1]user-interface vty 0 4
```

```
[sw1-line-vty0-4]authentication-mode scheme
[sw1-line-vty0-4]quit
[sw1]quit
<sw1>save
```

2）验证远程 telnet 远程访问

验证 1：

本机用户模式下输入 telnet 127.0.0.1，根据提示输入用户名以及密码。

```
<sw1>telnet 127.0.0.1
Trying 127.0.0.1 ...
Press CTRL+K to abort
Connected to 127.0.0.1 ...
Login: fsm
Password:
******************************************************************
* Copyright (c) 2004-2020 New H3C Technologies Co., Ltd. All rights reserved.*
* Without the owner's prior written consent,                      *
* no decompiling or reverse-engineering shall be allowed.         *
******************************************************************
%Jul 22 02:58:08:535 2021 sw1 SHELL/5/SHELL_LOGIN: fsm logged in from 127.0.0.1.
<sw1>
```

验证 2：

本机用户模式下输入 telnet 192.168.10.1，根据提示输入用户名以及密码。

```
<sw1>telnet 192.168.10.1
Trying 192.168.10.1 ...
Press CTRL+K to abort
Connected to 192.168.10.1 ...
Login: fsm
Password:
%Jul 22 03:06:06:080 2021 sw1 SHELL/5/SHELL_LOGIN: fsm logged in from 192.168.10.1.
******************************************************************
* Copyright (c) 2004-2020 New H3C Technologies Co., Ltd. All rights reserved.*
* Without the owner's prior written consent,                      *
* no decompiling or reverse-engineering shall be allowed.         *
******************************************************************
----------------------------------------------------------------
<sw1>
```

4. H3C 设备 -ssh 协议真机配置方法

1）基础配置　品牌：华三（H3C），型号：S5130S-28P-EI

```
<H3C>system-view
[H3C]sysname AR2
[AR2]ip route-static 0.0.0.0 0 192.168.10.254（根据实际情况配置）
[AR2]interface vlan 1
[AR2-Vlan-interface1]ip address 192.168.20.1 24
[AR2-Vlan-interface1]quit
[AR2]ssh server enable
[AR2]local-user fsm
New local user added.
[AR2-luser-manage-fsm]password simple a123456123456
```

```
[AR2-luser-manage-fsm]service-type ssh
[AR2-luser-manage-fsm]authorization-attribute user-role network-admin
[AR2-luser-manage-fsm]quit
[AR2]user-interface vty 0 4
[AR2-line-vty0-4]authentication-mode scheme
[AR2-line-vty0-4]protocol inbound ssh 新版本不需要配置；默认允许ssh
[AR2-line-vty0-4]quit
[AR2]sa f
Validating file. Please wait...
Saved the current configuration to mainboard device successfully.
[AR2]quit
<AR2>
```

2）验证远程 ssh 远程访问　本机用户模式下输入 ssh 127.0.0.1，根据提示输入用户名以及密码。

验证 1:

```
<AR2>ssh 127.0.0.1
Username: fsm
Press CTRL+C to abort.
Connecting to 127.0.0.1 port 22.
fsm@127.0.0.1's password:
Incorrect username or password, or server configuration error.
%Jul 22 03:37:16:969 2021 AR2 SSHS/6/SSHS_AUTH_PWD_FAIL: Authentication failed for user fsm from 127.0.0.1 port 64576 because of invalid username or wrong password.
fsm@127.0.0.1's password:
Enter a character ~ and a dot to abort.
%Jul 22 03:37:28:218 2021 AR2 SSHS/6/SSHS_AUTH_SUCCESS: SSH user fsm from 127.0.0.1 port 64576 passed password authentication.
%Jul 22 03:37:29:263 2021 AR2 SSHS/6/SSHS_CONNECT: SSH user fsm (IP: 127.0.0.1) connected to the server successfully.
******************************************************************************
* Copyright (c) 2004-2020 New H3C Technologies Co., Ltd. All rights reserved.*
* Without the owner's prior written consent,                                 *
* no decompiling or reverse-engineering shall be allowed.                    *
------------------------------------------------------------------------------
<AR2>
```

验证 2:

本机用户模式下输入 ssh 192.168.20.1，根据提示输入用户名以及密码。

```
<AR2>ssh 192.168.20.1
Username: fsm
Press CTRL+C to abort.
Connecting to 192.168.20.1 port 22.
The server is not authenticated. Continue? [Y/N]:y
Do you want to save the server public key? [Y/N]:y
fsm@192.168.20.1's password:
Enter a character ~ and a dot to abort.
%Jul 22 03:38:02:812 2021 AR2 SSHS/6/SSHS_AUTH_SUCCESS: SSH user fsm from 192.168.20.1 port 64577 passed password authentication.
%Jul 22 03:38:03:867 2021 AR2 SSHS/6/SSHS_CONNECT: SSH user fsm (IP: 192.168.20.1) connected to the server successfully.
******************************************************************************
```

```
* Copyright (c) 2004-2020 New H3C Technologies Co., Ltd. All rights reserved.*
* Without the owner's prior written consent,                                *
* no decompiling or reverse-engineering shall be allowed.                   *
*****************************************************************************
%Jul 22 03:38:04:967 2021 AR2 SHELL/5/SHELL_LOGIN: fsm logged in from
192.168.20.1.
<AR2>
```

3）查看登录用户情况

通过输入 dis users 命令查看正在访问交换机的信息，用户 fsm 是通过 vty 登录访问的。

```
[AR2]dis users
Idx     Line      Idle        Time              Pid     Type
0       AUX 0     00:00:00    Jul 22 05:22:59   484
+ 10    VTY 0     00:00:00    Jul 22 05:23:55   496     SSH
Following are more details.
AUX  0   :
         # 串口是管理员登录
         User role list: network-admin
VTY  0   :
         # 远程访问是 fsm 登录
         User name: fsm
         User role list: network-admin network-operator
         Location: 192.168.20.1
+        : Current operation user.
F        : Current operation user works in async mode.
```

项目 2　华为设备 telnet/ssh 远程访问配置

在企业中不可能只有一个品牌的网络设备，而目前华三与华为网络设备使用较多。所以本例介绍华为设备的 telnet/ssh 远程访问配置工作，配置步骤与华三类似，略有不同。

 项目简介

本项目介绍华为网络设备 telnet/ssh 远程访问配置方法，模拟与真机双重介绍，在本项目中，注意华为网络设备 ssh 配置与华三的区别，尤其是登录时，华为设备是在系统模式下输入 stelnet+IP，而华三网络设备是在用户模式下输入 ssh+IP。需要熟练掌握两个厂家的网络设备 telnet/ssh 配置方法。

一、telnet 远程访问协议配置

1. 通过 telnel 协议远程控制的拓扑图

实现功能：如图 2-2-1 所示，AR1 路由器模拟访问的主机（电脑），AR2 路由器通过配置 telnet 远程控制协议，实现远程控制。

网络设备接口以及 IP 地址规划参照表 2-2-1。

图2-2-1 telnet远程控制拓扑图

表 2-2-1 网络设备接口以及 IP 地址规划

设备	接口	IP
AR1	GigabitEthernet 0/0	192.168.10.1
AR2	GigabitEthernet 0/0	192.168.10.2

2. 项目配置步骤

步骤1　AR1 基本配置（客户端）

```
<Huawei>system-view
[Huawei]sysname AR1
[AR1]interface GigabitEthernet 0/0/0
[AR1-GigabitEthernet0/0/0]ip address 192.168.10.1 255.255.255.0
[AR1-GigabitEthernet0/0/0]quit
<AR1>save
```

步骤2　AR2 路由器配置 telnet 远程访问协议

```
<Huawei>system-view
[Huawei]sysname AR2
[AR2]interface GigabitEthernet 0/0/0
[AR2-GigabitEthernet0/0/0]ip address 192.168.10.2 255.255.255.0
# 进入 aaa 模式
[AR2]aaa
# 配置用户以及密码
[AR2-aaa]local-user fsm password cipher a123456
# 配置用户访问级别
[AR2-aaa]local-user fsm privilege level 15
# 配置用户 telnet 访问服务
[AR2-aaa]local-user fsm service-type telnet
[AR2-aaa]quit
# 配置远程服务数量
[AR2]user-interface vty 0 4
# 配置访问模式为 aaa
[AR2-ui-vty0-4]authentication-mode aaa
# 配置允许登录接入用户类型的协议，可以不配置
[AR2-ui-vty0-4]protocol inbound telnet
<AR2>save
```

> protocol inbound telnet 解释如下：这条在交换机 telnet 模式下可以不配置，命令 protocol inbound { all | ssh | telnet }用来配置允许登录接入用户类型的协议。protocol inbound telnet 为默认配置；如配置为 protocol inbound ssh 时，telnet 将无法登录；如果配置 protocol inbound all，则都可以登录。

步骤3 telnet 远程访问登录测试

在用户视图下输入 IP 地址：telnet 192.168.10.2，用户名 fsm，密码 a123456。

如图 2-2-2 中红色方框显示，远程访问到 AR2 设备界面。

图2-2-2　AR1远程通过Telnet登录AR2

二、ssh 远程访问协议配置

1. 通过 ssh 协议远程控制的拓扑图

实现功能：如图 2-2-3 所示，AR1 路由器模拟访问的主机（电脑），AR2 路由器通过配置 ssh 远程控制协议，实现远程控制。

网络设备接口以及 IP 地址规划参照表 2-2-2。

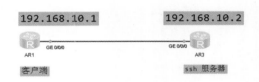

图2-2-3　ssh远程控制拓扑图

表 2-2-2　网络设备接口以及 IP 地址规划

设备	接口	IP
AR1	GigabitEthernet 0/0	192.168.10.1
AR2	GigabitEthernet 0/0	192.168.10.2

2. 项目配置

步骤1　AR1 交换机配置

```
<Huawei>system-view
[Huawei]sysname AR1
[AR1]interface GigabitEthernet 0/0/0
[AR1-GigabitEthernet0/0/0]ip address 192.168.10.1 255.255.255.0
[AR1-GigabitEthernet0/0/0]quit
[AR1]quit
<AR1>save
```

步骤2　AR2 交换机配置

```
<Huawei>system-view
[Huawei]sysname AR2
[AR2]interface GigabitEthernet 0/0/0
[AR2-GigabitEthernet0/0/0]ip address 192.168.10.2 255.255.255.0
```

```
[AR2-GigabitEthernet0/0/0]quit
# 启动ssh服务（注意是stelnet）
[AR2]stelnet server enable
# 创建用户以及密码
[AR2-aaa]local-user fsm password cipher a123456
Info: Add a new user.
# 允许ssh服务
[AR2-aaa]local-user fsm service-type ssh
# 配置访问级别
[AR2-aaa]local-user fsm privilege level 15
[AR2-aaa]quit
[AR2]user-interface vty 0 4
# 配置访问模式
[AR2-ui-vty0-4]authentication-mode aaa
# 配置VTY用户界面支持的协议是ssh（需要配置）
[AR2-ui-vty0-4] protocol inbound ssh
<AR2>save
```

步骤3 ssh远程访问登录测试

在AR1的系统视图下输入stelnet 192.168.20.2，但是报错无法访问，首次登录就是如此提示，如图2-2-4所示。按照提示，在AR1的系统模式下输入ssh client first-time enable，再次输入stelnet 192.168.20.2，输入用户名：fsm，密码：a123456，见图2-2-5，远程登录成功。

```
[AR1]stelnet 192.168.10.2
Please input the username:fsm
Trying 192.168.10.2 ...
Press CTRL+K to abort
Connected to 192.168.10.2 ...
Error: Failed to verify the server's public key.
Please run the command "ssh client first-time enable"to enable the first-time ac
cess function and try again.
[AR1]
```

图2-2-4 ssh访问报错提示

```
[AR1]stelnet 192.168.10.2
Please input the username:fsm
Trying 192.168.10.2 ...
Press CTRL+K to abort
Connected to 192.168.10.2 ...
The server is not authenticated. Continue to access it? (y/n)[n]:y
Apr 15 2021 20:12:42-08:00 AR1 %%01SSH/4/CONTINUE_KEYEXCHANGE(l)[0]:The server h
ad not been authenticated in the process of exchanging keys. When deciding wheth
er to continue, the user chose Y.
[AR1]
Save the server's public key? (y/n)[n]:y
The server's public key will be saved with the name 192.168.10.2. Please wait...

Apr 15 2021 20:12:44-08:00 AR1 %%01SSH/4/SAVE_PUBLICKEY(l)[1]:When deciding whet
her to save the server's public key 192.168.10.2, the user chose Y.
[AR1]
Enter password:
<AR2>
```

图2-2-5 远程登录成功

三、知识加油站

1. 华为真机 telnet 远程配置

品牌：华为，型号：S2700-9TP-SI-AC 交换机。

```
<Quidway>system-view
[Quidway]sysname sw1
[sw1]ip route-static 0.0.0.0 0 192.168.10.254（根据实际情况配置）
[sw1]interface vlan 1
# 在 vlan 1 视图下配置 IP 地址
[sw1-Vlanif1]ip address 192.168.10.1 24
[sw1-Vlanif1]quit
[sw1-aaa]local-user fsm password cipher a123456
Info: A new user added
[sw1-aaa]local-user fsm service-type telnet
[sw1-aaa]local-user fsm privilege level 15
[sw1-aaa]quit
[sw1]user-interface vty 0 4
[sw1-ui-vty0-4]authentication-mode aaa
[sw1-ui-vty0-4]quit
[sw1]quit
<sw1>save
```

验证：

本机用户模式下输入 ssh 127.0.0.1，根据提示输入用户名（fsm）以及密码 (a123456)。

```
<sw1>telnet 127.0.0.1
Trying 127.0.0.1 ...
Press CTRL+K to abort
Connected to 127.0.0.1 ...
Warning: Telnet is not a secure protocol, and it is recommended to use Stelnet.
Login authentication
Username:fsm
Password:
Info: The max number of VTY users is 5, and the number
      of current VTY users on line is 1.
<sw1>
```

2. 华为真机 ssh 配置

```
<Quidway>system-view
[Quidway]sysname sw1
[sw1]ip route-static 0.0.0.0 0 192.168.10.254（根据实际情况配置）
[sw1]stelnet server enable
[sw1]aaa
[sw1-aaa]local-user fsm password cipher a123456
Info: A new user added
[sw1-aaa]local-user fsm service-type ssh
[sw1-aaa]local-user fsm privilege level 15
[sw1-aaa]quit
[sw1]user-interface vty 0 4
[sw1-ui-vty0-4]authentication-mode aaa
[sw1-ui-vty0-4]protocol inbound ssh
[sw1-ui-vty0-4]quit
```

```
[sw1]quit
<sw1>save
```
验证：

本机用户模式下输入 ssh 127.0.0.1，根据提示输入用户名 (fsm) 以及密码（a123456）。

```
[sw1]stelnet 127.0.0.1
Please input the username:fsm
Trying 127.0.0.1 ...
Press CTRL+K to abort
Connected to 127.0.0.1 ...
Error: Failed to verify the server's public key.
Please run the command "ssh client first-time enable" to enable the first-
time access function and try again.
# 根据提示输入"ssh client first-time enable"
[sw1]ssh client first-time enable
[sw1]stelnet 127.0.0.1
Please input the username:fsm
Trying 127.0.0.1 ...
Press CTRL+K to abort
Connected to 127.0.0.1 ...
Enter password:
Info: The max number of VTY users is 5, and the number
      of current VTY users on line is 3.
<sw1>
```

项目3　华三设备 Console 认证配置

一般情况下，通过交换机的 Console 接口不需要认证就可以进入管理界面。在第一次配置交换机时，使用 Console 口登录，其他情况下都使用 telnet/ssh 远程访问来访问管理，对于重要的网络设备，为了防止非法人员通过 Console 接口修改配置，可以进行 Console 登录认证。

项目简介

登录网络设备：其一，远程通过 telnet/ssh 协议访问；其二，经过本机 Console 进行登录并管理，但是对于重要设备，比如核心、汇聚交换机，这些设备配置比较复杂，不希望非管理人员通过 Console 进入设备，就需要配置认证措施，可以配置用户加密码认证，也可以只配置密码认证。

一、Console 接口配置的几种方式

配置环境：华三设备，型号：H3C S5130S-28P-EI。

步骤 1 查看认证管理接口几种模式

```
# Console 登录配置需要选择 aux
[H3C]user-interface ?
INTEGER<0-73>  Number of the first line
aux            AUX line
class          Specify the line class to modify the default configuration
vty            Virtual type terminal (VTY) line
```

步骤 2 配置 Console 仅密码认证

```
# 进入串口配置视图
[H3C]user-interface aux 0
# 选择认证模式分别是空密码/密码认证/用户名与密码认证
[H3C-line-aux0]authentication-mode ?
# 空密码
none       Login without authentication
# 密码认证
password   Password authentication
# 用户名以及密码认证
scheme     Authentication use AAA
# 选择认证方式为密码认证
[H3C-line-aux0]authentication-mode password
# 设置密码
[H3C-line-aux0]set authentication password simple a123456
```

步骤 3 验证（设备重启后，当用 Console 连接时需要输入密码）

```
Press ENTER to get started.
# 输入密码（a123456）
Password:
# 进入交换机系统
******************************************************************
* Copyright (c) 2004-2020 New H3C Technologies Co., Ltd. All rights reserved.*
* Without the owner's prior written consent,                      *
* no decompiling or reverse-engineering shall be allowed.         *
******************************************************************
<H3C>%Jan  1 00:02:56:546 2013 H3C SHELL/5/SHELL_LOGIN: TTY logged in from aux0.
<H3C>
```

步骤 4 删除 Console 认证密码的方法

```
[H3C]user-interface aux 0
# 模式选择 none 表示空密码
[H3C-line-aux0]authentication-mode none
<H3C>save
```

再次重启交换机测试，不需要密码即可进入

```
<H3C>
<H3C>%Jan  1 00:01:22:457 2013 H3C SHELL/5/SHELL_LOGIN: TTY logged in from aux0.
```

步骤 5 配置 Console 用户名与密码认证

```
# 配置用户名
[H3C]local-user fsm
```

```
# 配置密码
[H3C-luser-manage-fsm]password simple a123456a123456
# 配置服务方式是 terminal（console 访问）
[H3C-luser-manage-fsm]service-type terminal
# 配置用户访问权限
[H3C-luser-manage-fsm]authorization-attribute user-role network-admin
# 进入 aux 0
[H3C]user-interface aux 0
# 配置认证模式为用户名与密码
[H3C-line-aux0]authentication-mode scheme
```

重启后根据提示输入用户名以及密码

```
Line aux0 is available.
Press ENTER to get started.
Login: fsm
Password:
******************************************************************
* Copyright (c) 2004-2020 New H3C Technologies Co., Ltd. All rights reserved.*
* Without the owner's prior written consent,                      *
* no decompiling or reverse-engineering shall be allowed.         *
******************************************************************
<H3C>%Jan  1 00:01:28:982 2013 H3C SHELL/5/SHELL_LOGIN: fsm logged in from aux0.
<H3C>
```

二、知识加油站

1. Console 线缆介绍

1）常见串口线　网络设备上都有一个类似网络的接口（RJ45），图2-3-1是H3C S10508X核心交换机，旁边备注Console接口，电脑通过Console进入管理界面，采用的是Console-com串口线，如图2-3-2所示。

因为目前大多数笔记本电脑无com接口，就需要用USB-com转换线，如图2-3-3所示。USB-com转换线与Console-com线缆配合使用，如图2-3-4所示。

技术更新换代，以上组合比较烦琐，市场上也早就有将它们融合在一起的配置线USB-RJ45，如图2-3-5所示。

图2-3-1　Console 接口

2）如何通过串口线登录网络设备的管理界面（软件安装方法不再累赘），如图2-3-6所示。

相关软件：一般使用 Xshell、SecureCRT

图2-3-2　Console-com线缆

图2-3-3　USB-com接口

图2-3-4　组合（Console-com＋USB-com）

采用串口线，在正确安装驱动后，设备管理器中查看虚拟com，如图2-3-7所示。

图2-3-6 交换机配置软件

图2-3-5 USB-RJ45配置线

图2-3-7 查看串口线com序号（本例是com5）

① SecureCRT 软件串口设置连接步骤。

步骤1 新建-快速连接，见图2-3-8。

步骤2 选择"Serial"，见图2-3-9。

步骤3 参数选择。端口：COM5，波特率：9600，数据位：8（你的电脑不一定是这个数字），奇偶校验：None，停止位：1，见图2-3-10。

通过以上步骤，进入网络管理界面，如图2-3-11所示。

② Xshell 软件串口设置连接步骤。

步骤1 新建-协议选择"SERIAKL"，见图2-3-12。

步骤2 类别中选择"SERIAKL"进行相关设置，见图2-3-13。输入用户名以及密码登录即可。

图2-3-8 快速连接

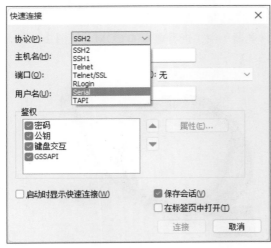

图2-3-9 选择"Serial"

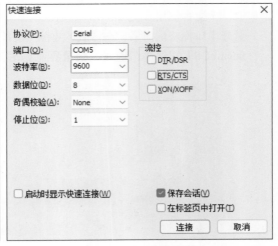

图2-3-10 参数配置

53

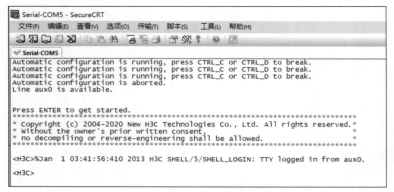

图2-3-11 进入管理界面

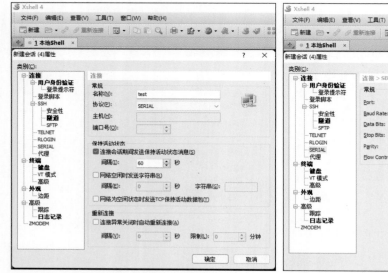

图2-3-12 新建-协议选择"SERIAKL"

图2-3-13 参数配置

3）无线蓝牙串口 近几年出现新型的无线蓝牙串口模块（图2-3-14）使用很方便，与串口RJ-45配合使用（图2-3-15）。笔记本需要有蓝牙功能，配对使用。

蓝牙串口模块配对后，在"我的电脑-设备管理器"中，查看虚拟的COM，见图2-3-16。SecureCRT软件串口连接，登录界面见图2-3-17。

图2-3-14 蓝牙com模块

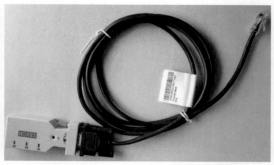

图2-3-15 蓝牙com模块与串口线连接

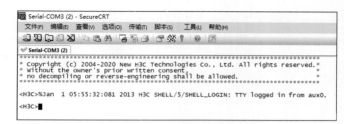

图2-3-16 无线蓝牙虚拟串口　　　　图2-3-17 无线蓝牙模块登录

2. 使用 SecureCRT、xshell 远程管理网络设备的方法（ssh/ telnet）

① SecureCRT 软件 ssh 远程管理网络设备。

步骤1　新建 - 快速连接。

步骤2　按照图 2-3-18 所示配置，协议选择 SSH2，主机名填写远程网络设备 IP 地址，端口 22，填写用户名，在弹出对话框中输入密码，即可进入登录界面（如第一次登录，需要接收公钥下载），见图2-3-19。

图2-3-18 设置界面　　　　图2-3-19 登录界面

② SecureCRT 软件 telnet 远程管理网络设备。

步骤1　新建快速连接。

步骤2　按照图 2-3-20 所示，协议选择 telnet，弹出界面输入主机名 IP 地址，端口 23，见图 2-3-21，填写用户名以及密码，即可进入登录界面。

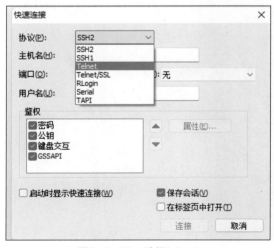
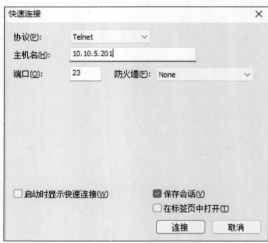

图2-3-20 选择telnet　　　　图2-3-21 IP地址以及端口配置

③ Xshell 软件 ssh 远程管理网络设备。

步骤 1　新建会话。

步骤 2　按照图 2-3-22 所示，协议选择 SSH，主机 IP 地址，端口 22，填写用户名，见图 2-3-23。弹出对话框输入密码，见图 2-3-24，即可进入登录界面，见图 2-3-25。如第一次登录，需要接收公钥下载。

图2-3-22　参数设置　　　　　　　　　　图2-3-23　输入用户名

　　　　　　　　　　　　　　　　　　　　图2-3-24　输入密码

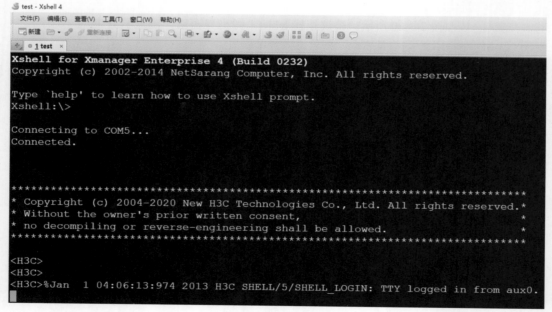

图2-3-25　登录界面

④ Xshell 软件 telnet 远程管理网络设备。

步骤1 新建会话。

步骤2 协议选择 TELNET，主机 IP 地址，端口 22，填写用户名，见图 2-3-26，弹出界面输入用户名以及密码，即可进入登录界面，见图 2-3-27。

图2-3-26 参数配置

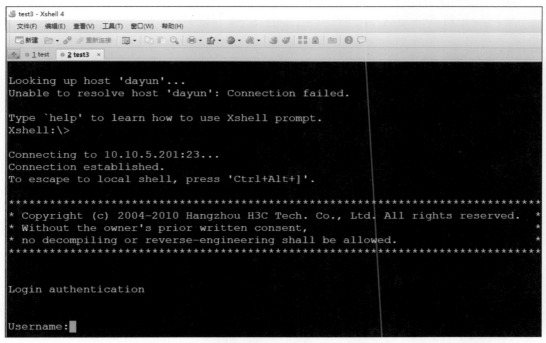

图2-3-27 输入用户名以及密码即可进入登录界面

项目 4 直连路由

企业中网络拓扑结构一般比较复杂，不同部门或者分支机构存在于不同的网段中（比如财务系统使用网段 192.168.10.0/24，而研发系统使用的是 192.168.20.0/24），就需要使用网络设备来连接不同的网络，实现网络互通。这种连接两个或多个网络的硬件设备，就是路由器，路由器在网络中起网关的作用，它读取每个数据包中的地址，并决定如何传送。

不同网段之间数据访问需要有路由信息（路由表），路由就是通往目标的路径。路由表中有三种路由：直连路由、静态路由和动态路由。

直连路由（Direct）一般只要在路由器接口配上 IP 地址，物理（Physical）与协议（Protocol）都是 up 状态就可建立，不需要网络管理人员维护，但是直连路由无法获取与其不直接相邻的路由信息。直连路由是所有路由协议当中优先级最高的。

如图 2-4-1 所示，对于路由器而言，只要接口配置 IP 地址，物理与协议均 up（图 2-4-2），就可以自动产生直连路由，对路由器 AR12 来说，有两条直连路由，红色箭头代表一条直连路由，网段是 192.168.10.0/24；绿色箭头代表另一条直连路由，网段是 192.168.20.0/24。这两条直连路由存在一个网络设备中，当 PC1 与 PC2 配置 IP 地址后，就可以互通。描述如下，详细讲解请看项目分析。

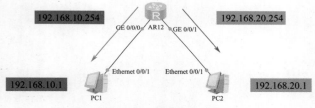

图 2-4-1 直连路由示意

图 2-4-1 中，按照图中 IP 地址配置后，在 AR2 路由器上，通过"display ip routing-table"命令查询路由条目如下：

```
[AR12]display ip routing-table
Destination/Mask    Proto    Pre  Cost   Flags  NextHop          Interface
192.168.10.0/24     Direct   0    0      D      192.168.10.254   GigabitEthernet0/0/0
192.168.20.0/24     Direct   0    0      D      192.168.20.254   GigabitEthernet0/0/1
     PC1（IP：192.168.10.1）-ping-PC2（IP：192.168.20.1）-通
PC>ping 192.168.20.1
Ping 192.168.20.1: 32 data bytes, Press Ctrl_C to break
From 192.168.20.1: bytes=32 seq=1 ttl=127 time=16 ms
……省略……
```

```
From 192.168.20.1: bytes=32 seq=4 ttl=127 time=16 ms
From 192.168.20.1: bytes=32 seq=5 ttl=127 time<1 ms
```

图2-4-2 通过命令查询接口物理与协议状态（绿框是接口名称，红框是物理up,蓝框是协议up）

一、直连路由

1. 直连路由项目拓扑图

实现功能：如图 2-4-3 所示，相邻路由器之间网络互通，但是不能跨越路由器实现网络通信，如要实现，需要配置相应的路由协议，比如静态路由（后面章节讲解）。

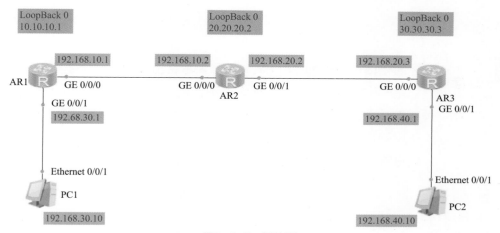

图2-4-3 拓扑图

网络设备接口以及 IP 地址规划参照表 2-4-1。

表 2-4-1 网络设备接口以及 IP 地址规划

序号	设备	接口	IP/Mask
1	AR1	GigabitEthernet 0/0/0	192.168.10.1/24
2	AR1	GigabitEthernet 0/0/1	192.168.30.1/24
3	AR1	LoopBack 0	10.10.10.1/32
4	AR2	GigabitEthernet 0/0/0	192.168.10.2/24
5	AR2	GigabitEthernet 0/0/1	192.168.20.2/24
6	AR2	LoopBack 0	20.20.20.2/32
7	AR3	GigabitEthernet 0/0/0	192.168.20.3
8	AR3	GigabitEthernet 0/0/1	192.168.40.1
9	AR3	LoopBack 0	30.30.30.3/32
10	PC1		192.168.30.10
11	PC2		192.168.40.10

2. 路由器接口 IP 地址基础配置

步骤 1　AR1 路由器接口 IP 配置

```
<Huawei>system-view
```

```
[Huawei]sysname AR1
# 进入接口视图
[AR1]interface GigabitEthernet 0/0/0
# 配置IP地址
[AR1-GigabitEthernet0/0/0]ip address 192.168.10.1 24
[AR1-GigabitEthernet0/0/0]quit
[AR1]interface GigabitEthernet 0/0/1
[AR1-GigabitEthernet0/0/1]ip address 192.168.30.1 24
[AR1]quit
# 进入环回接口视图
[AR1]interface LoopBack 0
# 环回接口配置IP地址
[AR1-LoopBack0]ip address 10.10.10.1 32
[AR1-LoopBack0]quit
[AR1]quit
<AR1>save
```

 环回接口一般称为 loopback 接口，是路由器上的一个逻辑接口，它是一个神奇的接口，只要设备不宕机，这个接口就一直处于 up 状态。环回接口在路由器上与物理接口类似，环回接口的 IP 地址，子网掩码一般设为 255.255.255.255（32 位）。环回接口网段与自身路由器其他网段属于直连路由。

通过命令"display ip interface brief"，查看 AR1 接口配置信息。

```
[AR1]display ip interface brief
*down: administratively down
^down: standby
(l): loopback
(s): spoofing
The number of interface that is up in Physical is 4
The number of interface that is down in Physical is 1
The number of interface that is up in Protocol is 4
The number of interface that is down in Protocol is 1
Interface                 IP Address/Mask      Physical    Protocol
GigabitEthernet0/0/0      192.168.10.1/24      up          up
GigabitEthernet0/0/1      192.168.30.1/24      up          up
GigabitEthernet0/0/2      unassigned           down        down
LoopBack0                 10.10.10.1/32        up          up(s)
NULL0                     unassigned           up          up(s)
```

AR1 路由器接口状态信息见表 2-4-2。

表 2-4-2 AR1 路由器接口状态信息

Interface	IP Address/Mask	Physical	Protocol
GigabitEthernet0/0/0	192.168.10.1/24	up	up
GigabitEthernet0/0/1	192.168.30.1/24	up	up
LoopBack0	10.10.10.1/32	up	up

> **注意**
>
> 环回接口是无法手动关闭的,但是物理接口可以关闭。具体如下:
>
> 1) 物理接口关闭与打开
>
> \# 进入接口
>
> `[AR1]interface GigabitEthernet 0/0/0`
>
> \# 输入 shutdown 命令,关闭接口
>
> `[AR1-GigabitEthernet0/0/0]shutdown`
>
> \# 输入 display this,查看接口状态,红色字体显示 shutdown,代表该接口已经关闭。
>
> ```
> [AR1-GigabitEthernet0/0/0]display this
> [V200R003C00]
> #
> interface GigabitEthernet0/0/0
> shutdown
> ip address 192.168.10.1 255.255.255.0
> #
> Return
> ```
>
> \# 重新打开接口,输入命令"undo shutdown"
>
> ```
> [AR1-GigabitEthernet0/0/0]undo shutdown
> [AR1-GigabitEthernet0/0/0]display this
> [V200R003C00]
> #
> interface GigabitEthernet0/0/0
> ip address 192.168.10.1 255.255.255.0
> #
> return
> ```
>
> 2) 环回接口不能关闭
>
> ```
> [AR1]interface LoopBack 0
> [AR1-LoopBack0]shutdown
> ```
>
> \# 当环回接口输入 shutdown,提示报错,是无法关闭的。
>
> `Error: Unrecognized command found at '^' position.`

步骤 2 AR2 路由器接口 IP 配置

```
<Huawei>system-view
[Huawei]sysname AR2
[AR2]interface GigabitEthernet 0/0/0
[AR2-GigabitEthernet0/0/0]ip address 192.168.10.2 24
[AR2-GigabitEthernet0/0/0]quit
[AR2]interface GigabitEthernet 0/0/1
[AR2-GigabitEthernet0/0/1]ip address 192.168.20.2 24
[AR2-GigabitEthernet0/0/1]quit
[AR2]interface LoopBack 0
[AR2-LoopBack0]ip address 20.20.20.2 32
[AR2-LoopBack0]quit
[AR2]quit
<AR2>save
```

查看 AR2 接口配置信息

```
[AR2]display ip interface brief
*down: administratively down
```

```
^down: standby
(l): loopback
(s): spoofing
The number of interface that is up in Physical is 4
The number of interface that is down in Physical is 1
The number of interface that is up in Protocol is 4
The number of interface that is down in Protocol is 1
Interface                     IP Address/Mask       Physical    Protocol
GigabitEthernet0/0/0          192.168.10.2/24       up          up
GigabitEthernet0/0/1          192.168.20.2/24       up          up
GigabitEthernet0/0/2          unassigned            down        down
LoopBack0                     20.20.20.2/32         up          up(s)
NULL0                         unassigned            up          up(s)
```

AR2 路由器接口状态信息如表 2-4-3 所示。

表 2-4-3　AR2 路由器接口状态信息

Interface	IP Address/Mask	Physical	Protocol
GigabitEthernet0/0/0	192.168.10.2/24	up	up
GigabitEthernet0/0/1	192.168.20.2/24	up	up
LoopBack0	20.20.20.2/32	up	up

步骤 3　AR3 路由器接口 IP 配置

```
<Huawei>system-view
[Huawei]sysname AR3
[AR3]interface GigabitEthernet 0/0/0
[AR3-GigabitEthernet0/0/0]ip address 192.168.20.3 24
[AR3-GigabitEthernet0/0/0]quit
[AR3]interface GigabitEthernet 0/0/1
[AR3-GigabitEthernet0/0/1]ip address 192.168.40.1 24
[AR3-GigabitEthernet0/0/1]quit
[AR3]interface LoopBack 0
[AR3-LoopBack0]ip address 30.30.30.3 32
[AR3-LoopBack0]quit
[AR3]quit
<AR3>save
```

查看 AR3 接口配置信息

```
[AR3]display ip interface brief
*down: administratively down
^down: standby
(l): loopback
(s): spoofing
The number of interface that is up in Physical is 4
The number of interface that is down in Physical is 1
The number of interface that is up in Protocol is 4
The number of interface that is down in Protocol is 1
Interface                     IP Address/Mask       Physical    Protocol
GigabitEthernet0/0/0          192.168.20.3/24       up          up
GigabitEthernet0/0/1          192.168.40.1/24       up          up
GigabitEthernet0/0/2          unassigned            down        down
LoopBack0                     30.30.30.3/32         up          up(s)
```

| | NULL0 | | unassigned | up | up(s) |

AR3 路由器接口状态信息见表 2-4-4。

表 2-4-4 AR3 路由器接口状态信息

Interface	IP Address/Mask	Physical	Protocol
GigabitEthernet0/0/0	192.168.20.3/24	up	up
GigabitEthernet0/0/1	192.168.40.1/24	up	up
LoopBack0	30.30.30.3/32	up	up

步骤 4 PC1 电脑配置基本参数

IP：192.168.30.10，子网掩码：255.255.255.0，网关：192.168.30.1。

配置如图 2-4-4 所示。

图2-4-4 PC1配置

步骤 5 PC2 电脑配置基本参数

IP：192.168.40.10，子网掩码：255.255.255.0，网关：192.168.40.1

配置如图 2-4-5 所示。

图2-4-5 PC2配置

步骤6 直连路由通信测试

AR1（IP：192.168.10.1）- ping- AR2（IP：192.168.10.2）- 通
```
<AR1>ping 192.168.10.2
  PING 192.168.10.2: 56  data bytes, press CTRL_C to break
    Reply from 192.168.10.2: bytes=56 Sequence=1 ttl=255 time=120 ms
    Reply from 192.168.10.2: bytes=56 Sequence=2 ttl=255 time=20 ms
……省略……
--- 192.168.10.2 ping statistics ---
  5 packet(s) transmitted
  5 packet(s) received
  0.00% packet loss
  round-trip min/avg/max = 10/38/120 ms
```
AR1（IP：192.168.10.1）- ping- PC1（192.168.30.10）- 通
```
<AR1>ping 192.168.30.10
  PING 192.168.30.10: 56  data bytes, press CTRL_C to break
    Reply from 192.168.30.10: bytes=56 Sequence=1 ttl=128 time=40 ms
    Reply from 192.168.30.10: bytes=56 Sequence=2 ttl=128 time=20 ms
……省略……
--- 192.168.30.10 ping statistics ---
  5 packet(s) transmitted
  5 packet(s) received
  0.00% packet loss
  round-trip min/avg/max = 10/22/40 ms
```
AR2（IP：192.168.20.2）- ping- AR3（IP：192.168.20.3）- 通
```
<AR2>ping 192.168.20.3
PING 192.168.20.3: 56  data bytes, press CTRL_C to break
Reply from 192.168.20.3: bytes=56 Sequence=1 ttl=255 time=60 ms
Reply from 192.168.20.3: bytes=56 Sequence=2 ttl=255 time=20 ms
……省略……
--- 192.168.20.3 ping statistics ---
5 packet(s) transmitted
5 packet(s) received
0.00% packet loss
round-trip min/avg/max = 20/30/60 ms
```
AR3（IP：192.168.40.1）- ping- PC2（IP：192.168.40.1）- 通
```
<AR3>ping 192.168.40.10
PING 192.168.40.10: 56  data bytes, press CTRL_C to break
Reply from 192.168.40.10: bytes=56 Sequence=1 ttl=128 time=20 ms
Reply from 192.168.40.10: bytes=56 Sequence=2 ttl=128 time=20 ms
……省略……
--- 192.168.40.10 ping statistics ---
5 packet(s) transmitted
5 packet(s) received
0.00% packet loss
round-trip min/avg/max = 10/18/20 ms
```

步骤7 查询直连路由条目

华为路由在没有配置任何信息的情况下默认有4条直连路由信息。

打开一个全新路由器，查询路由表信息。

查看路由表命令：`display ip routing-table`
```
[Huawei] display ip routing-table
```

```
Route Flags: R - relay, D - download to fib
------------------------------------------------------------------------
Routing Tables: Public
         Destinations : 4        Routes : 4
Destination/Mask    Proto   Pre  Cost      Flags NextHop      Interface
      127.0.0.0/8   Direct  0    0          D    127.0.0.1    InLoopBack0
      127.0.0.1/32  Direct  0    0          D    127.0.0.1    InLoopBack0
127.255.255.255/32  Direct  0    0          D    127.0.0.1    InLoopBack0
255.255.255.255/32  Direct  0    0          D    127.0.0.1    InLoopBack0
```

查看 AR1 的直连路由信息，其他路由器自行查看。

```
[AR1]display ip routing-table protocol direct
Route Flags: R - relay, D - download to fib
------------------------------------------------------------------------
Public routing table : Direct
         Destinations : 11       Routes : 11
Direct routing table status : <Active>
         Destinations : 11       Routes : 11
Destination/Mask    Proto   Pre  Cost      Flags NextHop      Interface
   10.10.10.1/32    Direct  0    0          D    127.0.0.1    LoopBack0
    127.0.0.0/8     Direct  0    0          D    127.0.0.1    InLoopBack0
    127.0.0.1/32    Direct  0    0          D    127.0.0.1    InLoopBack0
127.255.255.255/32  Direct  0    0          D    127.0.0.1    InLoopBack0
192.168.10.0/24     Direct  0    0          D    192.168.10.1 GigabitEthernet0/0/0
192.168.10.1/32     Direct  0    0          D    127.0.0.1    GigabitEthernet0/0/0
192.168.10.255/32   Direct  0    0          D    127.0.0.1    GigabitEthernet0/0/0
192.168.30.0/24     Direct  0    0          D    192.168.30.1 GigabitEthernet0/0/1
192.168.30.1/32     Direct  0    0          D    127.0.0.1    GigabitEthernet0/0/1
192.168.30.255/32   Direct  0    0          D    127.0.0.1    GigabitEthernet0/0/1
255.255.255.255/32  Direct  0    0          D    127.0.0.1    InLoopBack0
Direct routing table status : <Inactive>
         Destinations : 0        Routes : 0
```

路由表中各个字段的含义见表 2-4-5。

表 2-4-5 路由表中各字段的含义

Destination/Mask	目标网段/子网掩码	Flags	标志位
Proto	协议	NextHop	下一跳
Pre	优先级	Interface	出接口
Cost	开销值		

Destination/Mask：目标网段以及子网掩码，例如 192.168.50.0/24

Proto：显示路由条目的类型。直连路由：Direct，静态路由：Static，动态路由：RIP、OSPF。

Pre：数值越小，优先级别越高。默认情况下，直连路由是 0，静态路由是 60。

Cost：去往同一网络不同的路由，路由器会根据 cost 判断选出最优的路径，cost 越小，优先级越高。

Flags：路由标记。

NextHop：下一跳 IP 地址，相邻设备接口的 IP 地址。

Interface：出接口，从设备的该接口转发数据。

> 用PC1（IP：192.168.30.10）- ping - PC2（IP：192.168.40.10）能通吗？不通！为什么？因为直连路由无法跨设备通信。要实现全网互通就需要手工配置静态以及动态路由协议。
>
> PC1（IP：192.168.30.10）- ping - PC2（IP：192.168.40.10）- 不通
>
> ```
> PC>ping 192.168.40.10
> Ping 192.168.40.10: 32 data bytes, Press Ctrl_C to break
> Request timeout!
> Request timeout!
> --- 192.168.40.10 ping statistics ---
>   3 packet(s) transmitted
>   0 packet(s) received
>   100.00% packet loss
> ```

二、知识加油站

为什么ssh协议更安全？

由于telnet采用明文传输，无安全认证，存在很大安全隐患，出于安全方面考虑，尽量采用ssh加密协议进行远程控制。因为ssh采用非对称加密算法，需要两个秘钥：公开秘钥（publickey）和私有秘钥（privatekey），采用公开秘钥加密后的密文，只能通过对应的私有秘钥进行解密。

ssh协议分为（ssh1.5）与（ssh2.0），ssh2.0协议优于ssh1.5，在使用中使用ssh2.0协议。

使用非对称加密协议（ssh）登录流程步骤如下。

步骤1　通过ssh协议远程登录交换机10.10.4.30，交换机接收到电脑端访问登录请求，见图2-4-6，交换机将自己的公钥发给用户，用户访问需要接受公钥，见图2-4-7。

图2-4-6　登录界面

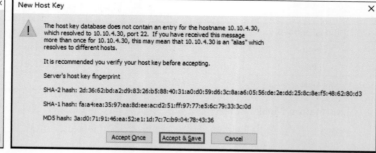

图2-4-7　接受公钥

步骤2　电脑端使用上述公钥，将数据进行加密发送给远程交换机。

步骤3　远程交换机用自己的私钥解密，然后验证其合法性。若密码正确，允许远程登录。因为私钥是交换机自己拥有，公钥与密码在网络中传输，如果泄密，因为没有私钥，也是无能为力，确保网络数据访问安全。

项目5　静态路由

静态路由（Static）配置简单，广泛应用于网络中，静态路由无需消耗网络设备的CPU资源来计算分析，但是静态路由需要网络管理员手动配置和维护，同时静态路由不能动态的根据现网拓扑的改变而改变。

静态路由格式：

[Huawei]ip route-static 目标网段 目标网段掩码 下一跳 IP 地址

项目简介

通过上节分析可知，数据不能全网互通，是因为路由表中缺少路由信息，通过配置静态路由或者动态路由（比如OSPF）都可以实现，本节重点学习如何配置静态路由。配置静态路由时，需要清楚目标网段以及下一跳地址，在路由器系统模式下进行配置。配置完毕，通过"dis ip routing-table"命令查看路由信息，检查配置是否正确。当路由器部署企业网络边界时，要通过配置默认路由，匹配"全世界"路由信息。在重要场合中，网络必须考虑链路冗余，在本节拓扑中，当AR2与AR3之间连接两条链路时，采用浮动路由（主备路由）实现数据通信，当其中一条链路故障时，数据会自动切换到另一条链路。

一、静态路由

1. 静态路由拓扑图

实现功能：如图2-5-1所示，通过路由器配置IP地址以及静态路由，实现各个网段之间网络互通，各个设备能互相ping通测试，网络通信正常。

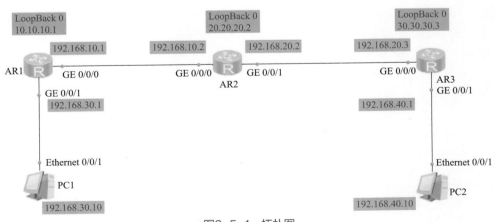

图2-5-1　拓扑图

67

2. 项目配置步骤

各个路由器的接口 IP 地址已经配置正确，物理以及协议全部 up，再进行静态路由配置。

步骤 1　AR1 静态路由配置

```
<AR1>system-view
# 以下四条是静态路由配置信息
[AR1]ip route-static 20.20.20.2 32 192.168.10.2
[AR1]ip route-static 192.168.20.0 24 192.168.10.2
[AR1]ip route-static 30.30.30.3 32 192.168.10.2
[AR1]ip route-static 192.168.40.0 24 192.168.10.2
[AR1]quit
<AR1>save
```

解释：

如果 AR1 路由器去往 AR2 的 192.168.20.0/24 网段，从拓扑图中可以看出，AR1 通往 AR2 路由器的下一跳是 AR2 g 0/0/0，这个接口的地址是 192.168.10.2。在路由器中配置如下：

```
[AR1]ip route-static 192.168.20.0 24 192.168.10.2。
```

步骤 2　AR2 静态路由配置

```
<AR2>system-view
[AR2]ip route-static 192.168.30.0 24 192.168.10.1
[AR2]ip route-static 10.10.10.1 32 192.168.10.1
[AR2]ip route-static 30.30.30.3 32 192.168.20.3
[AR2]ip route-static 192.168.40.0 24 192.168.20.3
[AR2]quit
<AR2>save
```

步骤 3　AR3 静态路由配置

```
<AR3>system-view
[AR3]ip route-static 192.168.10.0 24 192.168.20.2
[AR3]ip route-static 192.168.30.0 24 192.168.20.2
[AR3]ip route-static 20.20.20.2 32 192.168.20.2
[AR3]ip route-static 10.10.10.1 32 192.168.20.2
[AR3]quit
<AR3>save
```

步骤 4　再次测试通信

PC1（IP：192.168.30.10）- ping- PC2（IP：192.168.40.10）通了，就这么神奇？因为是在配置了静态路由信息后，各个路由器中都存在其他路由器中的路由信息，形成完整路由表，实现了全网互通。

```
PC>ping 192.168.40.10
Ping 192.168.40.10: 32 data bytes, Press Ctrl_C to break
From 192.168.40.10: bytes=32 seq=1 ttl=125 time=32 ms
……省略……
--- 192.168.40.10 ping statistics ---
  5 packet(s) transmitted
  5 packet(s) received
  0.00% packet loss
  round-trip min/avg/max = 31/31/32 ms
```

通过配置直连路由与静态路由实现了全网互通，各个路由器直连与静态路由信息见表 2-5-1。

表 2-5-1 路由信息

序号	设备	直连路由条目	静态路由条目（Static）
1	AR1	10.10.10.1/32 192.168.10.0/24 192.168.30.0/24	20.20.20.2/32 30.30.30.3/32 192.168.20.0/24 192.168.40.0/24
2	AR2	20.20.20.2/32 192.168.10.0/24 192.168.20.0/24	10.10.10.1/32 30.30.30.3/32 192.168.30.0/24 192.168.40.0/24
3	AR3	30.30.30.3/32 192.168.20.0/24 192.168.40.0/24	10.10.10.1/32 20.20.20.2/32 192.168.10.0/24 192.168.30.0/24

二、默认路由（又称缺省路由）

通过前面的表格分析，管理员配置的路由条目比较多，当网络环境比较大时，路由条目就非常繁杂，尤其是部署在企业出口的路由器，不可能明细化配置复杂的静态路由信息，一般在边界路由器配置默认路由，可以理解为默认路由可以通往"全世界"。默认路由也是静态路由的一种方式。

默认路由格式：

`[Huawei]ip route-static 0.0.0.0 0.0.0.0 下一跳 IP 地址`

目标网段与掩码中的"0"，表示任何值都可以，所以配置 0.0.0.0 能与任何目的地址匹配成功，形成默认路由要求的效果。

或者

`[Huawei]ip route-static 0.0.0.0 0 下一跳 IP 地址`

配置默认静态路由，掩码 0.0.0.0 可以简写为 0。

步骤 1 删除 AR1 路由器中原来配置的静态路由，重新配置默认路由。

```
# 查询现有的静态路由信息
[AR1]display current-configuration | include ip rou
ip route-static 20.20.20.2 255.255.255.255 192.168.10.2
ip route-static 30.30.30.3 255.255.255.255 192.168.10.2
ip route-static 192.168.20.0 255.255.255.0 192.168.10.2
ip route-static 192.168.40.0 255.255.255.0 192.168.10.2
```

"|"是重定向符，可以通过后面的命令过滤前面命令的输出。快速找到配置的静态路由。

`display current-configuration | include ip rou`

```
# 删除原来的静态路由，通过 undo 命令删除配置
[AR1]undo ip route-static 20.20.20.2 255.255.255.255 192.168.10.2
[AR1]undo ip route-static 30.30.30.3 255.255.255.255 192.168.10.2
[AR1]undo ip route-static 192.168.20.0 255.255.255.0 192.168.10.2
[AR1]undo ip route-static 192.168.40.0 255.255.255.0 192.168.10.2
# 再次查询静态路由是否删除完毕，查询静态路由删除完毕
[AR1]display current-configuration | include ip rou
[AR1]
# 配置默认路由
[AR1]ip route-static 0.0.0.0 0 192.168.10.2
[AR1]quit
<AR1>save
```

步骤 2　测试通信。

```
PC1（IP：192.168.30.10）- ping- PC2（IP：192.168.40.10）- 通
PC>ping 192.168.40.10
Ping 192.168.40.10: 32 data bytes, Press Ctrl_C to break
From 192.168.40.10: bytes=32 seq=1 ttl=125 time=16 ms
……省略……
--- 192.168.40.10 ping statistics ---
  5 packet(s) transmitted
  5 packet(s) received
  0.00% packet loss
  round-trip min/avg/max = 15/21/32 ms
```

同样的方法在 AR3 配置默认路由，但是不可以在 AR2 上配置。

为什么 AR2 上无法配置默认路由？因为 AR2 上配置默认路由有两条（指向两个地址），当数据包到达 AR2 的时候，AR2 并不清楚到底如何转发，会导致通信中断。

当路由器中默认路由与明细路由同时存在时，优先选择明细路由。

三、浮动路由（又称主备理由）

　　当 AR2 与 AR3 之间的两台路由配置的优先级不一样时，路由器选择优先级最高（Pre 数值越小越优先）的路由来转发报文，当路由优先级最高主路断开后，则备用线路激活，不影响数据转发。正常情况下高优先级的状态是 active（激活状态），其他处于 inactive（非激活状态），当高优先级的路由断开后，处于非激活状态就自动变为激活 active，当高优先级的路由恢复后，依然处于主路，转发数据。拓扑见图 2-5-2。

步骤 1　配置接口地址
AR2 路由器
```
<AR2>system-view
Enter system view, return user view with Ctrl+Z.
[AR2]interface GigabitEthernet 0/0/2
# 配置接口地址
[AR2-GigabitEthernet0/0/2]ip address 192.168.50.2 24
[AR2-GigabitEthernet0/0/2]quit
```

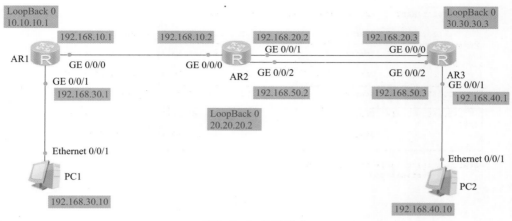

图2-5-2 浮动路由

```
[AR2]quit
<AR2>save
AR3 路由器
<AR3>system-view
Enter system view, return user view with Ctrl+Z.
[AR3]interface GigabitEthernet 0/0/2
[AR3-GigabitEthernet0/0/2]ip address 192.168.50.3 24
[AR3-GigabitEthernet0/0/2]quit
[AR3]quit
<AR3>save
```

步骤2　增加静态路由（6条）

AR2 路由器

```
[AR2]ip route-static 30.30.30.3 32 192.168.50.3
[AR2]ip route-static 192.168.40.0 24 192.168.50.3
```

AR3 路由器

```
[AR3]ip route-static 20.20.20.2 32 192.168.50.2
[AR3]ip route-static 10.10.10.1 32 192.168.50.2
[AR3]ip route-static 192.168.10.0 24 192.168.50.2
[AR3]ip route-static 192.168.30.0 24 192.168.50.2
```

步骤3　查看路由表

[AR2]dis ip routing-table protocol static

从图 2-5-3 中可以看出，通往 30.30.30.3/32 以及 192.168.40.0/24 都是两条路径。

[AR3]dis ip routing-table protocol static

从图 2-5-4 中可以看出通往 10.10.10.1/32、20.20.20.2/32、192.168.10.0/24、192.168.30.0/24 都是两条路径。

```
[AR2]dis ip routing-table protocol static
Route Flags: R - relay, D - download to fib
------------------------------------------------------------
Public routing table : Static
         Destinations : 4        Routes : 6       Configured Routes : 6

Static routing table status : <Active>
         Destinations : 4        Routes : 6
```

图2-5-3

```
Destination/Mask    Proto   Pre  Cost    Flags  NextHop         Interface
    10.10.10.1/32   Static  60   0         RD   192.168.10.1    GigabitEthernet
0/0/0
    30.30.30.3/32   Static  60   0         RD   192.168.20.3    GigabitEthernet
0/0/1
                    Static  60   0         RD   192.168.50.3    GigabitEthernet
0/0/2
    192.168.30.0/24 Static  60   0         RD   192.168.10.1    GigabitEthernet
0/0/0
    192.168.40.0/24 Static  60   0         RD   192.168.20.3    GigabitEthernet
0/0/1
                    Static  60   0         RD   192.168.50.3    GigabitEthernet
0/0/2

Static routing table status : <Inactive>
         Destinations : 0        Routes : 0
```

图2-5-3 链路冗余路由信息

```
[AR3]dis ip routing-table protocol static
Route Flags: R - relay, D - download to fib
------------------------------------------
Public routing table : Static
         Destinations : 4        Routes : 8        Configured Routes : 8

Static routing table status : <Active>
         Destinations : 4        Routes : 8

Destination/Mask    Proto   Pre  Cost    Flags  NextHop         Interface
    10.10.10.1/32   Static  60   0         RD   192.168.20.2    GigabitEthernet
0/0/0
                    Static  60   0         RD   192.168.50.2    GigabitEthernet
0/0/2
    20.20.20.2/32   Static  60   0         RD   192.168.20.2    GigabitEthernet
0/0/0
                    Static  60   0         RD   192.168.50.2    GigabitEthernet
0/0/2
    192.168.10.0/24 Static  60   0         RD   192.168.20.2    GigabitEthernet
0/0/0
                    Static  60   0         RD   192.168.50.2    GigabitEthernet
0/0/2
    192.168.30.0/24 Static  60   0         RD   192.168.20.2    GigabitEthernet
0/0/0
                    Static  60   0         RD   192.168.50.2    GigabitEthernet
0/0/2

Static routing table status : <Inactive>
         Destinations : 0        Routes : 0
```

图2-5-4 链路冗余路由信息

链路冗余路由信息见表 2-5-2。

表 2-5-2 链路冗余路由信息

设备	目标网段/掩码	下一跳地址	优先级	备注
AR2	10.10.10.1/32	192.168.10.1	60	
AR2	192.168.30.0/24	192.168.10.1	60	
AR2	30.30.30.3/24	192.168.20.3	60	
AR2	30.30.30.3/24	192.168.50.3	60	第二条支路路由
AR2	192.168.40.0/24	192.168.20.3	60	
AR2	192.168.40.0/24	192.168.50.3	60	第二条支路路由
AR3	10.10.10.1/32	192.168.20.2	60	
AR3	10.10.10.1/32	192.168.50.2	60	第二条支路路由

续表

设备	目标网段/掩码	下一跳地址	优先级	备注
AR3	20.20.20.2/32	192.168.20.2	60	
AR3	20.20.20.2/32	192.168.50.2	60	第二条支路路由
AR3	192.168.10.0/24	192.168.20.2	60	
AR3	192.168.10.0/24	192.168.50.2	60	第二条支路路由
AR3	192.168.30.0/24	192.168.20.2	60	
AR3	192.168.30.0/24	192.168.50.2	60	第二条支路路由

步骤 4 修改优先级别

通过修改 AR2 与 AR3 之间的静态路由的优先级别（数值越小越优先）来区分主线路与备用线路。默认静态路由优先级别是 60，将主链路静态路由优先级手动修改为 50。

AR2 路由器
```
[AR2]ip route-static 30.30.30.3 32 192.168.50.3 preference 50
[AR2]ip route-static 192.168.40.0 24 192.168.50.3 preference 50
```
AR3 路由器
```
[AR3]ip route-static 20.20.20.2 32 192.168.50.2 preference 50
[AR3]ip route-static 10.10.10.1 32 192.168.50.2 preference 50
[AR3]ip route-static 192.168.10.0 24 192.168.50.2 preference 50
[AR3]ip route-static 192.168.30.0 24 192.168.50.2 preference 50
```

步骤 5 重新查看路由表

AR2 路由表
```
[AR2]dis ip routing-table protocol static
```

通往 30.30.30.3/32、192.168.40.0/24，下一跳地址都是 192.168.50.3，处于 active 激活状态，见图 2-5-5。

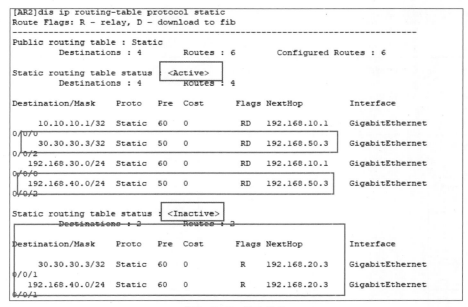

图2-5-5 路由信息

AR3 路由器
[AR3]display ip routing-table protocol static

通往 10.10.10.1/32、20.20.20.2/32、192.168.10.0/24、192.168.30.0/24 下一跳地址都是 192.168.50.2，处于 active 激活状态，见图 2-5-6。

```
[AR3]display ip routing-table protocol static
Route Flags: R - relay, D - download to fib
------------------------------------------------------------------------
Public routing table : Static
         Destinations : 4        Routes : 8        Configured Routes : 8

Static routing table status : <Active>
         Destinations : 4        Routes : 4

Destination/Mask    Proto   Pre  Cost      Flags NextHop         Interface
     10.10.10.1/32  Static  50   0          RD   192.168.50.2    GigabitEthernet0/0/2
     20.20.20.2/32  Static  50   0          RD   192.168.50.2    GigabitEthernet0/0/2
   192.168.10.0/24  Static  50   0          RD   192.168.50.2    GigabitEthernet0/0/2
   192.168.30.0/24  Static  50   0          RD   192.168.50.2    GigabitEthernet0/0/2

Static routing table status : <Inactive>
         Destinations : 4        Routes : 4

Destination/Mask    Proto   Pre  Cost      Flags NextHop         Interface
     10.10.10.1/32  Static  60   0          R    192.168.20.2    GigabitEthernet0/0/0
     20.20.20.2/32  Static  60   0          R    192.168.20.2    GigabitEthernet0/0/0
   192.168.10.0/24  Static  60   0          R    192.168.20.2    GigabitEthernet0/0/0
   192.168.30.0/24  Static  60   0          R    192.168.20.2    GigabitEthernet0/0/0

[AR3]
```

图 2-5-6　路由信息

目标网段优先级以及 IP 地址见表 2-5-3。

表 2-5-3　目标网段优先级以及 IP 地址

设备	目标网段/掩码	下一跳地址	优先级	备注
AR2	10.10.10.1/32	192.168.10.1	60	active
AR2	192.168.30.0/24	192.168.10.1	60	active
AR2	30.30.30.3/24	192.168.20.3	60	inactive
AR2	30.30.30.3/24	192.168.50.3	50	active
AR2	192.168.40.0/24	192.168.20.3	60	inactive
AR2	192.168.40.0/24	192.168.50.3	50	active
AR3	10.10.10.1/32	192.168.20.2	60	inactive
AR3	10.10.10.1/32	192.168.50.2	50	active
AR3	20.20.20.2/32	192.168.20.2	60	inactive

续表

设备	目标网段/掩码	下一跳地址	优先级	备注
AR3	20.20.20.2/32	192.168.50.2	50	active
AR3	192.168.10.0/24	192.168.20.2	60	inactive
AR3	192.168.10.0/24	192.168.50.2	50	active
AR3	192.168.30.0/24	192.168.20.2	60	inactive
AR3	192.168.30.0/24	192.168.50.2	50	active

从表 2-5-3 中可以看出，AR2- GigabitEthernet 0/0/2- AR3- GigabitEthernet 0/0/2 是主线路。依然测试 PC1-ping-PC2，当将主线路 AR2- GigabitEthernet 0/0/2 shutdown（关闭），观察通信情况，以及能否切换到备用线路 AR2- GigabitEthernet 0/0/1- AR3- GigabitEthernet 0/0/1。

AR2- GigabitEthernet 0/0/2，没有 shutdown 时，PC2-ping-PC1（持续 ping），见图 2-5-7。

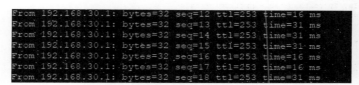

图 2-5-7　数据通信

```
# 进入 AR2 GigabitEthernet 0/0/2 接口
[AR2]interface GigabitEthernet 0/0/2
# 关闭接口（查看拓扑图，接口状态指示从绿色变为红色），见图 2-5-8。
[AR2-GigabitEthernet0/0/2]shutdown
```

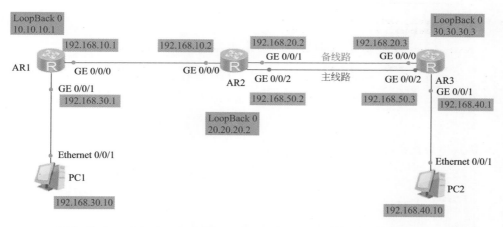

图 2-5-8　AR2-GigabitEthernet0/0/2 shutdown（接口状态由绿色变为红色）

AR2-GigabitEthernet0/0/2 shutdown 时，PC2-ping-PC1，从主线路切换到备用线路期间，数据通信，中间丢了几个数据包后，数据恢复正常通信，见图 2-5-9。

首先，观察 AR2-GigabitEthernet0/0/2 shutdown 后，主线路失效，观察 AR2 路由器的路由表，路由信息是 active 的，下一跳地址是 192.168.20.3，参照图 2-5-8，说明已经切换到备用线路，见图 2-5-10。

```
[AR2]display ip routing-table protocol static
```

```
From 192.168.30.1: bytes=32 seq=15 ttl=253 time=31 ms
From 192.168.30.1: bytes=32 seq=16 ttl=253 time=16 ms
From 192.168.30.1: bytes=32 seq=17 ttl=253 time=16 ms
From 192.168.30.1: bytes=32 seq=18 ttl=253 time=31 ms
From 192.168.30.1: bytes=32 seq=19 ttl=253 time=31 ms
From 192.168.30.1: bytes=32 seq=20 ttl=253 time=31 ms
From 192.168.30.1: bytes=32 seq=21 ttl=253 time=16 ms
From 192.168.30.1: bytes=32 seq=22 ttl=253 time=31 ms
From 192.168.30.1: bytes=32 seq=23 ttl=253 time=16 ms
From 192.168.30.1: bytes=32 seq=24 ttl=253 time=15 ms
From 192.168.30.1: bytes=32 seq=25 ttl=253 time=32 ms
From 192.168.30.1: bytes=32 seq=26 ttl=253 time=31 ms
Request timeout!
From 192.168.30.1: bytes=32 seq=28 ttl=253 time=16 ms
Request timeout!
From 192.168.30.1: bytes=32 seq=31 ttl=253 time=31 ms
From 192.168.30.1: bytes=32 seq=32 ttl=253 time=31 ms
From 192.168.30.1: bytes=32 seq=33 ttl=253 time=16 ms
From 192.168.30.1: bytes=32 seq=34 ttl=253 time=31 ms
From 192.168.30.1: bytes=32 seq=35 ttl=253 time=31 ms
From 192.168.30.1: bytes=32 seq=36 ttl=253 time=
```

图2-5-9 主备线路切换,丢掉几个数据包,然后数据通信正常

```
[AR2]display ip routing-table protocol static
Route Flags: R - relay, D - download to fib
------------------------------------------------------------
Public routing table : Static
         Destinations : 4      Routes : 6       Configured Routes : 6

Static routing table status : <Active>
         Destinations : 4      Routes : 4

Destination/Mask    Proto   Pre  Cost       Flags NextHop         Interface

     10.10.10.1/32  Static  60   0           RD   192.168.10.1    GigabitEthernet0/0/0
     30.30.30.3/32  Static  60   0           RD   192.168.20.3    GigabitEthernet0/0/1
    192.168.30.0/24 Static  60   0           RD   192.168.10.1    GigabitEthernet0/0/0
    192.168.40.0/24 Static  60   0           RD   192.168.20.3    GigabitEthernet0/0/1

Static routing table status : <Inactive>
         Destinations : 2      Routes : 2

Destination/Mask    Proto   Pre  Cost       Flags NextHop         Interface

     30.30.30.3/32  Static  50   0                192.168.50.3    Unknown
    192.168.40.0/24 Static  50   0                192.168.50.3    Unknown
```

图2-5-10 路由信息

[AR3]display ip routing-table protocol static

观察 AR3 路由器路由表,路由信息是 active 的,下一跳地址是 192.168.20.2。如图 2-5-11 所示。

```
[AR3]display ip routing-table protocol static
Route Flags: R - relay, D - download to fib
------------------------------------------------------------
Public routing table : Static
         Destinations : 4      Routes : 8       Configured Routes : 8

Static routing table status : <Active>
         Destinations : 4      Routes : 4

Destination/Mask    Proto   Pre  Cost       Flags NextHop         Interface

     10.10.10.1/32  Static  60   0           RD   192.168.20.2    GigabitEthernet0/0/0
     20.20.20.2/32  Static  60   0           RD   192.168.20.2    GigabitEthernet0/0/0
    192.168.10.0/24 Static  60   0           RD   192.168.20.2    GigabitEthernet0/0/0
```

```
     192.168.30.0/24  Static   60   0           RD   192.168.20.2   GigabitEthernet
0/0/0
Static routing table status : <Inactive>
         Destinations : 4        Routes : 4
Destination/Mask      Proto   Pre  Cost      Flags  NextHop       Interface
     10.10.10.1/32    Static   50   0               192.168.50.2  Unknown
     20.20.20.2/32    Static   50   0               192.168.50.2  Unknown
   192.168.10.0/24    Static   50   0               192.168.50.2  Unknown
   192.168.30.0/24    Static   50   0               192.168.50.2  Unknown
```

图2-5-11　路由信息

其次，tracert 命令路由追踪（AR2-GigabitEthernet0/0/2 shutdown）- 走备用线路
[AR3]tracert 192.168.30.1
traceroute to 192.168.30.1(192.168.30.1), max hops: 30 ,packet length: 40,press CTRL_C to break
1 192.168.20.2 20 ms   20 ms   20 ms
2 192.168.10.1 20 ms   20 ms   30 ms

从路由追踪上看，数据通信走的是备用线路。
当 AR2- GigabitEthernet0/0/2 接口恢复正常后。
[AR2-GigabitEthernet0/0/2]undo shutdown
再次观察 tracert 路由追踪情况。
[AR3]tracert 192.168.30.1
traceroute to 192.168.30.1(192.168.30.1), max hops: 30 ,packet length: 40,press CTRL_C to break
1 192.168.50.2 20 ms   20 ms   20 ms
2 192.168.10.1 20 ms   30 ms   20 ms

从路由追踪上看，数据通信走的是主线路。

项目6　交换机 MAC 地址表

在网络中建设中，交换机是最基础的网络设备，它基于 MAC 地址进行数据转发，交换机中 MAC 地址表并非一成不变，MAC 地址表不断更新，每一条 MAC 地址条目，生存时间默认是300s，到达时间没有刷新（通信）将删除。交换机使用时间越长，MAC 地址表条目越多，不需要发送的广播报文就越少，数据转发也就越迅速。

交换机的每个接口属于不同的冲突域，在网络中起到隔离冲突域。

交换机工作时，在 MAC 地址表中查找数据帧中的目标 MAC 地址，如果有，就将该数据帧发送到相应的交换机端口，数据转发；如果没有，就发送广播报文，

交换机向入端口以外的所有端口发送广播，查询目标 MAC 地址，找出对应关系，然后源 MAC 地址与交换机源端口、目的 MAC 与交换机出端口，在 MAC 地址表中形成对应关系。

一、神奇的 MAC 地址表

1. 拓扑图

如图 2-6-1 所示。

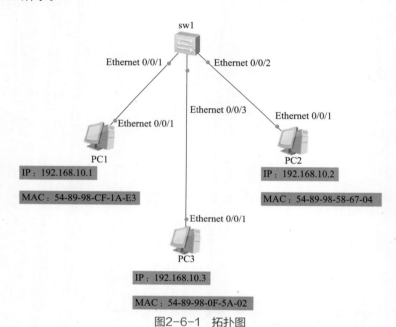

图2-6-1 拓扑图

地址规划如表 2-6-1 所示。

表 2-6-1 地址规划表

PC	IP 地址	MAC 地址	子网掩码
PC1	192.168.10.1	54-89-98-CF-1A-E3	255.255.255.0
PC2	192.168.10.2	54-89-98-58-67-04	255.255.255.0
PC3	192.168.10.3	54-89-98-0F-5A-02	255.255.255.0

2. 项目基本配置与查询 MAC 地址表

步骤1　查询 MAC 地址表

```
<Huawei>system-view
Enter system view, return user view with Ctrl+Z.
[Huawei]sysname sw1
[sw1]
```

```
# 输入"display mac-address"查询 mac 地址表
[sw1]display mac-address
# 目前没有任何信息
[sw1]
```

步骤2 配置 PC1 与 PC2、PC3 电脑 IP 地址等信息，参照图 2-6-2。

(a) PC1 IP地址配置

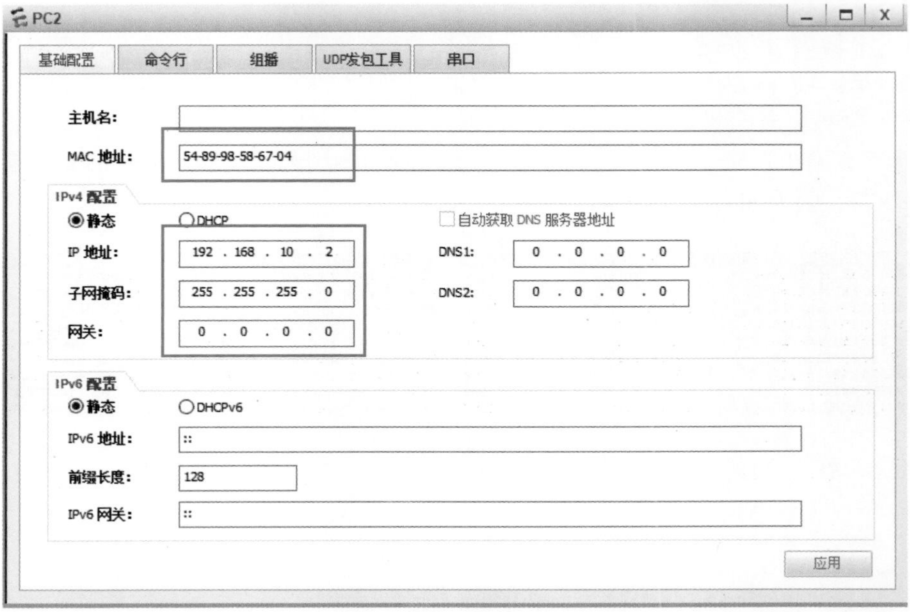

(b) PC2 IP地址配置

图2-6-2

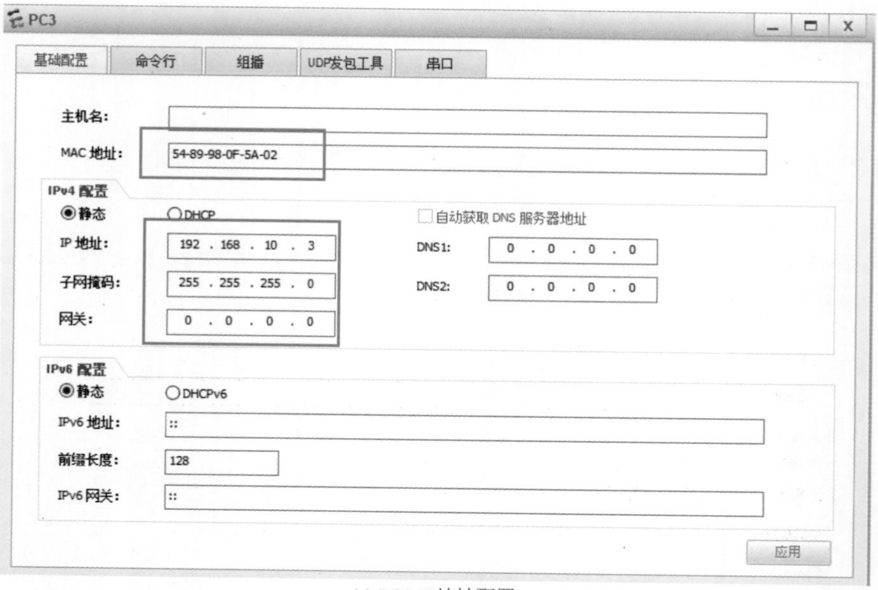

(c) PC3 IP地址配置

图2-6-2　IP地址

步骤3　ping 测试网络

PC1（IP：192.168.10.1）-ping-PC2（IP：192.168.10.2）-通
只有数据通信后，交换机才能建立 MAC 地址表，所以先执行 ping 命令。
```
PC>ping 192.168.10.2
From 192.168.10.2: bytes=32 seq=1 ttl=128 time=46 ms
From 192.168.10.2: bytes=32 seq=2 ttl=128 time=32 ms
……省略……

--- 192.168.10.2 ping statistics ---
  5 packet(s) transmitted
  5 packet(s) received
  0.00% packet loss
round-trip min/avg/max = 31/37/47 ms
```

以上可以看出，ping 通了 5 次，在 sw1-Ethernet 0/0/1 接口，通过 Wireshark 抓包软件看到 10 条报文，因为 ping 通 1 次包含来回两个报文。

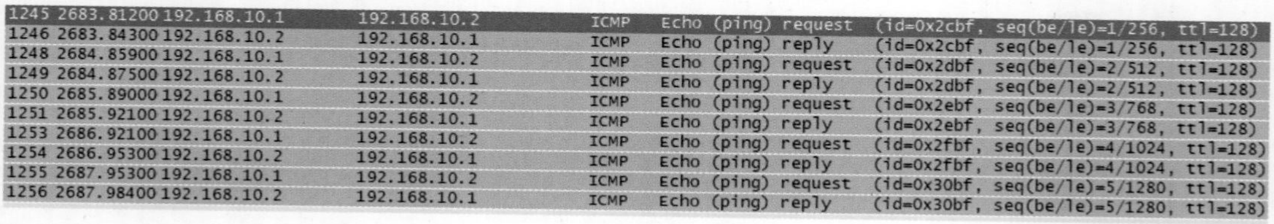

```
1245 2683.81200 192.168.10.1    192.168.10.2    ICMP   Echo (ping) request  (id=0x2cbf, seq(be/le)=1/256, ttl=128)
1246 2683.84300 192.168.10.2    192.168.10.1    ICMP   Echo (ping) reply    (id=0x2cbf, seq(be/le)=1/256, ttl=128)
1248 2684.85900 192.168.10.1    192.168.10.2    ICMP   Echo (ping) request  (id=0x2dbf, seq(be/le)=2/512, ttl=128)
1249 2684.87500 192.168.10.2    192.168.10.1    ICMP   Echo (ping) reply    (id=0x2dbf, seq(be/le)=2/512, ttl=128)
1250 2685.89000 192.168.10.1    192.168.10.2    ICMP   Echo (ping) request  (id=0x2ebf, seq(be/le)=3/768, ttl=128)
1251 2685.92100 192.168.10.2    192.168.10.1    ICMP   Echo (ping) reply    (id=0x2ebf, seq(be/le)=3/768, ttl=128)
1253 2686.92100 192.168.10.1    192.168.10.2    ICMP   Echo (ping) request  (id=0x2fbf, seq(be/le)=4/1024, ttl=128)
1254 2686.95300 192.168.10.2    192.168.10.1    ICMP   Echo (ping) reply    (id=0x2fbf, seq(be/le)=4/1024, ttl=128)
1255 2687.95300 192.168.10.1    192.168.10.2    ICMP   Echo (ping) request  (id=0x30bf, seq(be/le)=5/1280, ttl=128)
1256 2687.98400 192.168.10.2    192.168.10.1    ICMP   Echo (ping) reply    (id=0x30bf, seq(be/le)=5/1280, ttl=128)
```

步骤4　输入"display mac-address"命令，再次查询 MAC 表。

```
[sw1]display mac-address
MAC address table of slot 0:
-------------------------------------------------------------------------------
MAC Address      VLAN/         PEVLAN CEVLAN Port          Type       LSP/LSR-ID
                 VSI/SI                                               MAC-Tunnel
-------------------------------------------------------------------------------
```

```
5489-98cf-1ae3 1          -         -       Eth0/0/1        dynamic    0/-
5489-9858-6704 1          -         -       Eth0/0/2        dynamic    0/-
------------------------------------------------------------------------
Total matching items on slot 0 displayed = 2
```

步骤 5　抓包分析 MAC 地址建立过程

交换机 sw1 的 Ethernet 0/0/1 抓包观察，在 Wireshark 软件中输入 arp or icmp，过滤只显示相应的报文，如图 2-6-3 所示。

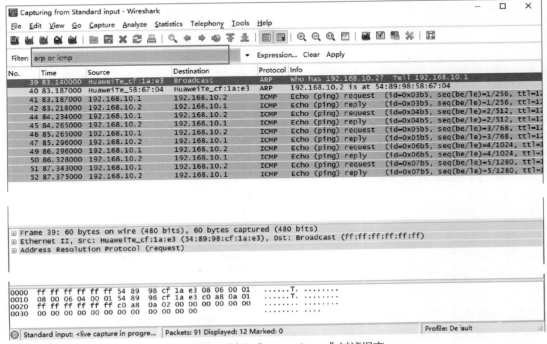

图2-6-3　输入"arp or icmp"过滤报文

第 39 条报文分析：PC1（IP:192.168.10.1、Mac: 5489-98cf-1ae3）与 PC2（IP：192.168.10.2）通信，因不清楚 PC2 的 MAC 地址，因此发送广播报文，如图 2-6-4 所示。

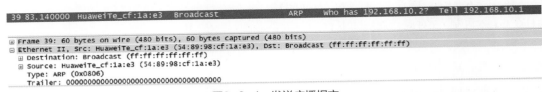

图2-6-4　发送广播报文

  谁是 192.168.10.2，请告诉 192.168.10.1，方式是通过广播形式发出去的（MAC 地址：ff:ff:ff:ff:ff:ff），当 PC2 接收到广播"通知"后，核对自己的 IP 地址就是 192.168.10.2。

40 条报文分析：如图 2-6-5 所示。

```
40 83.187000  HuaweiTe_58:67:04      HuaweiTe_cf:1a:e3    ARP    192.168.10.2 is at 54:89:98:58:67:04
⊞ Frame 40: 60 bytes on wire (480 bits), 60 bytes captured (480 bits)
⊟ Ethernet II, Src: HuaweiTe_58:67:04 (54:89:98:58:67:04), Dst: HuaweiTe_cf:1a:e3 (54:89:98:cf:1a:e3)
  ⊞ Destination: HuaweiTe_cf:1a:e3 (54:89:98:cf:1a:e3)
  ⊞ Source: HuaweiTe_58:67:04 (54:89:98:58:67:04)
    Type: ARP (0x0806)
    Trailer: 000000000000000000000000000000000000
```

图2-6-5　PC2的MAC地址

 192.168.10.2 的 MAC 地址：5489-9858-6704。

```
# 查询默认老化时间
[sw1]display mac-address aging-time
# 默认 300s
Aging time: 300 seconds
# 修改老化时间，默认单位是 s（秒）
[sw1]mac-address aging-time 100
# 验证查询是否修改正确
[sw1]display mac-address aging-time
  Aging time: 100 seconds
```

二、MAC 地址

MAC 地址又名物理地址，联网设备在网络中传输信息的时候，包含源地址与目的地址。每台物理设备上都有一个全球唯一的物理地址（MAC），相当于身份证，对应 TCP/IP 协议中的数据链路层。它的长度是 48bit，十六进制数表示，前 24 位是组织唯一标识符，由 IEEE 统一分配给生产商，后 24 位生产厂家自行分配给产品网卡的唯一数值。网卡的物理地址通常由生产厂家烧录到芯片中。

在图 2-6-6 中，物理地址为 6C4B-909D-41F6（十六进制），对应二进制如下。

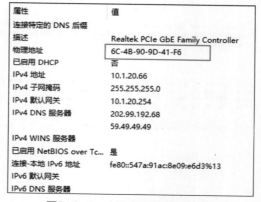

图2-6-6　查询网卡中MAC地址

0110 1100 0100 1011 1001 0000 1001 1101 0100 0001 1111 0110（二进制）

交换机中常用的几个命令介绍如下。

1. 查看交换机全局下 mac-address

命令：display mac-address

```
[Access]display mac-address
MAC ADDR          VLAN ID    STATE      PORT INDEX                 AGING TIME(s)
0003-0f02-daf0    1          LEARNED    GigabitEthernet1/0/24      AGING
0003-0f02-dc58    1          LEARNED    GigabitEthernet1/0/24      AGING
0003-0f02-ddae    1          LEARNED    GigabitEthernet1/0/24      AGING
```

2. 查看交换机某个 vlan 对应的 mac-address

命令：mac-address vlan XXX

```
[Access]display mac-address vlan 4053
MAC ADDR          VLAN ID   STATE      PORT INDEX               AGING TIME(s)
3cd2-e5a8-3630    4053      LEARNED    GigabitEthernet1/0/44    AGING
3cd2-e5a1-57c0    4053      LEARNED    GigabitEthernet3/0/24    AGING
```

3. 查询指定 mac-address 相关信息（在运维中经常用到该命令，进行 port 端口 vlan 修改）

命令：`display mac-address XXXX- XXXX - XXXX`

```
[Access]display mac-address 6C4B-909D-41F6
MAC ADDR          VLAN ID   STATE      PORT INDEX               AGING TIME(s)
6c4b-909d-41f6    135       LEARNED    GigabitEthernet1/0/35    AGING
```

通过查询，6C4B-909D-41F6 信息为 vlan: 135 port: GigabitEthernet1/0/35。

项目 7　vlan 与交换机端口模式 Access、Trunk

　　交换机每个接口处于不同的冲突域中，但是还是在一个广播域中，当网络中发送一个广播报文后，网络内的终端都能收到，这样消耗网络宽带资源。vlan（Virtual Local Area Network）的中文名为"虚拟局域网"，vlan 将局域网设备从逻辑上划分成一个个网段，vlan 技术主要解决了交换机无法隔离广播域的问题，主要原理是从逻辑上建立多个虚拟局域网，相同 vlan 用户互相通信正常，不同则 vlan 用户之间无法直接通信，其中一个 vlan 就是一个广播域。

　　交换机端口工作模式：如果交换机接口只允许一个 vlan 通过，则交换机接口是 Access 模式，一般用于连接计算机的端口；如果交换机接口允许多个 vlan 通过，则交换机接口模式是 Trunk。交换机与交换机之间连接的链路，称为主干链路。

　　交换机中配置 interface Vlanif，并且在 Vlanif 配置 IP 地址，作为网段网关地址，各个 PC 在配置 IP 地址时，应配置相应网关地址，实现数据互通。

　　当交换机接口连接 PC 终端时，接口配置为 Access 模式，只允许一个 vlan 通过。

　　当交换机之间数据通信，互联接口配置为 Trunk 模式，并且允许相关 vlan 通过。

一、vlan 与交换机端口模式 Access、Trunk

1. 控制拓扑图

实现功能：如图 2-7-1 所示。

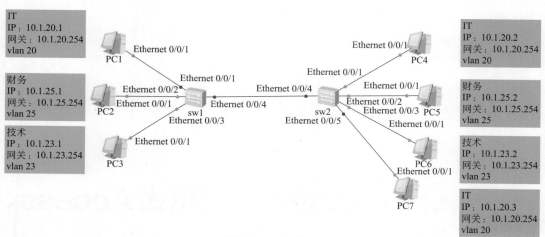

图2-7-1　拓扑图

IP 地址以及网关、vlan 如表 2-7-1 所示，子网掩码 255.255.255.0。

表 2-7-1　IP 地址以及网关、vlan

序号	PC	IP	网关	vlan
1	PC1	10.1.20.1	10.1.20.254	vlan20
2	PC2	10.1.25.1	10.1.25.254	vlan25
3	PC3	10.1.23.1	10.1.23.254	vlan23
4	PC4	10.1.20.2	10.1.20.254	vlan20
5	PC5	10.1.25.2	10.1.25.254	vlan25
6	PC6	10.1.23.2	10.1.23.254	vlan23
7	PC7	10.1.20.3	10.1.20.254	vlan20

2. 项目配置步骤

步骤 1　PC1 ～ PC7 的配置见图 2-7-2。

(a) PC1基本配置

图2-7-2

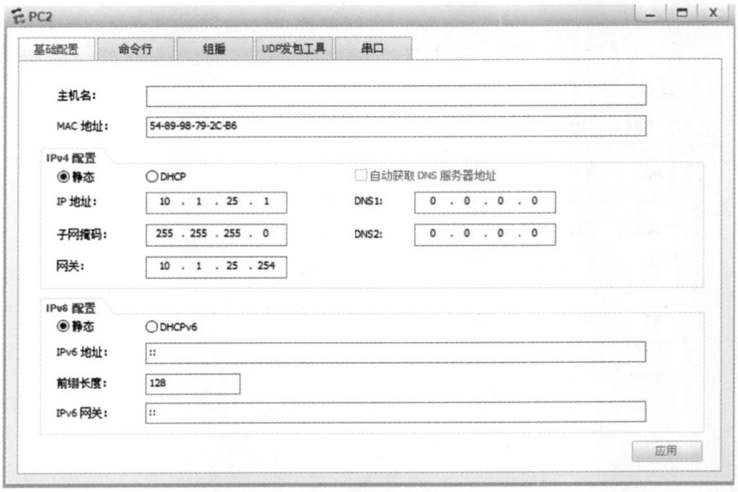

(b) PC2基本配置

(c) PC3基本配置

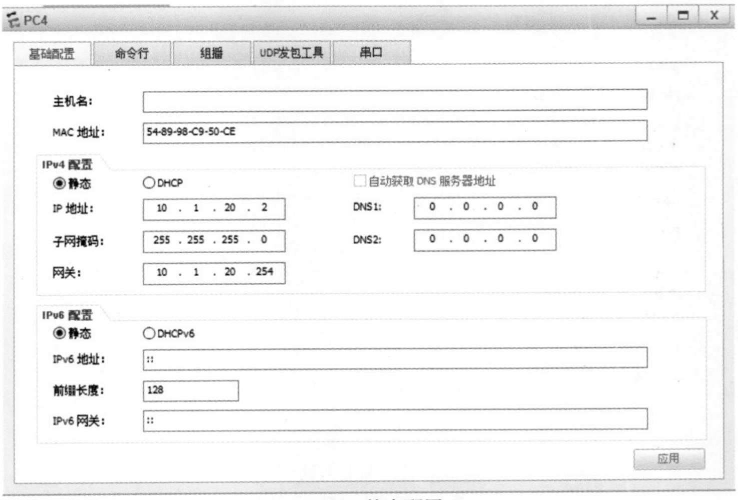

(d) PC4基本配置

图2-7-2

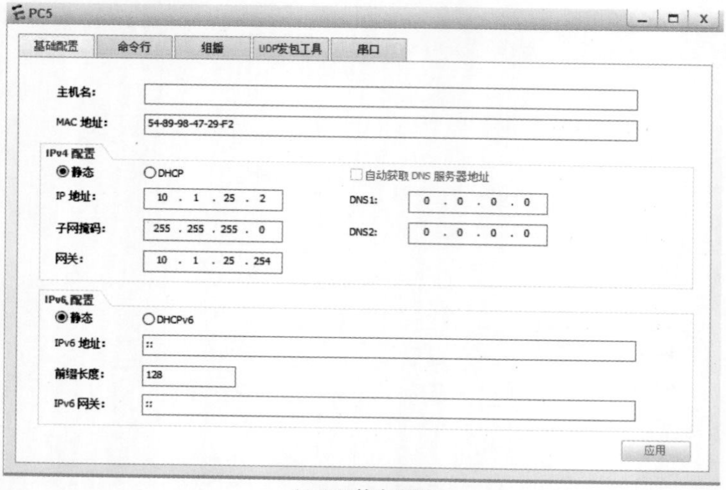

(e) PC5基本配置

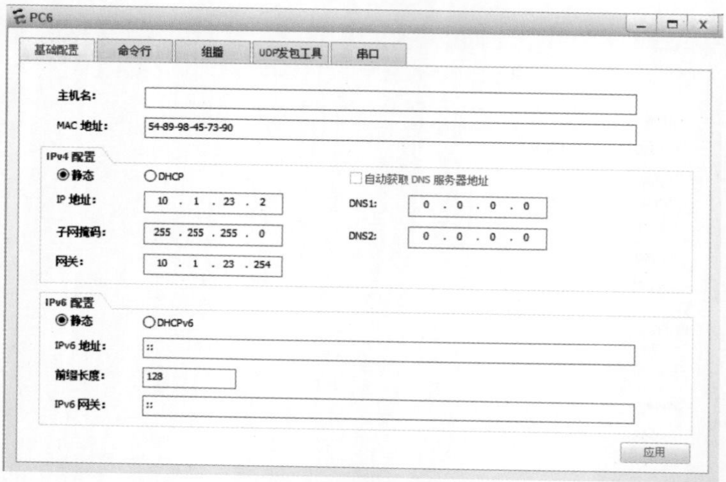

(f) PC6基本配置

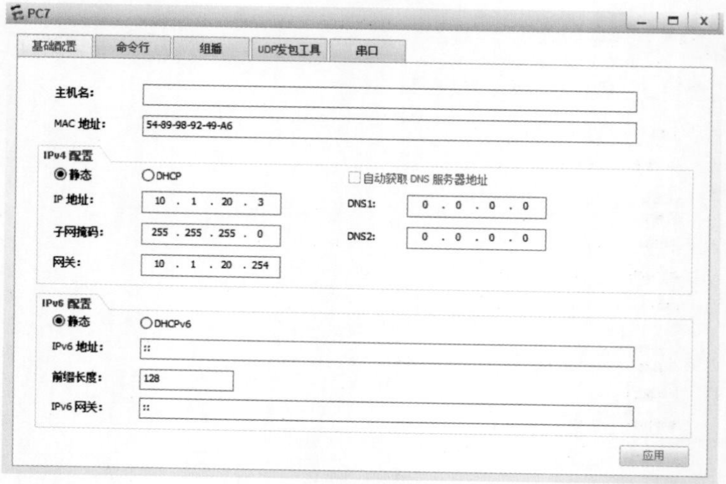

(g) PC7基本配置

图2-7-2　PC1～PC7的配置

步骤2 交换机配置。

sw1 基本配置

```
<Huawei>system-view
Enter system view, return user view with Ctrl+Z.
[Huawei]sysname sw1
# 创建vlan，如创建多个vlan，基本格式为vlan batch xx xx xx
[sw1]vlan batch 20 23 25
Info: This operation may take a few seconds. Please wait for a moment...done.
[sw1]interface Ethernet 0/0/1
# 因为接口连接是PC，接口类型修改为Access
[sw1-Ethernet0/0/1]port link-type access
# 允许vlan 20 通过
[sw1-Ethernet0/0/1]port default vlan 20
[sw1-Ethernet0/0/1]quit
# 以下是配置Ethernet 0/0/2，接口类型是Access，允许vlan 25通过
[sw1]interface Ethernet 0/0/2
[sw1-Ethernet0/0/2]port link-type access
[sw1-Ethernet0/0/2]port default vlan 25
[sw1-Ethernet0/0/2]quit
# 以下是配置Ethernet 0/0/3，接口类型是Access，允许vlan 23通过
[sw1]interface Ethernet 0/0/3
[sw1-Ethernet0/0/3]port link-type access
[sw1-Ethernet0/0/3]port default vlan 23
[sw1-Ethernet0/0/3]quit
# 以下是配置Ethernet 0/0/4，接口类型是Trunk，允许多个vlan 20 23 25通过
[sw1]interface Ethernet 0/0/4
[sw1-Ethernet0/0/4]port link-type trunk
[sw1-Ethernet0/0/4]port trunk allow-pass vlan 20 23 25
[sw1-Ethernet0/0/4]quit
[sw1]quit
<sw1>save
```

sw2 基本配置

```
<Huawei>system-view
Enter system view, return user view with Ctrl+Z.
[Huawei]sysname sw2
[sw2]
[sw2]vlan batch 20 23 25
Info: This operation may take a few seconds. Please wait for a moment...done.
[sw2]interface Ethernet 0/0/1
[sw2-Ethernet0/0/1]port link-type access
[sw2-Ethernet0/0/1]port default vlan 20
[sw2-Ethernet0/0/1]quit
[sw2]interface Ethernet 0/0/2
[sw2-Ethernet0/0/2]port link-type access
[sw2-Ethernet0/0/2]port default vlan 25
[sw2-Ethernet0/0/2]quit
[sw2]interface Ethernet 0/0/3
[sw2-Ethernet0/0/3]port link-type access
[sw2-Ethernet0/0/3]port default vlan 23
[sw2-Ethernet0/0/3]quit
[sw2]interface Ethernet 0/0/4
```

```
[sw2-Ethernet0/0/4]port link-type trunk
[sw2-Ethernet0/0/4]port trunk allow-pass vlan 20 23 25
[sw2-Ethernet0/0/4]quit
[sw2]quit
[sw2]interface Ethernet 0/0/5
[sw2-Ethernet0/0/5]port link-type access
[sw2-Ethernet0/0/5]port default vlan 20
[sw2-Ethernet0/0/5]quit
<sw2>save
```

步骤3　通过抓取报文观察相同 vlan 之间的通信过程。

观察报文之前，先执行 PC7（IP:10.1.20.3）-ping-PC1（IP:10.1.20.1）- 通。

① 在 sw2-Ethernet 0/0/5，抓取报文（过滤 arp or icmp 报文），见图2-7-3。

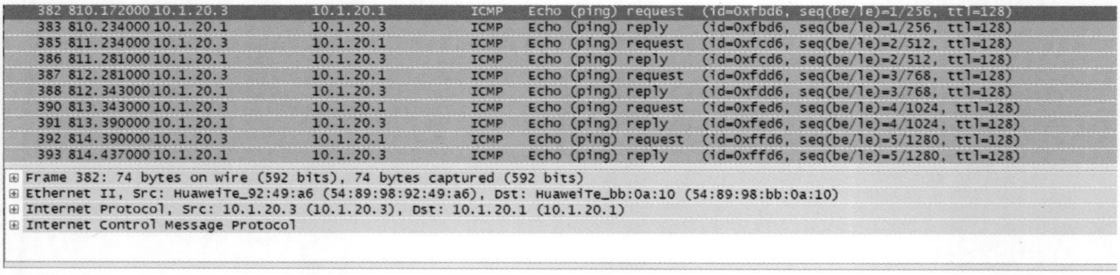

图2-7-3　sw2- Ethernet 0/0/5 报文信息

从报文中看出不带任何标签。

② sw2- Ethernet 0/0/4 抓取报文（arp or icmp），标签是 20。见图 2-7-4。

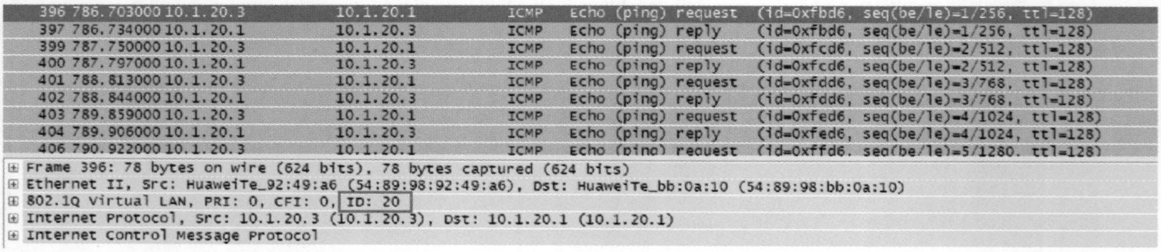

图2-7-4　sw2- Ethernet 0/0/4 报文信息

③ sw1- Ethernet 0/0/4 抓取报文（arp or icmp），标签是 20。见图 2-7-5。

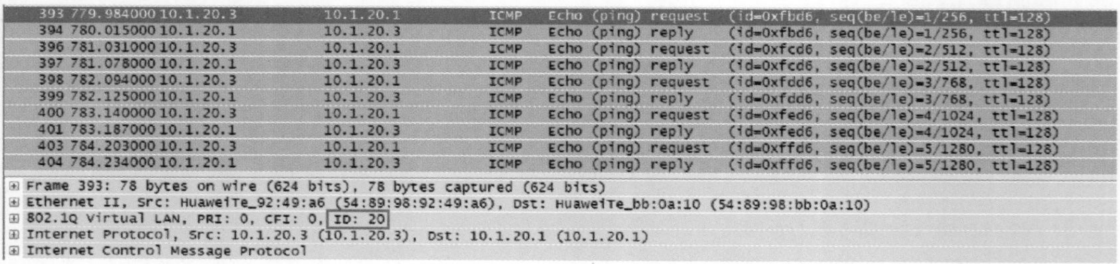

图2-7-5　sw1- Ethernet 0/0/4 抓取报文

④ sw1- Ethernet 0/0/1 抓取报文（arp or icmp），从报文中看出不带任何标签。见图 2-7-6。

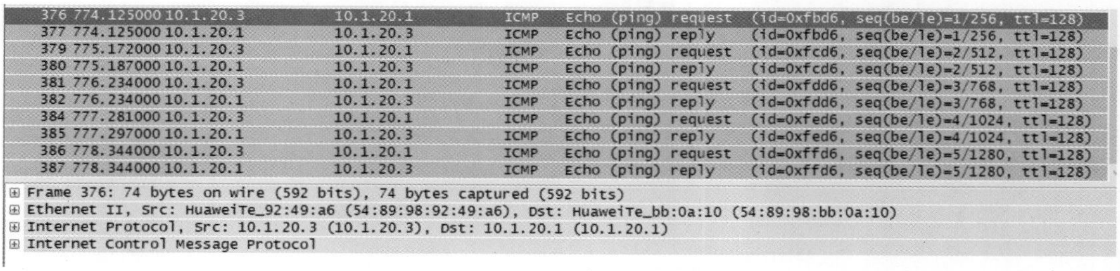

```
376 774.125000 10.1.20.3      10.1.20.1      ICMP    Echo (ping) request (id=0xfbd6, seq(be/le)=1/256, ttl=128)
377 774.125000 10.1.20.1      10.1.20.3      ICMP    Echo (ping) reply   (id=0xfbd6, seq(be/le)=1/256, ttl=128)
379 775.172000 10.1.20.3      10.1.20.1      ICMP    Echo (ping) request (id=0xfcd6, seq(be/le)=2/512, ttl=128)
380 775.187000 10.1.20.1      10.1.20.3      ICMP    Echo (ping) reply   (id=0xfcd6, seq(be/le)=2/512, ttl=128)
381 776.234000 10.1.20.3      10.1.20.1      ICMP    Echo (ping) request (id=0xfdd6, seq(be/le)=3/768, ttl=128)
382 776.234000 10.1.20.1      10.1.20.3      ICMP    Echo (ping) reply   (id=0xfdd6, seq(be/le)=3/768, ttl=128)
384 777.281000 10.1.20.3      10.1.20.1      ICMP    Echo (ping) request (id=0xfed6, seq(be/le)=4/1024, ttl=128)
385 777.297000 10.1.20.1      10.1.20.3      ICMP    Echo (ping) reply   (id=0xfed6, seq(be/le)=4/1024, ttl=128)
386 778.344000 10.1.20.3      10.1.20.1      ICMP    Echo (ping) request (id=0xffd6, seq(be/le)=5/1280, ttl=128)
387 778.344000 10.1.20.1      10.1.20.3      ICMP    Echo (ping) reply   (id=0xffd6, seq(be/le)=5/1280, ttl=128)
⊞ Frame 376: 74 bytes on wire (592 bits), 74 bytes captured (592 bits)
⊞ Ethernet II, Src: HuaweiTe_92:49:a6 (54:89:98:92:49:a6), Dst: HuaweiTe_bb:0a:10 (54:89:98:bb:0a:10)
⊞ Internet Protocol, Src: 10.1.20.3 (10.1.20.3), Dst: 10.1.20.1 (10.1.20.1)
⊞ Internet Control Message Protocol
```

图2-7-6　sw1- Ethernet 0/0/1 抓取报文

⑤ PC1-ping-PC2、PC1-ping-PC3 情况。

```
PC1（IP：10.1.20.1）-ping-PC2（IP：10.1.25.1）-不通
PC>ping 10.1.25.1
Ping 10.1.25.1: 32 data bytes, Press Ctrl_C to break
From 10.1.20.1: Destination host unreachable
……省略……

--- 10.1.20.254 ping statistics ---
   5 packet(s) transmitted
   0 packet(s) received
   100.00% packet loss
PC>ping 10.1.23.1
PC1（IP：10.1.20.1）-ping-PC3(IP：10.1.23.1)-不通
Ping 10.1.23.1: 32 data bytes, Press Ctrl_C to break
From 10.1.20.1: Destination host unreachable
……省略……

--- 10.1.20.254 ping statistics ---
   5 packet(s) transmitted
   0 packet(s) received
   100.00% packet loss
```

为什么呢？因为没有配置网关地址，请看下面步骤。

> 在进行操作步骤之前，首先了解网关是什么。
> 当两个不同网段的设备之间需要进行数据相互访问的时候，数据都是交给网关来处理，这个数据会从网关发出去。如果还是不理解，打个比方，两个大办公室，人员互相走动到彼此办公室，门口就相当于是个网关。网络中三层（理解为 IP 地址通信）数据转发，需要配置网关地址。比如 192.168.10.0/24 网段，当配置 IP：192.168.10.1 为该网段的网关地址时，IP 192.168.10.1 就不允许分给其他设备再使用。

步骤4　配置网关。

企业网络中网络架构一般分为接入层、汇聚层、核心层。在图 2-7-1 中增加汇聚层，交换机的虚拟接口 (interface Vlanif) 配置 IP 地址，作为各个网段对应 IP 地址的网关，当不同网段之间终端通信的时候，首先将数据发送给相应的网关。见图 2-7-7。

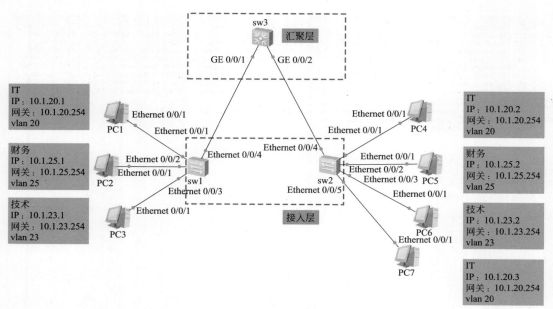

图2-7-7 网络架构

sw3-汇聚交换机配置

```
<Huawei>system-view
Enter system view, return user view with Ctrl+Z.
[Huawei]sysname sw3
[sw3]vlan batch 20 23 25
Info: This operation may take a few seconds. Please wait for a moment...done.
# 以下是配置Ethernet 0/0/1，接口类型是Trunk，允许多个vlan 20 23 25通过
[sw3]interface GigabitEthernet 0/0/1
[sw3-GigabitEthernet0/0/1]port link-type trunk
[sw3-GigabitEthernet0/0/1]port trunk allow-pass vlan 20 23 25
[sw3-GigabitEthernet0/0/1]quit
# 关闭信息提示。
[sw3]undo info-center enable
Info: Information center is disabled.
# 以下是配置Ethernet 0/0/2，接口类型是Trunk，允许多个vlan 20 23 25通过
[sw3]interface GigabitEthernet 0/0/2
[sw3-GigabitEthernet0/0/2]port link-type trunk
[sw3-GigabitEthernet0/0/2]port trunk allow-pass vlan 20 23 25
[sw3-GigabitEthernet0/0/2]quit
# 创建Vlan虚拟接口IP地址（作为Vlan20对应网段的网关）
[sw3]interface Vlanif 20
[sw3-Vlanif20]ip address 10.1.20.254 24
[sw3-Vlanif20]quit
# 创建Vlan虚拟接口IP地址（作为Vlan25对应网段的网关）
[sw3]interface Vlanif 25
[sw3-Vlanif25]ip address 10.1.25.254 24
[sw3-Vlanif25]quit
# 创建Vlan虚拟接口IP地址（作为Vlan23对应网段的网关）
[sw3]interface Vlanif 23
[sw3-Vlanif23]ip address 10.1.23.254 24
```

```
[sw3-Vlanif23]quit
[sw3]quit
<sw3>save
```

步骤5 通信测试。

PC1（IP：10.1.20.1）-ping-PC3(IP：10.1.23.1) - 通
```
PC>ping 10.1.23.1
Ping 10.1.23.1: 32 data bytes, Press Ctrl_C to break
From 10.1.23.1: bytes=32 seq=1 ttl=127 time=79 ms
……省略……

--- 10.1.23.1 ping statistics ---
  5 packet(s) transmitted
  5 packet(s) received
  0.00% packet loss
  round-trip min/avg/max = 62/75/79 ms
```
PC1（IP：10.1.20.1）-ping-PC2(IP：10.1.25.1) - 通
```
PC>ping 10.1.25.1
Ping 10.1.25.1: 32 data bytes, Press Ctrl_C to break
From 10.1.25.1: bytes=32 seq=1 ttl=127 time=79 ms
……省略……

--- 10.1.25.1 ping statistics ---
  5 packet(s) transmitted
  5 packet(s) received
  0.00% packet loss
  round-trip min/avg/max = 62/71/79 ms
```
PC1（IP：10.1.20.1）-ping-PC5(IP：10.1.25.2)-通
```
PC>ping 10.1.25.2
Ping 10.1.25.2: 32 data bytes, Press Ctrl_C to break
From 10.1.25.2: bytes=32 seq=1 ttl=127 time=78 ms
……省略……

--- 10.1.25.2 ping statistics ---
  5 packet(s) transmitted
  5 packet(s) received
  0.00% packet loss
  round-trip min/avg/max = 63/75/78 ms
```

步骤6 报文分析。

首先执行 PC1（IP：10.1.20.1）-ping-PC5(IP：10.1.25.2)，产生流量，见图2-7-8。

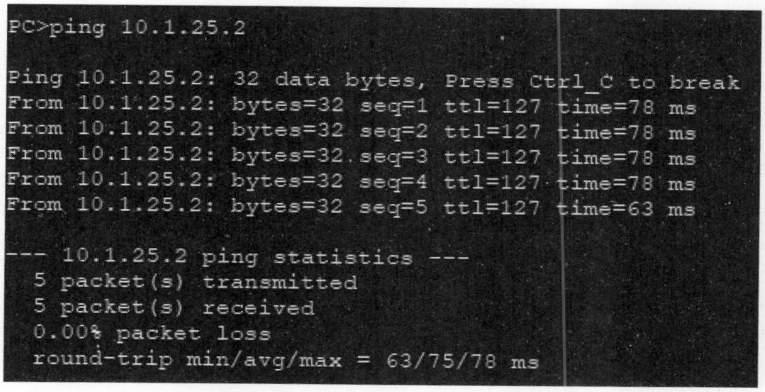

```
PC>ping 10.1.25.2
Ping 10.1.25.2: 32 data bytes, Press Ctrl_C to break
From 10.1.25.2: bytes=32 seq=1 ttl=127 time=78 ms
From 10.1.25.2: bytes=32 seq=2 ttl=127 time=78 ms
From 10.1.25.2: bytes=32 seq=3 ttl=127 time=78 ms
From 10.1.25.2: bytes=32 seq=4 ttl=127 time=78 ms
From 10.1.25.2: bytes=32 seq=5 ttl=127 time=63 ms

--- 10.1.25.2 ping statistics ---
  5 packet(s) transmitted
  5 packet(s) received
  0.00% packet loss
  round-trip min/avg/max = 63/75/78 ms
```

图2-7-8 PC1（IP: 10.1.20.1）-ping-PC5(IP: 10.1.25.2)

sw1- Ethernet 0/0/1 抓取报文（arp or icmp），不带标签，见图 2-7-9。

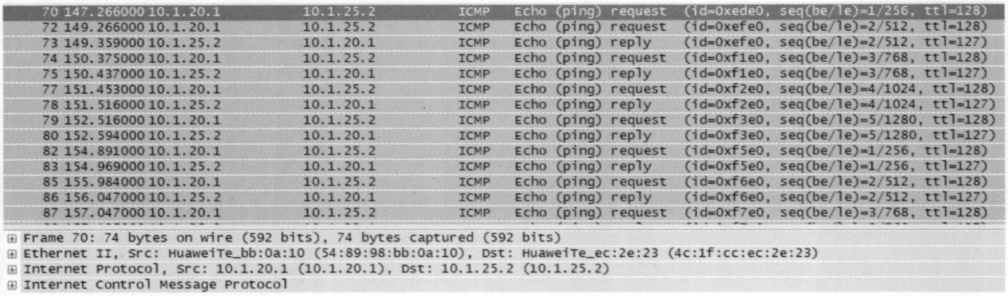

图2-7-9　sw1- Ethernet 0/0/1报文

sw1- Ethernet 0/0/4 抓取报文（arp or icmp），带标签 20，见图 2-7-10。

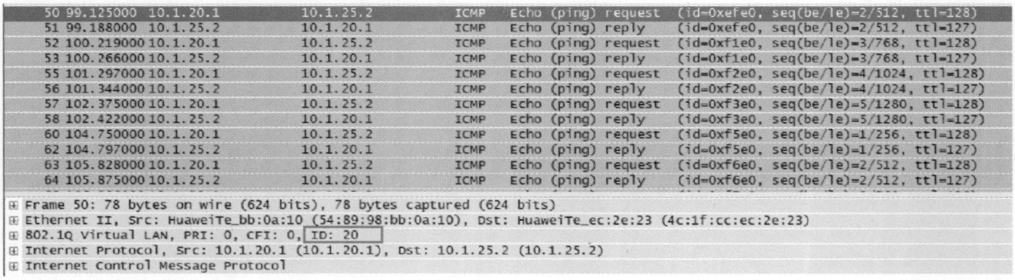

图2-7-10　sw1- Ethernet 0/0/4报文

sw3- GigabitEthernet 0/0/2 抓取报文（arp or icmp），带标签 25，见图 2-7-11。

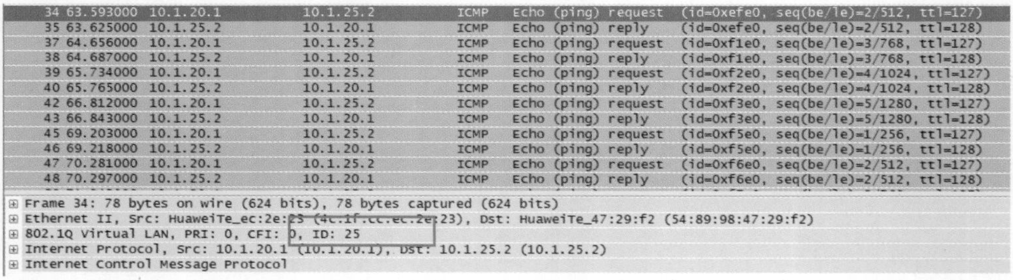

图2-7-11　sw3- GigabitEthernet 0/0/2报文

sw2- Ethernet 0/0/2 抓取报文（arp or icmp），不带标签，见图 2-7-12。

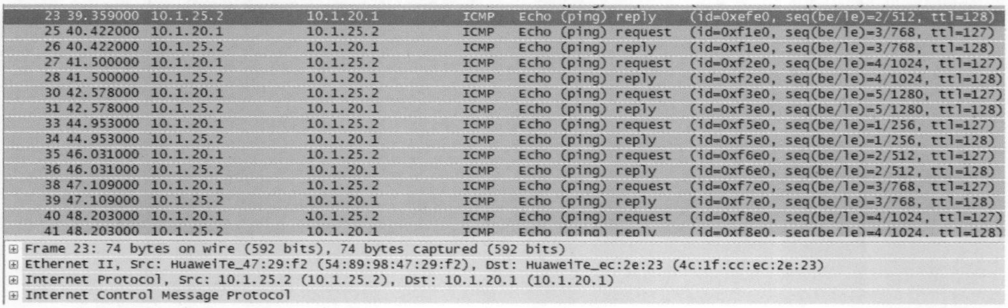

图2-7-12　sw2- Ethernet 0/0/2报文

二、知识加油站

交换机中常用的几个命令介绍如下。

1. 查看交换机 arp 信息

命令：display arp

```
[Access]display arp
Type: S-Static     D-Dynamic     O-Openflow     R-Rule     M-Multiport     I-Invalid
IP address         MAC address         VLAN/VSI name    Interface        Aging Type
10.10.40.1         1019-65d4-2601      4094             GE1/0/8          753   D
10.10.40.36        0c3a-fab7-6394      4094             GE1/0/7          302   D
```

2. 查询相应 vlan 下的 IP 地址

命令：display arp vlan XXXX

```
[Access] display arp vlan 2060
Type: S-Static     D-Dynamic     O-Openflow     R-Rule     M-Multiport     I-Invalid
IP address         MAC address         VLAN/VSI    Interface        Aging  Type
10.2.69.31         0017-6110-5911      2006        GE1/0/3          39     D
10.2.69.32         0017-6112-68e8      2006        GE1/0/3          39     D
10.2.69.34         0017-6112-68a0      2006        XGE1/0/26        39     D
```

对于 SDN 网络就需要按照下面的方法查询

命令：dis arp interface Vsi-interface XXX

```
[Access]dis arp interface Vsi-interface 10
  Type: S-Static   D-Dynamic    O-Openflow    R-Rule    M-Multiport    I-Invalid
IP address         MAC address         VLAN/VSI name    Interface        Aging Type
10.1.18.36         3cd2-e5a1-5480      vsi10            BAGG15           374   D
192.168.18.6       50bd-5f20-67c5      vsi10            BAGG12           321   D
192.168.18.101     3cd2-e5a1-4200      vsi10            BAGG13           352   D
192.168.18.102     3cd2-e5a1-5640      vsi10            BAGG15           374   D
192.168.18.103     3cd2-e5a1-58c0      vsi10            BAGG16           374   D
```

项目 8　H3C 链路聚合

链路聚合（Aggregation）可以实现流量在各端口之间进行负载分担，增加链路带宽，同一聚合组内各个成员端口之间冗余备份，提高了链路运行的可靠性，一般用于高可靠性网络中。链路聚合增加带宽的同时实现链路备份（即当一条链路出现故障，另一条链路仍能继续工作），保障业务访问网络不中断。

链路聚合是将交换机多个接口捆绑后形成一个逻辑聚合组，汇聚组内的所有物理链路作为一条逻辑链路来传送通信数据。见图 2-8-1，红框内是链路聚合，编号 10，蓝框内是聚合组成员，共计 6 个接口。

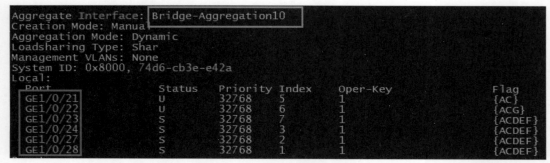

图2-8-1 交换机聚合口实例

本项目通过对交换机的两个 10G 接口配置链路聚合，实现两条链路捆绑，使带宽变为 20G。在高可靠网络架构中，一般采用链路聚合与堆叠技术配合使用。

一、链路聚合

1. 拓扑图

如图 2-8-2 所示，实现功能为链路备份与提高带宽容量。

网络连通情况可以用终端 PC 连接交换机进行测试，本节没有配置相关内容。

交换机接口见表 2-8-1。

图2-8-2 拓扑图

表 2-8-1 交换机接口

交换机名	接口	交换机名	接口
sw1	Ten-GigabitEthernet 1/0/51	sw2	Ten-GigabitEthernet 1/0/51
sw1	Ten-GigabitEthernet 1/0/52	sw2	Ten-GigabitEthernet 1/0/52

2. 项目配置步骤

步骤1 sw1 交换机基本配置

```
<H3C>system-view
[H3C]sysname sw1
# 批量创建 vlan
[sw1]vlan 100 to 200
# 新建二层聚合组 1（Aggregation 1）
[sw1]interface Bridge-Aggregation 1
[sw1-Bridge-Aggregation1]quit
# 批量进入交换机接口
[sw1]interface range Ten-GigabitEthernet 1/0/51 to Ten-GigabitEthernet 1/0/52
```

```
#将交换机接口加入聚合组1
[sw1-if-range]port link-aggregation group 1
[sw1-if-range]quit
#进入聚合组
[sw1]interface Bridge-Aggregation 1
#配置聚合链路为trunk类型
[sw1-Bridge-Aggregation1]port link-type trunk
#允许多个vlan通过
[sw1-Bridge-Aggregation1]port trunk permit vlan 100 200
[sw1-Bridge-Aggregation1]
[sw1-Bridge-Aggregation1]quit
[sw1]quit
<sw1>save
```

步骤2 sw2交换机基本配置

```
<H3C>system-view
[H3C]sysname sw2
#批量创建vlan
[sw2]vlan 100 to 200
[sw2]interface Bridge-Aggregation 1
[sw2-Bridge-Aggregation1]quit
[sw2]interface range Ten-GigabitEthernet 1/0/51 to Ten-GigabitEthernet 1/0/52
[sw2-if-range]port link-aggregation group 1
#将交换机接口加入聚合组1
[sw2-if-range]quit
#进入聚合组
[sw2]interface Bridge-Aggregation 1
#配置聚合链路为trunk类型
[sw2-Bridge-Aggregation1]port link-type trunk
#允许多个vlan通过
[sw2-Bridge-Aggregation1]port trunk permit vlan 100 200
[sw2-Bridge-Aggregation1]
[sw2-Bridge-Aggregation1]quit
[sw2]quit
<sw2>save
```

步骤3 验证

1）查看sw1聚合接口带宽是否变化，见图2-8-3。

命令：display interface brief<用户视图与系统视图均可>

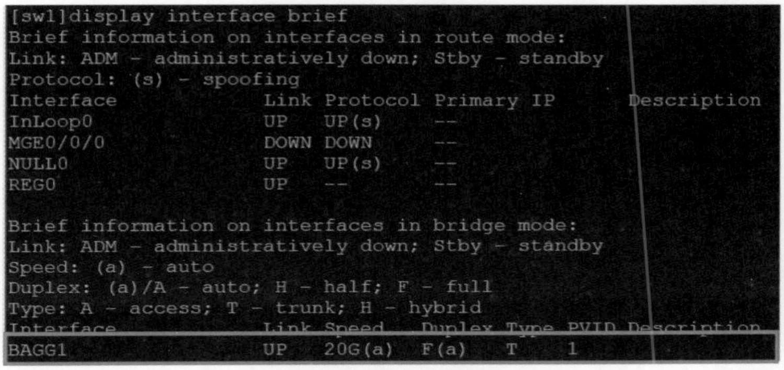

图2-8-3 sw1聚合接口带宽

从图中可以看出两个10G带宽的接口通过聚合技术变为20G。

2）查看sw2聚合接口带宽是否变化，见图2-8-4。

命令：`display interface brief`<用户视图与系统视图均可>

```
[sw2]display interface brief
Brief information on interfaces in route mode:
Link: ADM - administratively down; Stby - standby
Protocol: (s) - spoofing
Interface            Link Protocol Primary IP      Description
InLoop0              UP   UP(s)     --
MGE0/0/0             DOWN DOWN      --
NULL0                UP   UP(s)     --
REG0                 UP   --        --

Brief information on interfaces in bridge mode:
Link: ADM - administratively down; Stby - standby
Speed: (a) - auto
Duplex: (a)/A - auto; H - half; F - full
Type: A - access; T - trunk; H - hybrid
Interface            Link Speed    Duplex Type PVID Description
BAGG1                UP   20G(a)   F(a)   T    1
FGE1/0/53            DOWN 40G      A      A    1
FGE1/0/54            DOWN 40G      A      A    1
GE1/0/1              DOWN auto     A      A    1
GE1/0/2              DOWN auto     A      A    1
GE1/0/3              DOWN auto     A      A    1
```

图2-8-4 sw2聚合接口带宽

3）查看sw1聚合状态（sw2自行查看，不再列出），见图2-8-5。

① 查看所有聚合链路的详细状态信息。

命令：`display link-aggregation verbose`（用户视图与系统视图均可）

```
[sw1]display link-aggregation verbose
Loadsharing Type: Shar -- Loadsharing, NonS -- Non-Loadsharing
Port: A -- Auto
Port Status: S -- Selected, U -- Unselected, I -- Individual
Flags:  A -- LACP_Activity, B -- LACP_Timeout, C -- Aggregation,
        D -- Synchronization, E -- Collecting, F -- Distributing,
        G -- Defaulted, H -- Expired

Aggregate Interface: Bridge-Aggregation1
Aggregation Mode: Static
Loadsharing Type: Shar
  Port             Status  Priority  Oper-Key
--------------------------------------------------------------------
  XGE1/0/51          S       32768     1
  XGE1/0/52          S       32768     1
```

图2-8-5 sw1聚合状态

图2-8-5中看到接口的信息是S状态，表示选中。

Selected：处于此状态的成员端口称为选中端口。一般交换机两端聚合链路正常连接和启用的话，成员端口都处于选中状态，显示S。

② 查看聚合组内指定成员端口信息，见图2-8-6。

命令：`display link-aggregation member-port Ten-GigabitEthernet 1/0/51`

```
[sw1]display link-aggregation member-port Ten-GigabitEthernet 1/0/51
Flags: A -- LACP_Activity, B -- LACP_Timeout, C -- Aggregation,
       D -- Synchronization, E -- Collecting, F -- Distributing,
       G -- Defaulted, H -- Expired

Ten-GigabitEthernet1/0/51:
Aggregate Interface: Bridge-Aggregation1
Port Number: 52
Port Priority: 32768
Oper-Key: 1
```

图2-8-6 sw1聚合组内指定成员端口信息

二、知识加油站

1. 二层与三层聚合命令

二层聚合组配置命令：`[sw1]interface Bridge-Aggregation 1`

三层聚合组配置命令：[sw1]interface Route-Aggregation 1

2. range 命令

作用：在批量加入接口的时候使用。

示例：批量修改接口 vlan。

```
[sw1]vlan 100
[sw1-vlan100]quit
# 批量加入接口
[sw1]interface range GigabitEthernet 1/0/1 GigabitEthernet 1/0/2 GigabitEthernet 1/0/3
# 修改接口类型
[sw1-if-range]port link-type access
# 批量修改接口vlan
[sw1-if-range]port access vlan 100
```

查看效果，见图 2-8-7。

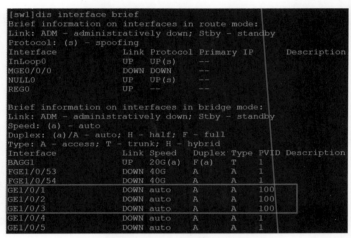

图2-8-7　批量修改接口效果

GE1/0/1 GE1/0/2 GE1/0/3 三个接口修改为 vlan 100。

3. 华为链路聚合配置

拓扑图参见图 2-8-8。

图2-8-8　华为交换机聚合拓扑图

与 H3C 交换机聚合类似，简要介绍如下。

sw1 配置

```
<Huawei>system-view
Enter system view, return user view with Ctrl+Z.
[Huawei]sysname sw1
```

```
[sw1]
# 批量创建vlan
[sw1]vlan batch 100 200
# 创建逻辑聚合组1
[sw1]interface Eth-Trunk 1
# 将物理端口加入逻辑聚合组
[sw1-Eth-Trunk1]trunkport GigabitEthernet 0/0/23 to 0/0/24
Info: This operation may take a few seconds. Please wait for a moment...done.
# 修改逻辑口类型为trunk
[sw1-Eth-Trunk1]port link-type trunk
# 允许通过Vlan通过
[sw1-Eth-Trunk1]port trunk allow-pass vlan 100 200
```

sw2 自行完成配置。

华为交换机聚合查看聚合状态

命令：display eth-trunk 1

```
[sw1-Eth-Trunk1]display eth-trunk 1
Eth-Trunk1's state information is:
WorkingMode: NORMAL              Hash arithmetic: According to SIP-XOR-DIP
Least Active-linknumber: 1       Max Bandwidth-affected-linknumber: 8
Operate status: up               Number Of Up Port In Trunk: 2
--------------------------------------------------------------------------
PortName                         Status         Weight
GigabitEthernet0/0/23            Up             1
GigabitEthernet0/0/24            Up             1
```

项目9 华三 irf 堆叠技术

irf 堆叠，是指将多台设备通过线缆连接组合在一起，虚拟化成一台设备，可以简化网络配置与运维管理，结合链路聚合技术，实现高可靠网络架构。堆叠使用时需要注意，各个厂家的设备不能混用。irf 堆叠由多台成员设备组成，一个 irf 中同时只能存在一台 Master，其他成员设备都是 Standy/Slave。Master 设备负责 irf 的运行管理和维护，Standy/Slave 设备作为备份的同时也处理业务。如 Master 设备故障，系统会迅速自动选举新的 Master，以保证通信业务正常。

设备堆叠时，采用专用堆叠线，见图 2-9-1，也可以用光模块 + 光纤组成。

堆叠主要命令：

```
[sw1]dis irf              <-- 查看设备的 irf 配置信息 -->
[sw1]display irf configuration   <-- 查看 irf 的端口信息 -->
```

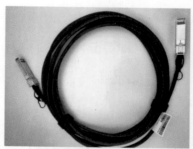

图 2-9-1 专用堆叠线以及现场使用

[sw1]display irf topology <-- 查看 irf 的拓扑信息 -->

堆叠中角色主要功能见表 2-9-1。

表 2-9-1 堆叠中角色

序号	角色	主要功能
1	Master	负责 irf 的运行管理和维护，同时刻堆叠中只有一台 Master
2	Standy	是 Master 的备设备，如 Master 故障，接替 Master，堆叠中只有一台 Standy 设备
3	Slave	业务转发

Master 设备竞选规则：优先级越高越优先；交换机 MAC 地址小的优先；在系统中运行时间长的优先，一般通过调整设备优先级来决定 Master 设备。

项目简介

本项目通过配置 irf 堆叠与链路聚合技术，实现了网络设备与链路冗余，提高了网络可靠性。项目中进行了各种网络模拟测试，比如堆叠中一台设备故障、链路聚合中一条线路故障等，结果显示网络通信没有受到影响。为了防止堆叠分裂，导致网络中断，必须进行 irf 冲突检测（MAD 功能）配置，在发生堆叠分裂时，数据包只通过 irf 堆叠中编号小的交换机。

一、堆叠技术 irf

1. 拓扑图

拓扑图如图 2-9-2 所示，实现功能：通过 irf 以及聚合技术，实现设备与线路冗余备份。

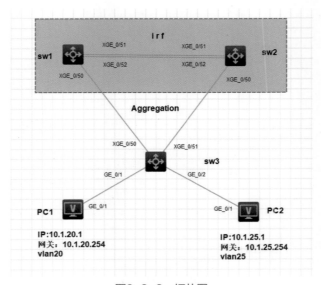

图 2-9-2 拓扑图

99

PC 参数如表 2-9-2 所示。

表 2-9-2　PC-IP 地址表

序号	PC	IP	掩码	网关	vlan
1	PC1	10.1.20.1	255.255.255.0	10.1.20.254	20
2	PC2	10.1.25.1	255.255.255.0	10.1.25.254	25

irf 堆叠接口如表 2-9-3 所示。

表 2-9-3　irf 堆叠接口

序号	设备	接口	
1	sw1	Ten-GigabitEthernet 1/0/51	Ten-GigabitEthernet 1/0/52
2	sw2	Ten-GigabitEthernet 1/0/51	Ten-GigabitEthernet 1/0/52

Aggregation（聚合）接口如表 2-9-4 所示。

表 2-9-4　Aggregation 接口

序号	设备	接口	
1	sw1	Ten-GigabitEthernet 1/0/50	Ten-GigabitEthernet 2/0/50
2	sw3	Ten-GigabitEthernet 1/0/50	Ten-GigabitEthernet 1/0/51

2. 项目配置步骤

步骤 1　sw1-irf 基本配置

```
<H3C>system-view
System View: return to User View with Ctrl+Z.
[H3C]sysname sw1
# 设置修改优先级别
```

member-id：表示设备在 IRF 中的成员编号，priority：表示优先级，取值范围为 1～32。优先级值越大表示优先级越高，优先级高的设备竞选时成为主设备的可能性越大。

```
[sw1]irf member 1 priority 5
# 批量进入交换机接口，使用"rang"命令
[sw1]interface range Ten-GigabitEthernet 1/0/51 Ten-GigabitEthernet 1/0/52
# 关闭交换机接口
[sw1-if-range]shutdown
[sw1-if-range]quit
# 创建堆叠逻辑接口
[sw1]irf-port 1/2
# 将堆叠接口加入到逻辑接口
[sw1-irf-port1/2]port group interface Ten-GigabitEthernet 1/0/51
# 提示信息是需要激活，暂时不执行激活命令
You must perform the following tasks for a successful I RF setup:
Save the configuration after completing irf configurati on.
Execute the "irf-port-configuration active" command to activate the irf ports.
# 将堆叠接口加入逻辑接口
[sw1-irf-port1/2]port group interface Ten-GigabitEthernet 1/0/52
[sw1-irf-port1/2]quit
# 批量进入交换机接口
[sw1]interface range Ten-GigabitEthernet 1/0/51 Ten-GigabitEthernet 1/0/52
# 打开交换机接口
[sw1-if-range]undo shutdown
```

```
[sw1-if-range]quit
[sw1]quit
<sw1>save
```

步骤 2　sw2-irf 基本配置

```
<H3C>system-view
System View: return to User View with Ctrl+Z.
[H3C]sysname sw2
# 修改交换机接口编号，由 1 修改为 2
[sw2]irf member 1 renumber 2
# 提示是否继续，选择 y
Renumbering the member ID may result in configuration change or loss. Continue?[Y/N]:y
[sw2]quit
# 这里一定不要忘记保存！！！
<sw2>save
# 重启交换机
<sw2>reboot
Start to check configuration with next startup configuration file, please wait.........DONE!
This command will reboot the device. Continue? [Y/N]:y
Now rebooting, please wait...
%Jul 31 16:53:16:892 2021 sw2 DEV/5/SYSTEM_REBOOT: System is rebooting now.
```

查询接口交换机接口编号是否变化？通过查看编号已经从 1 变为 2。

```
<sw2>display interface brief
Brief information on interfaces in route mode:
Link: ADM - administratively down; Stby - standby
Protocol: (s) - spoofing
Interface            Link Protocol Primary IP      Description
InLoop0              up   up(s)    --
MGE0/0/0             down down     --
NULL0                up   up(s)    --
REG0                 uP   --       --
Brief information on interfaces in bridge mode:
Link: ADM - administratively down; Stby - standby
Speed: (a) - auto
Duplex: (a)/A - auto; H - half; F - full
Type: A - access; T - trunk; H - hybrid
Interface            Link Speed   Duplex Type PVID Description
FGE2/0/53            DOWN 40G     A      A    1
FGE2/0/54            DOWN 40G     A      A    1
GE2/0/1              DOWN auto    A      A    1
GE2/0/2              DOWN auto    A      A    1
GE2/0/3              DOWN auto    A      A    1
……省略其他输出
```

解释：编号全部从"1"变为"2"，也就是 GE1/0/1 变为 GE2/0/1 ……

继续 sw2-irf 配置

```
[sw2]interface range Ten-GigabitEthernet 2/0/51 Ten-GigabitEthernet 2/0/52
[sw2-if-range]shutdown
[sw2]irf-port 2/1
[sw2-irf-port2/1]port group interface Ten-GigabitEthernet 2/0/51
# 提示激活，暂时不激活，放到最后
You must perform the following tasks for a successful IRF setup:
```

```
Save the configuration after completing irf configuration.
Execute the "irf-port-configuration active" command to activate the IRF ports.
[sw2-irf-port2/1]port group interface Ten-GigabitEthernet 2/0/52
[sw2-irf-port2/1]quit
[sw2]interface range Ten-GigabitEthernet 2/0/51 Ten-GigabitEthernet 2/0/52
[sw2-if-range]undo shutdown
[sw2-if-range]quit
[sw2]quit
<sw2>save
```

步骤 3 sw1 与 sw2 交换机配置 irf 激活命令

命令：irf-port-configuration active

sw1

```
[sw1]irf-port-configuration active
```
sw2

#sw2 交换机执行激活命令之前，之前的配置一定要保存，当输入这条命令后，交换机立刻重启，堆叠不成功的原因大多数是没有及时进行配置保存。

```
[sw2]irf-port-configuration active
```

重启完毕，两台交换机合体为一，交换机 sw2 的名字也变为 sw1。交换机的接口数量也变为 2 倍。通过"display interface brief"查看。

```
[sw1] display interface brief
Brief information on interfaces in route mode:
Link: ADM - administratively down; Stby - standby
Protocol: (s) - spoofing
Interface            Link Protocol Primary IP      Description
InLoop0              up   up(s)    --
MGE0/0/0             down down     --
NULL0                up   up(s)    --
REG0                 up   --       --
Brief information on interfaces in bridge mode:
Link: ADM - administratively down; Stby - standby
Speed: (a) - auto
Duplex: (a)/A - auto; H - half; F - full
Type: A - access; T - trunk; H - hybrid
Interface            Link Speed   Duplex Type PVID Description
FGE1/0/53            DOWN 40G     A      A    1
FGE1/0/54            DOWN 40G     A      A    1
FGE2/0/53            DOWN 40G     A      A    1
FGE2/0/54            DOWN 40G     A      A    1
……省略
```

步骤 4 查看 irf 相关信息

① 查看设备的 irf 配置信息。

命令: dis irf

在 Master 交换机上输入命令查看。

```
    [sw1]dis irf
MemberID    Role       Priority    CPU-Mac              Description
*+1         Master     5           b234-692e-0104       ---
  2         Standby    1           b234-7163-0204       ---
--------------------------------------------------------------
* indicates the device is the master.
```

```
+ indicates the device through which the user logs in.
The bridge MAC of the IRF is: b234-692e-0100
Auto upgrade                : yes
Mac persistent              : 6 min
Domain ID                   : 0
```

两个设备堆叠后，对外的地址是：b234-692e-0100。

* 标识该设备是 Master。

+ 代表用户登录的设备。

dis irf 命令查询结果字段解释如表 2-9-5 所示。

表 2-9-5　dis irf 命令查询结果字段解释

MemberID	Role	Priority	CPU-Mac	Description
成员设备编号	成员设备角色	优先级	设备 MAC	描述（无描述时，显示 ----）

Standby 交换机上输入命令"dis irf"查看，注意 MemberID 数字前"+"位置变化。

```
<sw1>sys
System View: return to User View with Ctrl+Z.
[sw1]dis irf
   MemberID     Role        Priority    CPU-Mac             Description
   *1           Master      5           b234-692e-0104      ---
   +2           Standby     1           b234-7163-0204      ---
--------------------------------------------------------------------------
* indicates the device is the master.
+ indicates the device through which the user logs in.
The bridge MAC of the IRF is: b234-692e-0100
Auto upgrade                : yes
Mac persistent              : 6 min
Domain ID                   : 0
```

② 查看 irf 的端口信息。

命令：display irf configuration

```
[sw1]display irf configuration
MemberID  NewID    IRF-Port1                        IRF-Port2
1         1        disable                          Ten-GigabitEthernet1/0/51
                                                    Ten-GigabitEthernet1/0/52
2         2        Ten-GigabitEthernet2/0/51        disable
                   Ten-GigabitEthernet2/0/52
```

③ 查看 irf 设备链路信息。

命令：display irf link

```
[sw1]display irf link
Member 1
IRF Port   Interface                       Status
1          disable                         --
2          Ten-GigabitEthernet1/0/51       up
           Ten-GigabitEthernet1/0/52       up
Member 2
IRF Port   Interface                       Status
1          Ten-GigabitEthernet2/0/51       up
           Ten-GigabitEthernet2/0/52       up
2          disable                         --
```

步骤5 创建 vlan 以及网关地址

因为两个交换机已经合体，在哪台交换机配置都可以。

```
# 批量创建 vlan
[sw1]vlan 20 to 25
# 配置 IP 地址，作为 vlan 20 对应网段的网关地址
[sw1]interface vlan-interface 20
[sw1-Vlan-interface20]ip address 10.1.20.254 24
[sw1-Vlan-interface20]quit
# 配置 IP 地址，作为 vlan 25 对应网段的网关地址
[sw1]interface vlan-interface 25
[sw1-vlan-interface25]ip address 10.1.25.254 24
[sw1-vlan-interface25]quit
[sw1]quit
<sw1>save
```

步骤6 配置聚合链路

```
# 创建聚合组 1
[sw1]interface Bridge-Aggregation 1
# 批量进入交换机接口
[sw1-Bridge-Aggregation1]quit
[sw1]interface range Ten-GigabitEthernet 1/0/50 Ten-GigabitEthernet 2/0/50
# 将接口加入聚合组
[sw1-if-range]port link-aggregation group 1
# 退出
[sw1-if-range]quit
```

查询聚合状态

命令：display link-aggregation verbose

```
[sw1]display link-aggregation verbose
Loadsharing Type: Shar -- Loadsharing, NonS -- Non-Loadsharing
Port: A -- Auto
Port Status: S -- Selected, U -- Unselected, I -- Individual
Flags:   A -- LACP_Activity, B -- LACP_Timeout, C -- Aggregation,
         D -- Synchronization, E -- Collecting, F -- Distributing,
         G -- Defaulted, H -- Expire
Aggregate Interface: Bridge-Aggregation1
Aggregation Mode: Static
Loadsharing Type: Shar
  Port            Status    Priority  Oper-Key
--------------------------------------------------------------------------------
  XGE1/0/50       S         32768     1
  XGE2/0/50       S         32768     1
[sw1]quit
<sw1>save
```

步骤7 配置聚合口类型以及允许通过 vlan

```
[sw1]interface Bridge-Aggregation 1
# 配置聚合口接口类型
[sw1-Bridge-Aggregation1]port link-type trunk
# 允许需求 vlan 通过
[sw1-Bridge-Aggregation1]port trunk permit vlan 20 25
[sw1-Bridge-Aggregation1]quit
[sw1]quit
<sw1>save
```

步骤8 sw3 基本配置

```
<H3C>system-view
System View: return to User View with Ctrl+Z.
[H3C]sysname sw3
[sw3]vlan 20 to 25
# 以下创建聚合组，并且将物理接口加入聚合组
[sw3]interface Bridge-Aggregation 1
[sw3-Bridge-Aggregation1]quit
[sw3]interface range Ten-GigabitEthernet 1/0/50 Ten-GigabitEthernet 1/0/51
[sw3-if-range]port link-aggregation group 1
[sw3-if-range]quit
# 配置接口类型，允许vlan通过
[sw3]interface Bridge-Aggregation 1
[sw3-Bridge-Aggregation1]port link-type trunk
[sw3-Bridge-Aggregation1]port trunk permit vlan 20 25
[sw3-Bridge-Aggregation1]quit
[sw3]
# 配置access类型，并且允许接口通过
[sw3]interface GigabitEthernet 1/0/1
[sw3-GigabitEthernet1/0/1]port link-type access
[sw3-GigabitEthernet1/0/1]port access vlan 20
[sw3-GigabitEthernet1/0/1]quit
# 配置access，并且允许接口通过
[sw3]interface GigabitEthernet 1/0/2
[sw3-GigabitEthernet1/0/2]port link-type access
[sw3-GigabitEthernet1/0/2]port access vlan 25
[sw3-GigabitEthernet1/0/2]quit
[sw3]quit
<sw3>save
```

步骤9 PC 配置

PC1 和 PC2 基础配置见图 2-9-3。

依次点击启用（接口管理）- 填写静态 IP 地址（IPV4 配置）- 启用。

(a) PC1 基础配置　　　　　　　　　　　(b) PC2 基础配置

图2-9-3 PC1和PC2的基础配置

步骤10 通信测试

```
PC1（IP:10.1.20.1）-ping-PC2(IP:10.1.25.1)-通
56 bytes from 10.1.25.1: icmp_seq=0 ttl=254 time=3.000 ms
……省略……
```

二、知识加油站

1. 初体验聚合链路冗余的魅力

sw1

通过："display interface brief"查询聚合组带宽是否变为20G。

```
[sw1]display interface brief
Brief information on interfaces in route mode:
Link: ADM - administratively down; Stby - standby
Protocol: (s) - spoofing
Interface            Link Protocol Primary IP      Description
InLoop0              up   up(s)    --
MGE0/0/0             down down     --
NULL0                up   up(s)    --
REG0                 up   --       --
vlan20               up   up       10.1.20.254
vlan25               up   up       10.1.25.254
Brief information on interfaces in bridge mode:
Link: ADM - administratively down; Stby - standby
Speed: (a) - auto
Duplex: (a)/A - auto; H - half; F - full
Type: A - access; T - trunk; H - hybrid
Interface            Link Speed   Duplex Type PVID Description
# 聚合带宽显示20G，聚合正确
BAGG1                up   20G(a)  F(a)   T    1
```

sw3

通过："display interface brief"查询聚合组带宽是否变为20G。

```
[sw3]display interface brief
Brief information on interfaces in route mode:
Link: ADM - administratively down; Stby - standby
Protocol: (s) - spoofing
Interface            Link Protocol Primary IP      Description
InLoop0              up   up(s)    --
MGE0/0/0             down down     --
NULL0                up   up(s)    --
REG0                 up   --       --
Brief information on interfaces in bridge mode:
Link: ADM - administratively down; Stby - standby
Speed: (a) - auto
Duplex: (a)/A - auto; H - half; F - full
Type: A - access; T - trunk; H - hybrid
Interface            Link Speed   Duplex Type PVID Description
BAGG1                up   20G(a)  F(a)   T    1
```

既然是聚合，首先测试PC1与PC2之间通信情况，在PC1上输入ping -c 5000 10.1.25.1，如图2-9-4所示，将L1链路断开，观察丢包延时情况，结果如下（通信正常）。

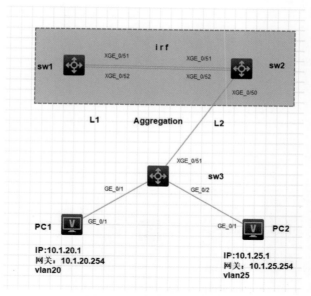

图2-9-4 断开L1测试通信

```
Ping 10.1.25.1 (10.1.25.1): 56 data bytes, press CTRL_C to break
56 bytes from 10.1.25.1: icmp_seq=0 ttl=254 time=3.000 ms
……省略……
```

以上说明通信正常无影响。

恢复L1，升级测试，关闭堆叠设备sw2，再次观察通信情况（丢了三个包），如图2-9-5所示。

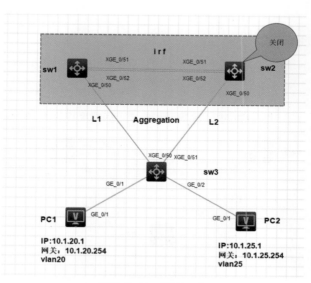

图2-9-5 关闭sw2

```
56 bytes from 10.1.25.1: icmp_seq=121 ttl=254 time=2.000 ms
Request time out
Request time out
Request time out
```

107

```
56 bytes from 10.1.25.1: icmp_seq=125 ttl=254 time=2.000 ms
56 bytes from 10.1.25.1: icmp_seq=126 ttl=254 time=2.000 ms
……省略……
56 bytes from 10.1.25.1: icmp_seq=126 ttl=254 time=2.000 ms
```

以上说明，当设备堆叠后，网络可靠性非常高，在链路聚合正常的情况下，网络几乎无影响。

恢复设备拓扑原始状态。请自行实验将两条堆叠线都断开，还能通信吗？

堆叠、聚合拓扑等效图如图2-9-6所示。

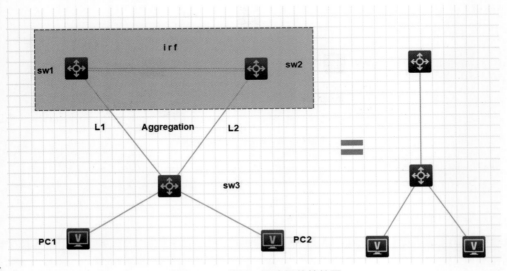

图2-9-6　堆叠、聚合拓扑等效图

2. 堆叠线断开的通信情况分析

步骤1　当断开一条堆叠线，见图2-9-7，测试网络正常，见图2-9-8。

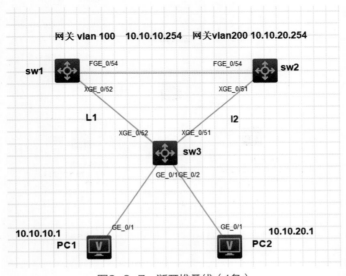

图2-9-7　断开堆叠线（1条）

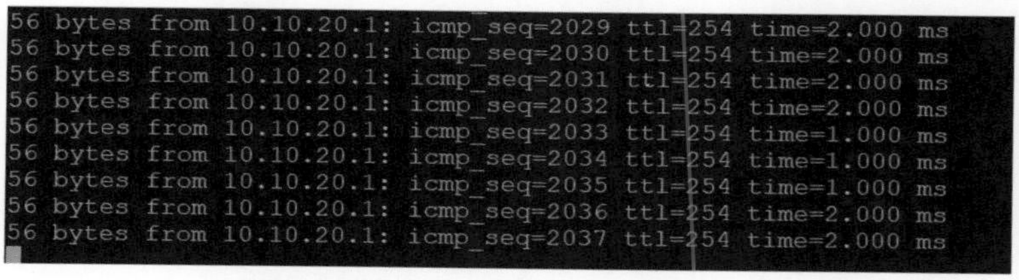

图2-9-8　断开一条堆叠线，网络正常

步骤2　当断开堆叠线2条，见图2-9-9，网络中断！见图2-9-10。

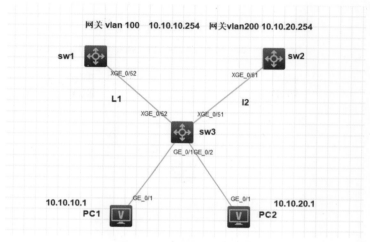

图2-9-9　堆叠线全部断开

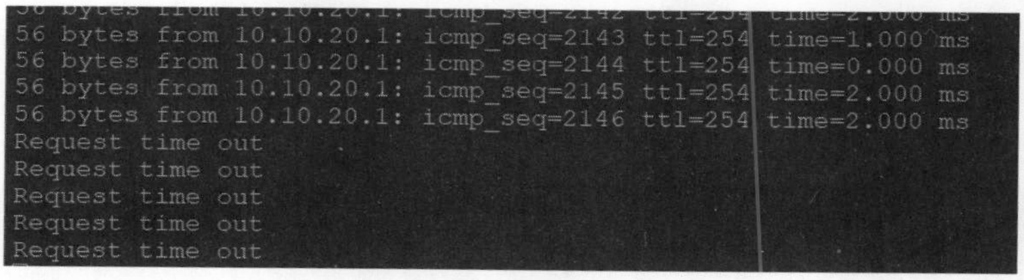

图2-9-10　网络中断

网络中断的原因：当堆叠线断开后，交换机sw1与sw2都认为自己是Master，当流量到达sw1与sw2的时候，无法选路而导致网络中断。解决办法，配置MAD检测，当出现堆叠线全部断开后，只允许一个设备转发流量。

步骤3　配置MAD-lacp检测

sw1交换机配置

```
[sw1]interface Bridge-Aggregation 1
# 修改聚合模式为动态
[sw1-Bridge-Aggregation1]link-aggregation mode dynamic
#mad 检测使能
[sw1-Bridge-Aggregation1]mad enable
You need to assign a domain ID (range: 0-4294967295)
[Current domain is: 0]: 0
The assigned domain ID is: 0
MAD LACP only enable on dynamic aggregation interface.
```

109

sw3 交换机配置

```
[sw3]interface Bridge-Aggregation 1
[sw3-Bridge-Aggregation1]link-aggregation mode dynamic
[sw3-Bridge-Aggregation1]mad enable
You need to assign a domain ID (range: 0-4294967295)
[Current domain is: 0]: 0
The assigned domain ID is: 0
MAD LACP only enable on dynamic aggregation interface.
```

irf domain 默认是 0。当整个系统中多个堆叠域的时候，最好配置堆叠域编号。

断开所有堆叠线，如图 2-9-11 所示，再进行网络测试（通），如图 2-9-12 所示。

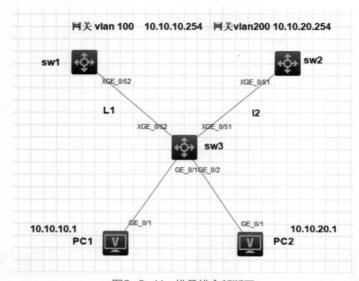

图2-9-11　堆叠线全部断开

图2-9-12　网络测试

查看 sw1 交换机三层接口信息，全部为 up 状态，见图 2-9-13。

图2-9-13　Master 设备vlan up

查看 sw2（实际显示的 sw1）交换机，全部为 down 状态，见图 2-9-14。不进行流量转发。

```
[sw1]dis ip interface brief
*down: administratively down
(s): spoofing  (l): loopback
Interface              Physical Protocol IP Address       Description
MGE0/0/0               down     down     --               --
Vlan100                down     down     10.10.10.254     --
Vlan200                down     down     10.10.20.254     --
```

图2-9-14　非Master设备vlan down

通过查看交换机三层接口信息，交换机 sw2 的 vlan100、vlan200 全部 down，流量只能从 sw1 经过，也就不存在流量无法选择的情况发生。当堆叠线全部断开后（也就是堆叠分裂），配置 mad 检查功能后，流量从 member 编号小的交换机通过。

步骤4　mad-bfd 检测

删除上面步骤中配置的 mad-lacp 的检测协议。

在 sw1 与 sw2 设备之间增加一条线路（sw1-g1/0/1，sw2-g1/0/1），用于 bfd 检测。见图 2-9-15。

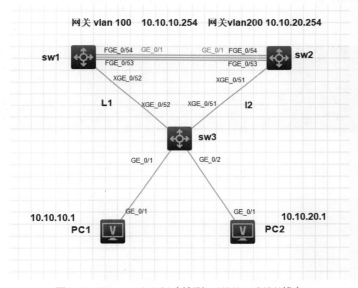

图2-9-15　mad-bfd（检测 g1/0/1-g2/0/1线）

```
# 创建 vlan 4094
[sw1]vlan 4094
# 将 GigabitEthernet 1/0/1 GigabitEthernet 2/0/1 划分到 vlan 4094
[sw1-vlan4094]port GigabitEthernet 1/0/1 GigabitEthernet 2/0/1
[sw1]interface Vlan-interface 4094
# 启用 mad bfd 使能功能
[sw1-Vlan-interface4094]mad bfd enable
# 配置 IP 地址
[sw1-Vlan-interface4094]mad ip address 192.168.2.1 24 member 1
[sw1-Vlan-interface4094]mad ip address 192.168.2.2 24 member 2
[sw1-Vlan-interface4094]quit
# 关闭 GigabitEthernet 1/0/1 接口 stp 功能（因为 bfd 检测与 stp 互斥，需要关闭 stp 功能）
[sw1]interface GigabitEthernet 1/0/1
[sw1-GigabitEthernet1/0/1]undo stp enable
[sw1-GigabitEthernet1/0/1]quit
# 关闭 GigabitEthernet 1/0/1 接口 stp 功能
```

```
[sw1]interface GigabitEthernet 2/0/1
[sw1-GigabitEthernet2/0/1]undo stp enable
```

Display mad 命令查看目前 mad 检测功能，mad bfd 已经启用，见图 2-9-16。

```
<sw1>display mad
<sw1>display mad
MAD ARP disabled.
MAD ND disabled.
MAD LACP disabled.
MAD BFD enabled.
```

图2-9-16 查看所有mad

当所有堆叠线断开（堆叠分裂）的时候系统启用 mad-bfd 检测，交换机 sw1 的三层接口处于 up 状态，进行数据转发，见图 2-9-17。而 sw2（显示 sw1），三层接口全部处于 down 状态，不进行数据转发，见图 2-9-18。

```
[sw1]display ip interface brief
*down: administratively down
(s): spoofing   (l): loopback
Interface            Physical Protocol IP Address      Description
MGE0/0/0             down     down     --              --
Vlan100              up       up       10.10.10.254    --
Vlan200              up       up       10.10.20.254    --
Vlan4094             down     down     192.168.2.1     --
```

图2-9-17 sw1 继续数据转发

```
[sw1]dis ip interface brief
*down: administratively down
(s): spoofing   (l): loopback
Interface            Physical Protocol IP Address      Description
MGE0/0/0             down     down     --              --
Vlan100              down     down     10.10.10.254    --
Vlan200              down     down     10.10.20.254    --
Vlan4094             down     down     192.168.2.2     --
```

图2-9-18 sw2（显示sw1）停止数据转发

当堆叠线全部断开的时候，启用 mad-bfd 检测，测试网络正常，见图 2-9-19。

```
<H3C>ping 10.10.20.1
Ping 10.10.20.1 (10.10.20.1): 56 data bytes, press CTRL_C to break
56 bytes from 10.10.20.1: icmp_seq=0 ttl=254 time=1.000 ms
56 bytes from 10.10.20.1: icmp_seq=1 ttl=254 time=2.000 ms
56 bytes from 10.10.20.1: icmp_seq=2 ttl=254 time=1.000 ms
56 bytes from 10.10.20.1: icmp_seq=3 ttl=254 time=3.000 ms
56 bytes from 10.10.20.1: icmp_seq=4 ttl=254 time=1.000 ms
```

图2-9-19 因mad-bfd启用，数据通信正常。

总结： 当系统检测到 irf 堆叠分裂时，两个 irf 中成员编号小的交换机继续正常运行，编号大的交换机接口转入 down，暂且不转发业务，当 irf 堆叠从分裂恢复正常后，编号大设备接口自动进入流量转发状态。

项目 10 H3C 端口镜像

通过端口镜像对网络的流量进行监控分析，可以对企业内部的网络流量进行监控管理，能快速地定位故障。在交换机或者其他网络设备上，将一个或多个端口的流量转发到某一个指定端口（镜像端口）来实现网络流量监控管理。

项目简介

通过将核心交换机三个万兆接口数据镜像到某个接口，利用全网流量分析设备进行数据分析，展示异常流量，可以帮助运维人员及时发现故障，进行相关处理，例如防火墙配置安全策略等。

一、端口镜像

1. 拓扑图

如图 2-10-1 所示，真实环境：要求对数据中心以及某厂区所有数据流量进行监控管理。因为所有的数据都需经过核心进行数据转发，所以在核心上进行端口镜像，采用全网流量监测设备进行网络流量实时检测。

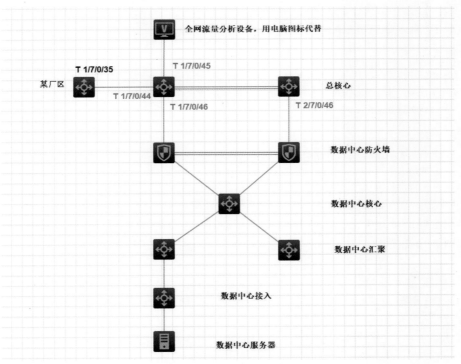

图2-10-1 拓扑图

交换机关于镜像端口有关如表 2-10-1 所示。

表 2-10-1 镜像源端口与目的端口

被镜像源端口	Ten-GigabitEthernet 1/7/0/44	Ten-GigabitEthernet 1/7/0/46	Ten-GigabitEthernet 2/7/0/46
端口镜像（目的端口）	Ten-GigabitEthernet 1/7/0/45		

2. 项目配置

步骤1　流量镜像基本配置

基本命令：

```
# 查看镜像组1
dis mirroring-group 1
# 新建镜像组1
mirroring-group 1 local
# 配置被监控源端口，可配置多个。both代表进出都监控，inbound代表监控入，outbound代表监控出
mirroring-group 1 mirroring-port GigabitEthernet 5/0/3 both（示例）
# 配置监控端口，一般与专门流量分析的设备连接。
mirroring-group 1 monitor-port GigabitEthernet 3/0/45（示例）
```

核心交换机端口镜像配置

```
[Access]mirroring-group 1 local
[Access]mirroring-group 1 mirroring-port Ten-GigabitEthernet 1/7/0/46 both
[Access]mirroring-group 1 mirroring-port Ten-GigabitEthernet 2/7/0/46 both
[Access]mirroring-group 1 mirroring-port Ten-GigabitEthernet 1/7/0/44 both
[Access]mirroring-group 1 monitor-port Ten-GigabitEthernet 1/7/0/45
```

步骤2　查询端口镜像配置

```
[Access]display mirroring-group 1
Mirroring group 1:
    Type: Local
    Status: Active
    Mirroring port:
        # 以下三个镜像源端口
        Ten-GigabitEthernet1/7/0/44   Both
        Ten-GigabitEthernet1/7/0/46   Both
        Ten-GigabitEthernet2/7/0/46   Both
        # 镜像目的端口
Monitor port: Ten-GigabitEthernet1/7/0/45
```

专用全网流量分析设备针对Ten-GigabitEthernet1/7/0/45数据进行分析，诊断结果实时显示，如图2-10-2所示。

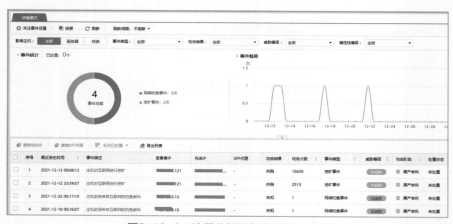

图2-10-2　流量分析设备实时报告截图

二、知识加油站

1. 诊断交换机光模块参数的当前测量值

命令：display transceiver diagnosis interface Ten-GigabitEthernet 1/7/0/1
[Access]display transceiver diagnosis interface Ten-GigabitEthernet 1/7/0/1
Ten-GigabitEthernet1/7/0/1 transceiver diagnostic information:
  Current diagnostic parameters:
     Temp.(Voltage(V)   Bias(mA)   RX power(dBm)   TX power(dBm)
     21   3.35   33.13   -4.50   -2.25
Alarm thresholds:
     Temp.(Voltage(V)   Bias(mA)   RX power(dBm)   TX power(dBm)
High 73   3.80   88.00   2.50   1.00
Low   -3   2.81   1.00   -16.40   -11.20

蓝色框内是参考数值，红框内数值含义解释：

Temp - 温度　显示 21℃

Voltage - 电压　显示 3.35V

Bias 偏置电流　显示 33.13mA

RX power 接收光功率 –4.50dBm

TX power 发射光功率 –2.25dBm

2. 查询光模块主要特征参数

命令：display transceiver interface Ten-GigabitEthernet 1/7/0/1
[Access]display transceiver interface Ten-GigabitEthernet 1/7/0/1
  Ten-GigabitEthernet1/7/0/1 transceiver information:
    Transceiver Type   : 10G_BASE_LR_SFP
    Connector Type   : LC
    Wavelength(nm)   : 1310
    Transfer Distance(km)   : 10(SMF)
    Digital Diagnostic Monitoring : YES
    Vendor Name   : H3C
    Ordering Name   : SFP-XG-LX-SM1310-D

解释：
Transceiver Type（收发器类型）-10G_BASE_LR_SFP（万兆单模光模块）
Connector Type（链接器类型）- LC
Wavelength(nm)（波长）-1310
Transfer Distance(km)（传输距离）-10km
Digital Diagnostic Monitoring（数字诊断检测）- 是
Vendor Name（厂商名称）-H3C
Ordering Name（光模块名称）- SFP-XG-LX-SM1310-D

3. 查光模块电子标签信息（验证真伪）

命令：display transceive manuinfo interface Ten-GigabitEthernet 1/7/0/1
[Access]display transceive manuinfo interface Ten-GigabitEthernet 1/7/0/1
Ten-GigabitEthernet1/7/0/1 transceiver manufacture information:
Manu. Serial Number : 210231A1RQN20A0000NY
Manufacturing Date : 2020-10-29
Vendor Name   : H3C

Manu. Serial Number- 序列号（查看是否与实物的序列号一致）
Manufacturing Date- 生产日期
Vendor Name- 生产厂家

4. 光接口检查

如光接口一直无法建立通信，可以检查光路内环是否正常。用一条尾纤将光模块 RX 与 TX 连接，插到交换机。如图 2-10-3 和图 2-10-4 所示。

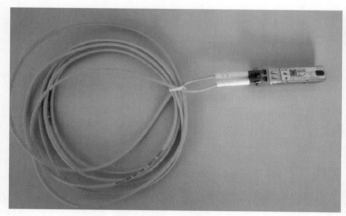

图2-10-3　光模块自环　　　　　　　　图2-10-4　交换机验证

操作步骤：

步骤1　清空接口统计信息

命令：`reset counters interface GigabitEthernet 1/0/27`

`<DYJT-BQ-Access_test.4.36>reset counters interface GigabitEthernet 1/0/27`

`<Access>reset counters interface GigabitEthernet 1/0/27`

步骤2　配置接口环回内部

命令：`loopback internal`

`[DYJT-BQ-Access_test.4.36-GigabitEthernet1/0/27]loopback internal`
`Loop internal succeeded!`

`[Access-GigabitEthernet1/0/27]loopback internal`
`Loop internal succeeded!`

步骤3　查看统计信息

命令：`display interface GigabitEthernet 1/0/27`

```
[Access]display interface GigabitEthernet 1/0/27
GigabitEthernet1/0/27
Current state: up
Line protocol state: up
IP packet frame type: Ethernet II, hardware address: 0c3a-fab7-63b0
Description: GigabitEthernet1/0/27 Interface
Bandwidth: 1000000 kbps
Loopback is not set
Media type is optical fiber, Port hardware type is 1000_BASE_LX_SFP
1000Mbps-speed mode, full-duplex mode
Link speed type is autonegotiation, link duplex type is autonegotiation
Flow-control is not enabled
Maximum frame length: 10240
Allow jumbo frames to pass
```

```
        Broadcast max-ratio: 100%
        Multicast max-ratio: 100%
        Unicast max-ratio: 100%
        PVID: 1
        MDI type: Automdix
        Port link-type: Access
         Tagged VLANs:    None
         Untagged VLANs: 1
        Port priority: 0
        Last link flapping: 0 hours 12 minutes 34 seconds
        Last clearing of counters: 05:45:36 Mon 07/12/2021
        Current system time:2021-07-12 05:53:12
        Last time when physical state changed to up:2021-07-12 05:40:38
        Last time when physical state changed to down:2021-07-12 05:40:36
        Peak input rate: 79 bytes/sec, at 2021-07-12 05:50:52
        Peak output rate: 79 bytes/sec, at 2021-07-12 05:50:52
        Last 300 seconds input: 0 packets/sec 75 bytes/sec 0%
        Last 300 seconds output: 0 packets/sec 75 bytes/sec 0%
        Input (total):   254 packets, 34217 bytes
                 0 unicasts, 10 broadcasts, 244 multicasts, 0 pauses
        Input (normal):   254 packets, - bytes
                 0 unicasts, 10 broadcasts, 244 multicasts, 0 pauses
        Input:  0 input errors, 0 runts, 0 giants, 0 throttles
                 0 CRC, 0 frame, - overruns, 0 aborts
                 - ignored, - parity errors
        Output (total): 254 packets, 34217 bytes
                 0 unicasts, 10 broadcasts, 244 multicasts, 0 pauses
        Output (normal): 254 packets, - bytes
                 0 unicasts, 10 broadcasts, 244 multicasts, 0 pauses
        Output: 0 output errors, - underruns, - buffer failures
                 0 aborts, 0 deferred, 0 collisions, 0 late collisions
                 0 lost carrier, - no carrier
```

如果接口能 up，同时检查 Input (total) 与 Output (total) 包数量一致，说明光模块与交换机没有问题。测试使用的是千兆模块。

项目 11 华为 ACL 访问控制列表

ACL 访问控制列表，就是定义不同的规则，设备根据规则进行数据包"允许通过"或者"拒绝"进行分类，从而实现网络访问行为的控制、流量限制等措施。

ACL 报文规则是顺序匹配，在创建基本 ACL 或者高级 ACL 时，根据项目需求，需要建立 1 条或者多条规则，网络设备接收到流量后，按照规则顺序进行匹配，如果匹配，执行定义规则的允许/拒绝流量；如果从开始到结尾都没有匹配，数据流量不作任何处理。华为设备默认流量是放行，与思科相反。

项目简介

ACL 访问控制在网络中经常使用。在项目中，通过配置 ACL 基本与高级策略，允许或者限制某些特定流量或者协议。在项目中通过多个例子，学习掌握各种 ACL 用法。在项目中调用 ACL 时特别要注意，华为设备在调用接口时，若没有匹配条目，则默认允许（permit）放行；在调用 vty 接口时，若没有匹配条目，默认是拒绝（deny）。

举例说明通配符如表 2-11-1 所示。

表 2-11-1　举例说明通配符

IP 地址以及网段	通配符
192.168.10.1	0.0.0.0
192.168.10.0	0.0.0.255
192．168.0.0	0.0.255.255

华为 ACL 分类见表 2-11-2。

表 2-11-2　华为 ACL 分类

序号	范围	参数	分类
1	2000～2999	源 IP 地址等	基本 ACL
2	3000～3999	源 IP、目的 IP、源 MAC、目的 MAC、端口等	高级 ACL

一、ACL 访问控制

1. 拓扑

如图 2-11-1 所示。

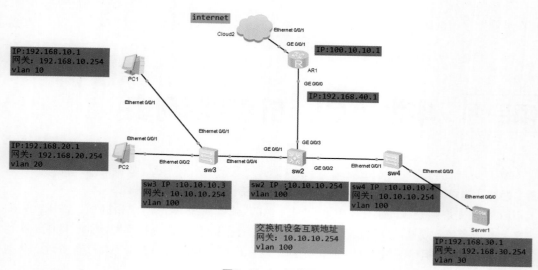

图2-11-1　拓扑图

IP 地址、网关等参数见表 2-11-3。

表 2-11-3　网络设备参数配置

序号	设备	IP	网关	vlan	接口
1	PC1	192.168.10.1	192.168.10.254	10	/
2	PC2	192.168.20.1	192.168.20.254	20	/
3	Server	192.168.30.1	192.168.30.254	30	/
4	sw3	/	/	10	Etherent 0/0/1
5	sw3	/	/	20	Etherent 0/0/2
6	sw3	/	/	10 20 30 40 100	Etherent 0/0/4
7	sw3	10.10.10.3	10.10.10.254	100	互联地址
8	sw2	/	/	10 20 30 40 100	Etherent 0/0/1
9	sw2	/	/	10 20 30 40 100	Etherent 0/0/2
10	sw2	/	/	40	Etherent 0/0/3
11	sw2	10.10.10.254	/	100	互联地址
12	sw4	/	/	10 20 30 40 100	Etherent 0/0/1
13	sw4	/	/	30	Etherent 0/0/3
14	sw4	10.10.10.4	10.10.10.254	100	互联地址
15	AR1	192.168.40.1	/	/	GE 0/0/0
16	AR1	100.10.10.1	/	/	GE 0/0/1

2．基本配置（实现网络互通）

步骤 1　PC 配置（PC1、PC2、Server1），如图 2-11-2 所示。

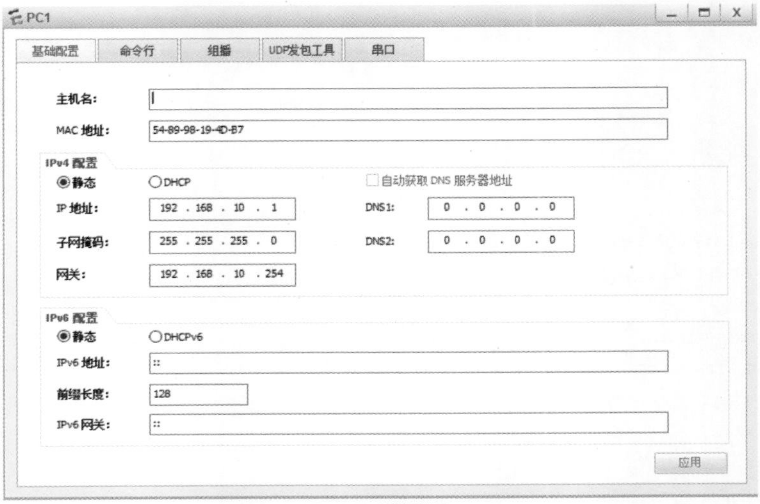

(a) PC1参数配置

图2-11-2

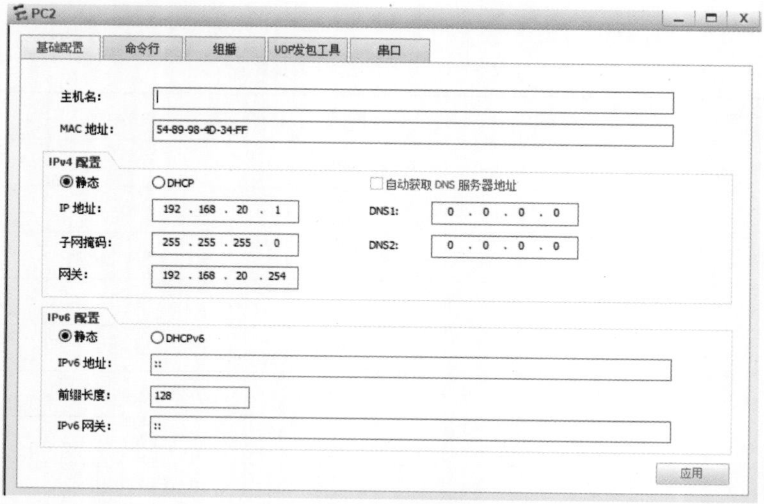

(b) PC2 参数配置

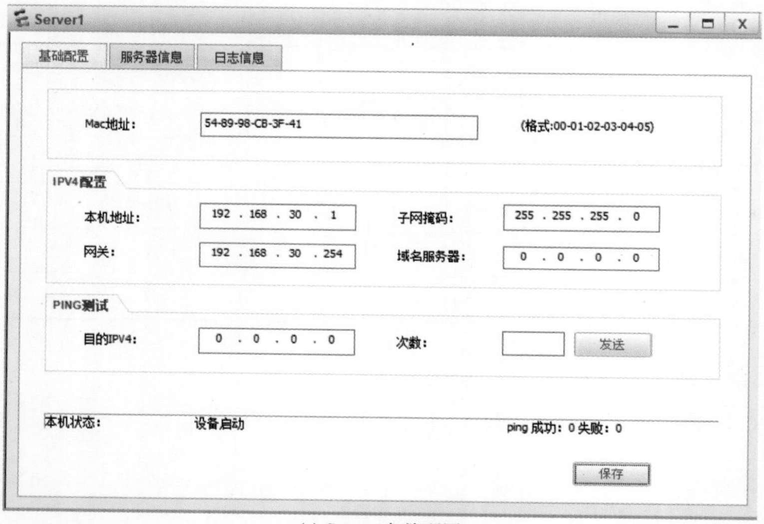

(c) Server 参数配置

图2-11-2　PC配置

步骤2　交换机、路由器配置。

sw3

```
<Huawei>system-view
Enter system view, return user view with Ctrl+Z.
# 关闭信息中心
[Huawei]un in en
Info: Information center is disabled.
[Huawei]sysname sw3
# 批量创建vlan
[sw3]vlan batch 10 20 30 40 100
# 以下是允许vlan 10 通过Ethernet0/0/1
[sw3]interface Ethernet 0/0/1
[sw3-Ethernet0/0/1]port link-type access
```

```
[sw3-Ethernet0/0/1]port default vlan 10
[sw3-Ethernet0/0/1]quit
# 以下是允许 vlan 20 通过 Ethernet0/0/2
[sw3-Ethernet0/0/2]port link-type access
[sw3-Ethernet0/0/2]port default vlan 20
[sw3-Ethernet0/0/2]quit
# 以下是允许 vlan 10 20 30 40 100 通过 Ethernet0/0/4
[sw3]interface Ethernet 0/0/4
[sw3-Ethernet0/0/4]port link-type trunk
[sw3-Ethernet0/0/4]port trunk allow-pass vlan 10 20 30 40
[sw3-Ethernet0/0/4]quit
# 配置交换机 sw3 的互联 IP 地址
[sw3]interface Vlanif 100
[sw3-Vlanif100] ip address 10.10.10.3 24
# 配置默认路由
[sw3] ip route-static 0.0.0.0 0.0.0.0 10.10.10.254
[sw3]quit
<sw3>save
sw2
<huawei>system-view
Enter system view, return user view with Ctrl+Z
[huawei]sysname sw2
[sw2]vlan batch 10 20 30 40 100
# 配置网关地址 192.168.10.254
[sw2]interface Vlanif 10
[sw2-Vlanif10]ip address 192.168.10.254 24
[sw2-Vlanif10]quit
# 配置网关地址 192.168.20.254
[sw2]interface Vlanif 20
[sw2-Vlanif20]ip address 192.168.20.254 24
[sw2-Vlanif20]quit
# 配置网关地址 192.168.30.254
[sw2]interface Vlanif 30
[sw2-Vlanif30]ip address 192.168.30.254 24
[sw2-Vlanif30]quit
# 配置网关地址 192.168.40.254
[sw2]interface Vlanif 40
[sw2-Vlanif40]ip address 192.168.40.254 24
[sw2-Vlanif40]quit
# 配置交换机互联网关地址 192.168.40.254
[sw2]interface Vlanif 100
[sw2-Vlanif100]ip address 10.10.10.254 24
[sw2-Vlanif100]quit
# 以下是允许 vlan 10 20 30 40 100 通过 GigabitEthernet0/0/1
[sw2]interface GigabitEthernet 0/0/1
[sw2-GigabitEthernet0/0/1]port link-type trunk
[sw2-GigabitEthernet0/0/1]port trunk allow-pass vlan 10 20 30 40 100
[sw2-GigabitEthernet0/0/1]quit
# 以下是允许 vlan 10 20 30 40 100 通过 GigabitEthernet0/0/2
[sw2]interface GigabitEthernet 0/0/2
[sw2-GigabitEthernet0/0/2]port link-type trunk
```

```
[sw2-GigabitEthernet0/0/2]port trunk allow-pass vlan 10 20 30 40 100
[sw2-GigabitEthernet0/0/2]quit
# 以下是允许 vlan 40 通过 GigabitEthernet 0/0/3
[sw2-GigabitEthernet0/0/3]port link-type access
[sw2-GigabitEthernet0/0/3]port default vlan 40
[sw2-GigabitEthernet0/0/3]quit
# 配置明细路由
[sw2]ip route-static 100.10.10.0 255.255.255.0 192.168.40.1
[sw2]quit
<sw2>save
```

sw4

```
<Huawei>system-view
[Huawei]sysname sw4
[sw4]vlan batch 10 20 30 40 100
[sw4]interface Ethernet 0/0/1
[sw4-Ethernet0/0/1]port link-type trunk
[sw4-Ethernet0/0/1]port trunk allow-pass vlan 10 20 30 40 100
[sw4-Ethernet0/0/1]quit
[sw4]interface Ethernet 0/0/3
[sw4-Ethernet0/0/3]port link-type access
[sw4-Ethernet0/0/3]port default vlan 30
[sw4-Ethernet0/0/3]quit
# 配置交换机 sw4 的互联 IP 地址
[sw4]interface Vlanif 100
[sw4-Vlanif100] ip address 10.10.10.4 255.255.255.0
# 配置默认路由
[sw4]ip route-static 0.0.0.0 0.0.0.0 10.10.10.254
[sw4]quit
<sw4>save
```

AR1

```
<Huawei>sys
Enter system view, return user view with Ctrl+Z.
[Huawei]sysname AR1
[AR1]interface GigabitEthernet 0/0/0
[AR1-GigabitEthernet0/0/0]ip address 192.168.40.1 24
[AR1]interface GigabitEthernet 0/0/1
# 创建模拟运营商公网地址
[AR1-GigabitEthernet0/0/1]ip address 100.10.10.1 24
# 配置默认路由
[AR1]ip route-static 0.0.0.0 0 192.168.40.254
```

步骤3 网络互通测试

```
PC1（IP:192.168.10.1）-ping-Server1(IP:192.168.30.1)- 通
PC>ping 192.168.30.1
Ping 192.168.30.1: 32 data bytes, Press Ctrl_C to break
From 192.168.30.1: bytes=32 seq=1 ttl=254 time=78 ms
From 192.168.30.1: bytes=32 seq=2 ttl=254 time=78 ms
……省略……
--- 192.168.30.1 ping statistics ---
  5 packet(s) transmitted
  5 packet(s) received
```

```
  0.00% packet loss
round-trip min/avg/max = 63/78/93 ms
```
PC1（IP:192.168.10.1）-ping-PC2（IP:192.168.20.1）-通
```
PC>ping 192.168.20.1
Ping 192.168.20.1: 32 data bytes, Press Ctrl_C to break
From 192.168.20.1: bytes=32 seq=1 ttl=127 time=94 ms
From 192.168.20.1: bytes=32 seq=2 ttl=127 time=78 ms
……省略……
--- 192.168.20.1 ping statistics ---
  5 packet(s) transmitted
  5 packet(s) received
```
PC1（IP:192.168.10.1）-ping-AR1（IP:192.168.40.1）-通
```
PC>ping 192.168.40.1
Ping 192.168.40.1: 32 data bytes, Press Ctrl_C to break
From 192.168.40.1: bytes=32 seq=1 ttl=254 time=93 ms
From 192.168.40.1: bytes=32 seq=2 ttl=254 time=63 ms
……省略……
--- 192.168.40.1 ping statistics ---
  5 packet(s) transmitted
  5 packet(s) received
```
PC2（IP:192.168.20.1）-ping-AR1（IP:192.168.40.1）-通
```
PC>ping 192.168.40.1
Ping 192.168.40.1: 32 data bytes, Press Ctrl_C to break
From 192.168.40.1: bytes=32 seq=1 ttl=254 time=63 ms
……省略……
--- 192.168.40.1 ping statistics ---
  5 packet(s) transmitted
  5 packet(s) received
```
Server1（IP:192.168.30.1）-ping-AR1（IP:192.168.40.1）-通，见图2-11-3。

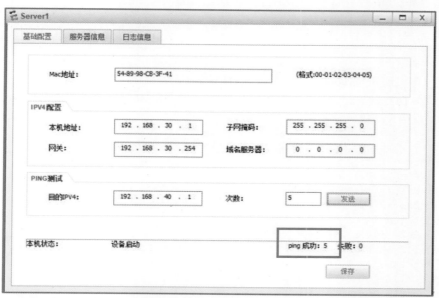

图2-11-3　Server1（IP:192.168.30.1）-ping-AR1（IP:192.168.40.1）-通

设备互通情况统计见表2-11-4。

表 2-11-4　设备互通情况统计

设备	IP	设备	IP	通信
PC1	192.168.10.1	PC2	192.168.20.1	通
PC1	192.168.10.1	Server1	192.168.30.1	通
PC1	192.168.10.1	sw3	10.10.10.3	通
PC1	192.168.10.1	sw2	10.10.10.254	通
PC1	192.168.10.1	sw4	10.10.10.4	通
PC1	192.168.10.1	AR1	192.168.40.1	通
PC1	192.168.10.1	AR1	100.10.10.1	通
PC2	192.168.20.1	PC3	192.168.30.1	通
PC3	192.168.30.1	AR1	192.168.40.1	通
PC3	192.168.30.1	AR1	100.10.10.1	通

步骤 4　ACL 控制列表项目规划

1）实现功能：PC1 不能通过 AR1 访问数据，但是能访问数据中心服务器。

sw2 交换机配置

```
# 创建基本 acl 2000
[sw2]acl 2000
# 定义规则 rule 5，拒绝来自源地址 192.168.10.1 的所有数据流量
[sw2-acl-basic-2000]rule 5 deny source 192.168.10.1 0
[sw2-acl-basic-2000]quit
[sw2]interface GigabitEthernet 0/0/3
# 在交换机 GigabitEthernet0/0/3 接口出方向上调用 ACL 策略
[sw2-GigabitEthernet0/0/3]traffic-filter outbound acl 2000
[sw2-GigabitEthernet0/0/3]quit
[sw2]quit
<sw2>save
```

测试（可以访问 Server1）

```
PC1（IP:192.168.10.1）-ping-Server1(IP:192.168.30.1)-通
PC>ping 192.168.30.1
Ping 192.168.30.1: 32 data bytes, Press Ctrl_C to break
From 192.168.30.1: bytes=32 seq=1 ttl=254 time=63 ms
……省略……

--- 192.168.30.1 ping statistics ---
  5 packet(s) transmitted
  5 packet(s) received
  0.00% packet loss
  round-trip min/avg/max = 47/72/94 ms
```

ping 路由器 AR1 不通（必经 AR1 的 GE 0/0/3，已经配置 ACL 策略限制）

```
PC>ping 192.168.40.1
Ping 192.168.40.1: 32 data bytes, Press Ctrl_C to break
Request timeout!
……省略……

--- 192.168.40.1 ping statistics ---
  5 packet(s) transmitted
```

```
  0 packet(s) received
  100.00% packet loss
```

2）通过 ACL 限制，网络打印机只允许某个网段打印。

实现功能：Server1 模拟为网络打印机，目前在企业中普遍使用网络打印机，规定 192.168.20.0 网段能打印，其他客户端不允许网络打印。

删除原来配置 ACL 配置

```
<sw2>system-view
[sw2]interface GigabitEthernet 0/0/3
# 进入交换机接口，删除 ACL 调用。
[sw2-GigabitEthernet0/0/3]undo traffic-filter outbound acl 2000
[sw2-GigabitEthernet0/0/3]quit
# 删除 ACL
[sw2]undo acl 2000
# 查看所有 ACL
[sw2]display acl all
Total nonempty ACL number is 0
```

继续该项目配置

```
<sw4>sys
Enter system view, return user view with Ctrl+Z.
[sw4]acl 2000
# 允许 192.168.20.0/24 数据通信
[sw4-acl-basic-2000]rule 5 permit source 192.168.20.0 0.0.0.255
# 拒绝其他网段通信
[sw4-acl-basic-2000]rule 10 deny source any
[sw4-acl-basic-2000]quit
[sw4]interface Ethernet 0/0/3
[sw4-Ethernet0/0/3]traffic-filter outbound acl 2000
[sw4-Ethernet0/0/3]quit
[sw4]quit
<sw4>save
```

测试 PC2（IP:192.168.20.1）-ping-Server1(IP:192.168.30.1)-通

```
PC>ping 192.168.30.1
Ping 192.168.30.1: 32 data bytes, Press Ctrl_C to break
From 192.168.30.1: bytes=32 seq=1 ttl=254 time=63 ms
……省略……

--- 192.168.30.1 ping statistics ---
  5 packet(s) transmitted
  5 packet(s) received
  0.00% packet loss
  round-trip min/avg/max = 62/75/94 ms
```

测试 PC1（IP:192.168.10.1）-ping-Server1(IP:192.168.30.1)-不通

```
PC>ping 192.168.30.1
Ping 192.168.30.1: 32 data bytes, Press Ctrl_C to break
Request timeout!
Request timeout!
……省略……

--- 192.168.30.1 ping statistics ---
  5 packet(s) transmitted
  0 packet(s) received
  100.00% packet loss
```

3）AR1 不允许 sw4 通过 ssh 访问 (AR1 接口下限制)

删除原来配置 ACL，并且查询所有 ACL。

```
[sw4]interface Ethernet 0/0/3
[sw4-Ethernet0/0/3]undo traffic-filter outbound acl 2000
[sw4-Ethernet0/0/3]quit
[sw4]undo acl 2000
[sw4]display acl all
Total nonempty ACL number is 0
```

AR1 的 ssh 远程访问配置

```
[AR1]aaa
[AR1-aaa]local-user fsm password cipher a123456
Info: Add a new user.
[AR1-aaa]local-user fsm service-type ssh
[AR1-aaa]local-user fsm privilege level 15
[AR1-aaa]quit
[AR1]stelnet server enable
[AR1]user-interface vty 0 4
[AR1-ui-vty0-4]authentication-mode aaa
[AR1-ui-vty0-4]protocol inbound ssh
[AR1-ui-vty0-4]quit
```

sw4-ssh-AR1　没有进行 ACL 限制，进行测试，可以 ssh 到 AR1

```
[sw4]stelnet 192.168.40.1
Please input the username:fsm
Trying 192.168.40.1 ...
Press CTRL+K to abort
Connected to 192.168.40.1 ...
Error: Failed to verify the server's public key.
Please run the command "ssh client first-time enable" to enable the first-time ac
cess function and try again.
[sw4]ssh client first-time enable
[sw4]stelnet 192.168.40.1
Please input the username:fsm
Trying 192.168.40.1 ...
Press CTRL+K to abort
Connected to 192.168.40.1 ...
The server is not authenticated. Continue to access it? [Y/N] :y
Save the server's public key? [Y/N] :y
The server's public key will be saved with the name 192.168.40.1. Please wait...
Enter password:
------------------------------------------------------------------------------
  User last login information:
------------------------------------------------------------------------------
  Access Type: SSH
  IP-Address : 192.168.40.254 ssh
  Time       : 2021-08-05 13:31:49-08:00
------------------------------------------------------------------------------
```

AR1-ACL 配置

```
[AR1]acl 3000
# 拒绝交换机 10.10.10.4 通过 22 端口（ssh）
[AR1-acl-adv-3000]rule 5 deny tcp source 10.10.10.4 0 destination-port eq 22
[AR1-acl-adv-3000]quit
```

```
[AR1]interface GigabitEthernet 0/0/0
# 在 GigabitEthernet0/0/0 inbound 调用 acl
[AR1-GigabitEthernet0/0/0]traffic-filter inbound acl 3000
[AR1-GigabitEthernet0/0/0]quit
[AR1]quit
<AR1>save
```

在 sw4 通过 ssh 协议访问 AR1，进行 ACL 限制之后，进行测试，已经无法通过 ssh 远程到 AR1。

```
[sw4]stelnet 192.168.40.1
Please input the username:fsm
Trying 192.168.40.1 ...
Press CTRL+K to abort
```

测试其他交换机 sw3 通过 ssh 协议访问 AR1，可以访问。

```
[sw3]stelnet 192.168.40.1
Please input the username:fsm
Trying 192.168.40.1 ...
Press CTRL+K to abort
Connected to 192.168.40.1 ...
Enter password:
    ------------------------------------------------------------
    User last login information:
    ------------------------------------------------------------
    Access Type: ssh
    IP-Address : 10.10.10.3 ssh
    Time       : 2021-08-05 13:46:28-08:00
    ------------------------------------------------------------
```

sw2 通过 ssh 协议访问 AR1，可以访问。

```
[sw2]stelnet 192.168.40.1
Please input the username:fsm
Trying 192.168.40.1 ...
Press CTRL+K to abort
Connected to 192.168.40.1 ...
Enter password:
    ------------------------------------------------------------
    User last login information:
    ------------------------------------------------------------
    Access Type: ssh
    IP-Address : 192.168.40.254 ssh
    Time       : 2021-08-05 23:42:46-08:00
    ------------------------------------------------------------
```

在 AR1 上查询 ACL 匹配条目

```
[AR1]display acl all
Total quantity of nonempty ACL number is 1
Advanced ACL 3000, 1 rule
Acl's step is 5
# 可以看到匹配条目 9 matches（9 条匹配策略），说明策略生效
 rule 5 deny tcp source 10.10.10.4 0 destination-port eq 22 (9 matches)
……
```

在 ACL3000 策略中，增加一条允许条目（在交换机接口下调用，ACL 默认最后一条就是允许，可以不加）

```
[AR1-acl-adv-3000]rule 10 permit tcp
```
再次查询
```
[AR1]display acl 3000
Advanced ACL 3000, 2 rules
Acl's step is 5
rule 5 deny tcp source 10.10.10.4 0 destination-port eq 22 (4 matches)
rule 10 permit tcp (39 matches)
```
……

查看调用信息
```
[AR1]display traffic-filter applied-record
--------------------------------------------------------------------
Interface                   Direction    AppliedRecord
--------------------------------------------------------------------
GigabitEthernet0/0/0        inbound      acl 3000
--------------------------------------------------------------------
```

4）AR1 不允许 sw4 通过 ssh 访问（在 AR1 的 vty 下进行限制）

删除接口下配置
```
[AR1-acl-adv-3000]undo rule 10
[AR1-GigabitEthernet0/0/0]undo traffic-filter inbound
[AR1]user-interface vty 0 4
# 用户接口调用 ACL
[AR1-ui-vty0-4]acl 3000 inbound
[AR1-ui-vty0-4]quit
[AR1]quit
<AR1>save
```

测试 sw4 无法访问通过 ssh-AR1
```
[sw4]stelnet 192.168.40.1
Please input the username:fsm
Trying 192.168.40.1 ...
Press CTRL+K to abort
```

测试 sw3 也无法访问，为什么？
```
[sw3]stelnet 192.168.40.1
Please input the username:fsm
Trying 192.168.40.1 ...
Press CTRL+K to abort
```

测试 sw2 也无法访问，为什么？
```
[sw2]stelnet 192.168.40.1
Please input the username:fsm
Trying 192.168.40.1 ...
Press CTRL+K to abort
```
……

非常重要，在 vty 下调用规则，最后默认是拒绝，而在交换机接口下调用最后默认是放行。所以说需要添加一条允许规则，如下：

```
[AR1-acl-adv-3000]rule 10 permit tcp
```

再次进行测试，就可以了。

```
[sw3]stelnet 192.168.40.1
Please input the username:fsm
Trying 192.168.40.1 ...
Press CTRL+K to abort
Connected to 192.168.40.1 ...
Enter password:
------------------------------------------------------------------------
  User last login information:
------------------------------------------------------------------------
  Access Type: ssh
  IP-Address : 192.168.40.254 ssh
  Time       : 2021-08-05 23:46:35-08:00
------------------------------------------------------------------------
<AR1>
[sw2]stelnet 192.168.40.1
Please input the username:fsm
Trying 192.168.40.1 ...
Press CTRL+K to abort
Connected to 192.168.40.1 ...
Enter password:
------------------------------------------------------------------------
  User last login information:
------------------------------------------------------------------------
  Access Type: ssh
  IP-Address : 10.10.10.3 ssh
  Time       : 2021-08-05 23:58:52-08:00
------------------------------------------------------------------------
<AR1>
```

查询匹配条目

```
[AR1]display acl all
Total quantity of nonempty ACL number is 1
Advanced ACL 3000, 2 rules
Acl's step is 5
rule 5 deny tcp source 10.10.10.4 0 destination-port eq 22 (8 matches)
rule 10 permit tcp (2 matches)
```

查看用户登录

```
[AR1]display users
  User-Intf    Delay     Type    Network Address         AuthenStatus    AuthorcmdFlag
  0   CON 0          00:00:00                                            pass
  Username : Unspecified
  129 VTY 0      00:00:00    ssh     100.10.10.1              pass
  Username : fsm
+ 130 VTY 1      00:00:00    ssh     192.168.40.1             pass
  Username : fsm
```

5）将 PC2 更换为 Client1，通过配置 ACL 控制策略，实现 Client1 能访问 ftp 服务器（Server1），但是无法 ping 通 Server1 服务器（IP:192.168.30.1），拓扑图见图 2-11-4。Server1 基础配置见图 2-11-5。如图 2-11-6 所示，在 Server1 菜单中选择"服务器信息"-"FtpServer"，配置文件目录，点击"启动"。

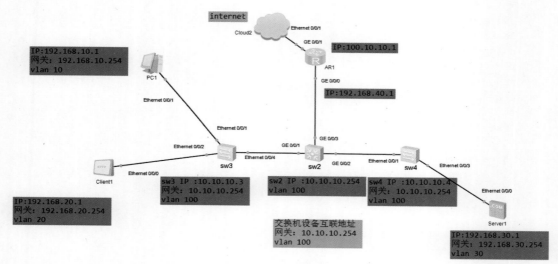

图2-11-4　拓扑图

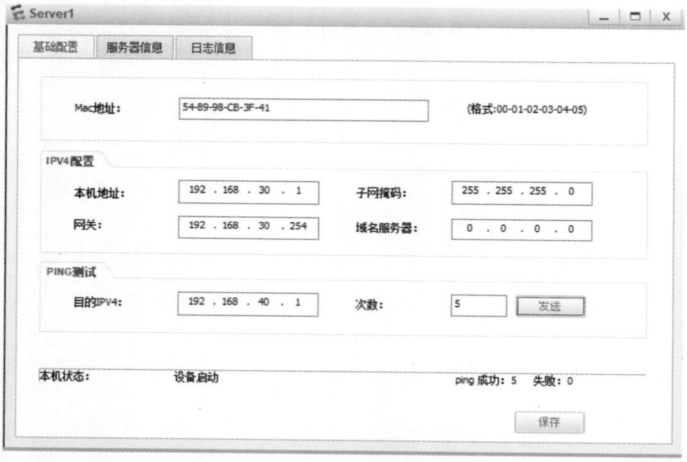

图2-11-5　Server 1 基础配置

图2-11-6　Server FtpServer 配置信息

Client1 基础配置如图 2-11-7 所示。

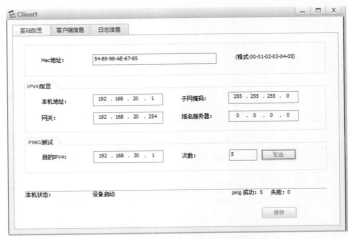

图2-11-7　Client1 基础配置

在图 2-11-7 中，Client1 选择"客户端信息"，选择"FtpClient"，填写服务器地址相关信息进行配置。如图 2-11-8 所示。

图2-11-8　FtpClient 配置信息

在图 2-11-8 中，点击"登录"，测试可以登录，并且将本地"fsm.txt"上传到服务器（Server 1）。如图 2-11-9 所示。

图2-11-9　登录测试

测试 Client1(IP:10.1.20.1)-Ping-Server 服务器 (IP: 192.168.30.1)- 通，如图 2-11-10 所示。

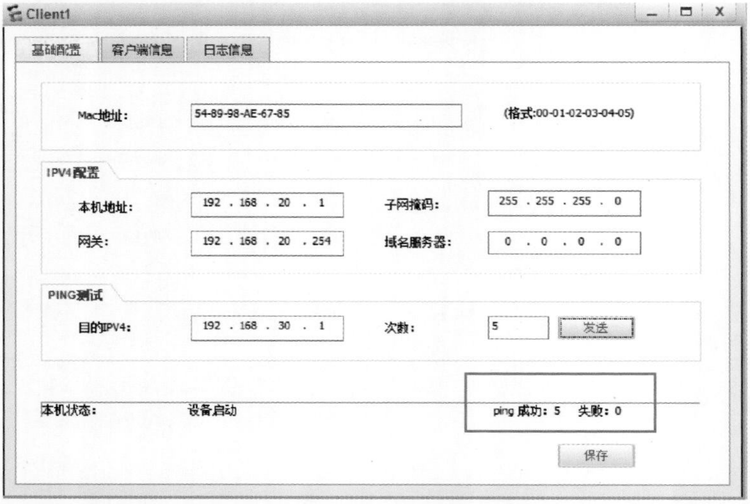

图2-11-10　测试Client1(IP:10.1.20.1)-Ping-Server 服务器(IP: 192.168.30.1)-通

在 sw3 配置 ACL 配置：

```
<sw3>sys
Enter system view, return user view with Ctrl+Z.
[sw3]acl 3000
# 允许 192.168.20.1 端口号 21（ftp）访问 Server (IP:192.168.30.1)
[sw3-acl-adv-3000]rule 5 permit tcp source 192.168.20.1 0 destination
192.168.30.1 0 destination-port eq 21
# 拒绝 192.168.20.1 ping Server (IP:192.168.30.1)
[sw3-acl-adv-3000]rule 10 deny icmp source 192.168.20.1 0 destination
 192.168.30.1 0
# 查询配置策略
[sw3-acl-adv-3000]dis this
#
acl number 3000
rule 5 permit tcp source 192.168.20.1 0 destination 192.168.30.1 0 destina
tion-port eq ftp
rule 10 deny icmp source 192.168.20.1 0 destination 192.168.30.1 0
#
return
[sw3-acl-adv-3000]quit
[sw3]quit
<sw3>save
<sw3>sys
[sw3]interface Ethernet 0/0/2
# 在 Ethernet0/0/2 接口 inbound 方向调用 acl 3000
[sw3-Ethernet0/0/2]traffic-filter inbound acl 3000
[sw3-Ethernet0/0/2]quit
[sw3]quit
<sw3>save
```

测试 Client(IP:192.168.20.1)- ping-Server(IP: 192.168.30.1)- 不通，如图 2-11-11 所示。

测试是否能登录 ftp 服务器，可以登录。如图 2-11-12 所示。

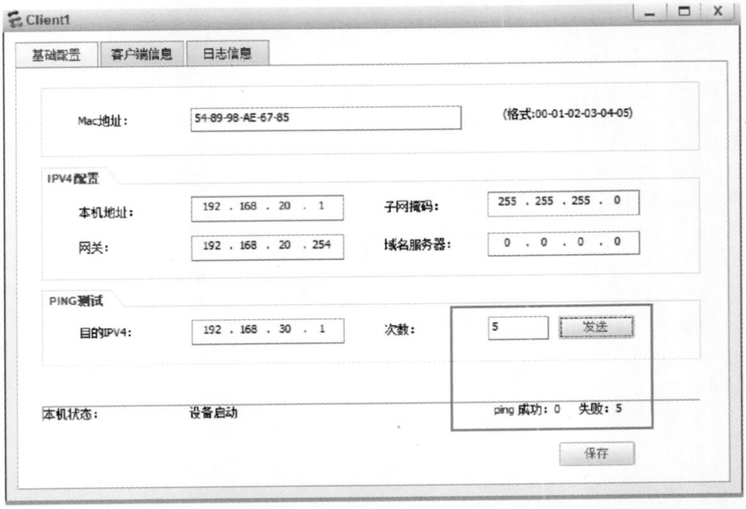

图2-11-11 测试 Client(IP:192.168.20.1)- ping-Server(IP: 192.168.30.1)- 不通

图2-11-12 登录ftp服务器

二、知识加油站

实例：网络打印机 IP 地址：10.1.20.31，要求设置：允许 10.1.20.0/24 访问打印，其他网段全部拒绝。分别使用 ACL 基本与高级策略实现。

交换机型号：S5130S-28P-EI，版本：version 7.1.070，Release 6320。

1. 基本 ACL 控制策略配置

```
[H3C.4.36]ACL number 2000
[H3C.4.36-acl-ipv4-basic-2000]rule 5 permit source 10.1.20.0 0.0.0.255
[H3C.4.36-acl-ipv4-basic-2000]rule 10 deny
[H3C.4.36-acl-ipv4-basic-2000]quit
[H3C.4.36]interface GigabitEthernet 1/0/10
[H3C.4.36-GigabitEthernet1/0/10]packet-filter 2000 outbound
```

133

```
[H3C.4.36-GigabitEthernet1/0/10]quit0
[H3C.4.36]save
```

2. 高级 ACL 控制

```
[H3C.4.36]acl number 3000
[H3C.4.36-acl-ipv4-adv-3000]rule 5 permit ip source 10.1.20.0 0.0.0.255 destination 10.1.20.31 0.0.0.0
[H3C.4.36-acl-ipv4-adv-3000]rule deny ip source any
[H3C.4.36-acl-ipv4-adv-3000]quit
[H3C.4.36]interface  GigabitEthernet 1/0/10
[H3C.4.36-GigabitEthernet1/0/10]packet-filter 3000 outbound
[H3C.4.36-GigabitEthernet1/0/10]quit
[H3C.4.36]save
```

项目 12　环回接口与 H3C 模拟器网络设备通信

在进行网络实验时，创建 loopback 接口（环回网卡）方便实验操作。本节详细分解具体配置以及操作。

项目简介

本项目通过物理主机配置环回适配器，与华三、华为模拟器网络设备互联通信。同时本项目还介绍了如何使用物理主机网卡与模拟中网络设备通信，项目中通过 SecureCRT 访问网络设备，进行测试工作。

一、在 Windows 10 中如何配置 loopback 网卡

步骤 1　电脑桌面，点击"此电脑"图标，右键选择"管理"。选择"设备管理器"，右侧"fsm"，fsm 是本项目中计算机的名字，如图 2-12-1 所示。再点击菜单中"操作"，选择"添加过时硬件"。见图 2-12-2。

步骤 2　选择"安装我手动从列表选择的硬件（M）"，然后点击"下一步"。如图 2-12-3 所示。

步骤 3　选择"网络适配器"。如图 2-12-4 所示。

步骤 4　厂商选择"Microsoft"，型号选择"Microsoft KM-TEST 环回适配器"。如图 2-12-5 所示。

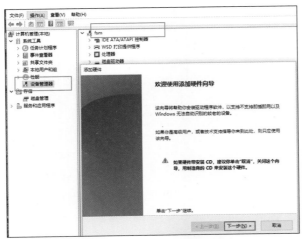

图2-12-1　选择设备管理器

图2-12-2　选择添加过时硬件

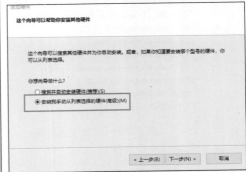

图2-12-3　选择"安装我手动从列表选择的硬件（M）"

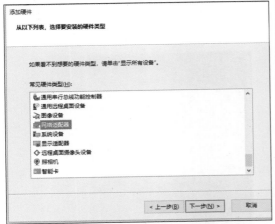

图2-12-4　选择"网络适配器"

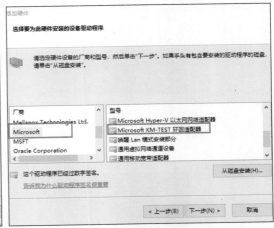

图2-12-5　选择"Microsoft KM-TEST环回适配器"

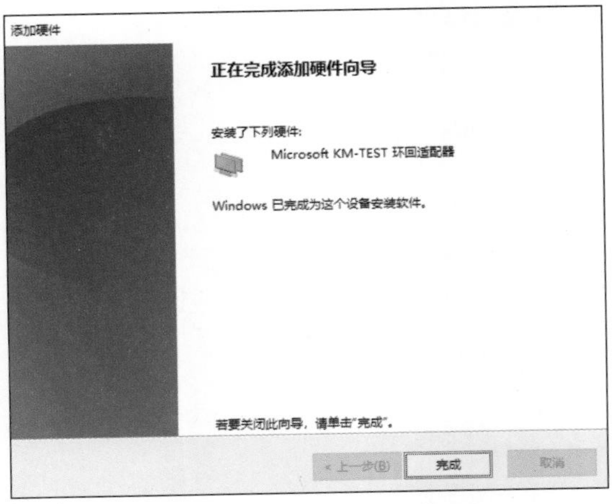

步骤5　安装完成。如图2-12-6所示。

步骤6　打开网络适配器，查看新增加适配器，如图2-12-7所示，然后打开"网络与共享中心"，选择"更改适配器设置"进行环回地址配置。如图2-12-8所示。

配置完毕，重启生效。

图2-12-6　安装完成

图2-12-7　增加适配器

图2-12-8　"更改适配器设置"进行环回地址配置

二、HCL 模拟器中交换机与环回接口通信

1. 创建交换机、Host、连线

如图 2-12-9 ～图 2-12-11 所示。

图2-12-9　新建交换机

图2-12-10　选择"Host"

图2-12-11　选择"Manual"

在图 2-12-12 中，鼠标放在 Host 上，左键选择图中的环回网卡。

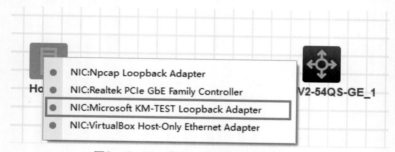
图2-12-12　"Host"选择环回接口网卡

IP 地址配置见图 2-12-13。

(a) 交换机IP地址配置　　　　　　　　(b) 主机IP地址配置

图2-2-13　IP地址配置

2. H3C 交换机配置

步骤 1　基本配置

```
<H3C>sys
[H3C]interface GigabitEthernet 1/0/1
# 接口切换到三层模式
H3C 三层端口切换命令
命令：port link-mode bridge /route
bridge 是桥接（也就是二层），route 是路由（也就是三层）
[H3C-GigabitEthernet1/0/1]port link-mode route
# 配置与环回在一个网段的 IP 地址
[H3C-GigabitEthernet1/0/1]ip address 172.99.0.2 24
```

步骤 2　测试与物理主机 ping 情况

1）交换机 H3C（172.99.0.2）-ping- 物理主机（172.99.0.1）- 通

```
<H3C>ping 172.99.0.1
Ping 172.99.0.1 (172.99.0.1): 56 data bytes, press CTRL_C to break
56 bytes from 172.99.0.1: icmp_seq=0 ttl=64 time=0.000 ms
……省略……
--- Ping statistics for 172.99.0.1 ---
5 packet(s) transmitted, 5 packet(s) received, 0.0% packet loss
round-trip min/avg/max/std-dev = 0.000/1.000/4.000/1.549 ms
```

2）物理主机（172.99.0.1）-ping- 交换机 H3C（172.99.0.2）- 通

```
C:\Users\fsm>ping 172.99.0.2
正在 Ping 172.99.0.2 具有 32 字节的数据：
来自 172.99.0.2 的回复：字节=32 时间<1ms TTL=255
来自 172.99.0.2 的回复：字节=32 时间<1ms TTL=255
……省略……
172.99.0.2 的 Ping 统计信息：
    数据包：已发送 = 4，已接收 = 4，丢失 = 0 (0% 丢失)，
往返行程的估计时间（以毫秒为单位）：
最短 = 0ms，最长 = 0ms，平均 = 0ms
```

步骤 3　测试通过物理主机 SecureCRT 软件 -ssh 协议 - 访问 H3C 交换机

1）交换机 ssh 远程访问配置

```
[H3C]ssh server enable
[H3C]local-user fsm
New local user added.
```

```
[H3C-luser-manage-fsm]password simple a123456
[H3C-luser-manage-fsm]service-type ssh
[H3C-luser-manage-fsm]authorization-attribute user-role network-admin
[H3C]user-interface vty 0 4
[H3C-line-vty0-4]authentication-mode scheme
[H3C-line-vty0-4]quit
[H3C]save
```

2）物理主机通过SecureCRT-ssh软件访问交换机设置，如图2-12-14和图2-12-15所示。

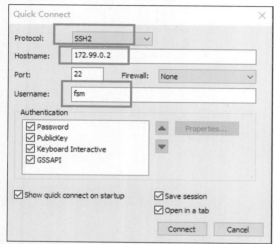

图2-12-14　SecureCRT-ssh软件访问交换机

图2-12-15　输入用户名以及密码

步骤4　通过"display users"，查询访问用户。如图2-12-16所示。

图2-12-16　访问用户是fsm，通过"vty"登录

三、知识加油站

1. H3C 模拟器通过选择本机网卡桥接通信

1）Host 选择物理主机网卡，如图2-12-17所示。

2）拓扑图，如图2-12-18所示。

交换机地址配置：10.1.20.5，物理主机地址：10.1.20.66。

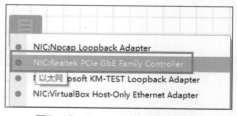

图2-12-17　Host物理主机网卡

3）交换机基础配置：

```
[H3C-GigabitEthernet1/0/1]port link-mode route
```

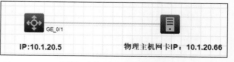

图2-12-18　拓扑图

```
[H3C-GigabitEthernet1/0/1]ip address 10.1.20.5 24
```

4）通信测试

交换机 H3C（10.1.20.5）-ping- 物理主机（10.1.20.66）- 通
```
[H3C]ping 10.1.20.66
Ping 10.1.20.66 (10.1.20.66): 56 data bytes, press CTRL_C to break
56 bytes from 10.1.20.66: icmp_seq=0 ttl=64 time=1.000 ms
56 bytes from 10.1.20.66: icmp_seq=1 ttl=64 time=0.000 ms
……省略……
--- Ping statistics for 10.1.20.66 ---
5 packet(s) transmitted, 5 packet(s) received, 0.0% packet loss
round-trip min/avg/max/std-dev = 0.000/0.600/1.000/0.490 ms
```
物理主机（IP: 10.1.20.66）-ping- 交换机 H3C（IP: 10.1.20.5）- 通
```
C:\Users\fsm>ping 10.1.20.5
正在 Ping 10.1.20.5 具有 32 字节的数据：
来自 10.1.20.5 的回复：字节=32 时间<1ms TTL=255
来自 10.1.20.5 的回复：字节=32 时间<1ms TTL=255
……省略……
10.1.20.5 的 Ping 统计信息：
    数据包：已发送 = 4，已接收 = 4，丢失 = 0 (0% 丢失)，
往返行程的估计时间（以毫秒为单位）：
    最短 = 0ms，最长 = 0ms，平均 = 0ms
```

5）测试通过物理主机 SecureCRT 软件 -telnet 协议 - 访问 H3C 交换机

H3C 交换机 telnet 协议 配置：
```
[H3C]telnet server enable
[H3C]local-user fsm
New local user added.
[H3C-luser-manage-fsm]password simple a12346
[H3C-luser-manage-fsm]service-type telnet
[H3C-luser-manage-fsm]authorization-attribute user-role network-admin
[H3C-luser-manage-fsm]quit
[H3C]user-interface vty 0 63
[H3C-line-vty0-63]authentication-mode scheme
[H3C-line-vty0-63]quit
[H3C]save force
```

物理主机 SecureCRT-telnet 访问交换机设置，见图 2-12-19。

输入用户名以及密码，登录信息如图 2-12-20 所示。

2. H3C 模拟器通过选择本虚拟网卡桥接通信

1）见图 2-12-21，网络适配器选择"VirtualBox"，并且配置地址。拓扑图见图 2-12-22。

2）交换机基础配置
```
[H3C-GigabitEthernet1/0/1]port link-mode route
[H3C-GigabitEthernet1/0/1]ip address 192.168.56.10 24
```

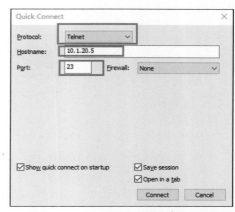

图2-12-19　SecureCRT-telnet访问交换机设置

```
********************************************************************
* Copyright (c) 2004-2017 New H3C Technologies Co., Ltd. All rights reserved.*
* Without the owner's prior written consent,                        *
* no decompiling or reverse-engineering shall be allowed.           *
********************************************************************

login: fsm
Password:
<H3C>sy
<H3C>system-view
System View: return to User View with Ctrl+Z.
[H3C]dis
[H3C]display users
 Idx  Line            Idle        Time              Pid    Type
  0   CON 0           00:00:45    Aug 06 22:47:40   225
+ 85  VTY 1           00:00:00    Aug 06 23:01:18   281    TEL

Following are more details.
VTY 1   :
        : User name: fsm
        : Location: 10.1.20.66
 +      : Current operation user.
 F      : Current operation user works in async mode.
[H3C]
```

图2-12-20　登录信息

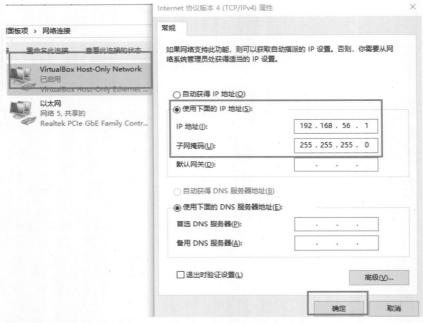

图2-12-21　配置"VirtualBox"虚拟网卡地址

图2-12-22　拓扑图

[H3C-GigabitEthernet1/0/1]quit

3）交换机主机（192.168.56.10）-ping- 物理主机（192.168.56.1）- 通

[H3C]ping 192.168.56.1
Ping 192.168.56.1 (192.168.56.1): 56 data bytes, press CTRL_C to break
56 bytes from 192.168.56.1: icmp_seq=0 ttl=64 time=1.000 ms
56 bytes from 192.168.56.1: icmp_seq=1 ttl=64 time=0.000 ms
……省略……

```
--- Ping statistics for 192.168.56.1 ---
5 packet(s) transmitted, 5 packet(s) received, 0.0% packet loss
round-trip min/avg/max/std-dev = 0.000/0.400/1.000/0.490 ms
```
物理主机（192.168.56.1）-ping- 交换机主机（192.168.56.10）- 通
```
C:\Users\fsm>ping 192.168.56.10
正在 Ping 192.168.56.10 具有 32 字节的数据：
来自 192.168.56.10 的回复：字节=32 时间<1ms TTL=255
……省略……
192.168.56.10 的 Ping 统计信息：
    数据包：已发送 = 4，已接收 = 4，丢失 = 0 (0% 丢失)，
    往返行程的估计时间（以毫秒为单位）：
    最短 = 0ms，最长 = 0ms，平均 = 0ms
C:\Users\fsm>
```

4）ssh 协议远程访问测试

H3C 交换机 ssh 配置忽略，仅测试，见图 2-12-23。登录信息见图 2-12-24。

图2-12-23　配置信息　　　　　　　　图2-12-24　登录信息

3. 华为 eNSP 环回接口与物理主机通信

1）拓扑图见图 2-12-25。

Cloud2 配置见图 2-12-26，绑定环回地址（IP：172.168.16.1），出端口编号选择 2，选中双向通道。

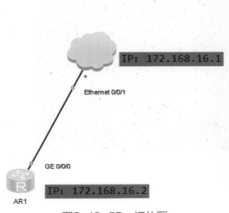

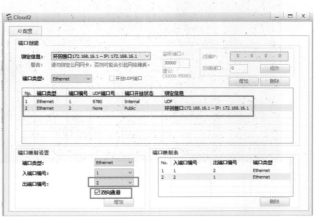

图2-12-25　拓扑图　　　　　　　　图2-12-26　Cloud2 配置

2）华为路由器配置

```
<Huawei>system-view
Enter system view, return user view with Ctrl+Z.
[Huawei]sysname AR1
[AR1]interface GigabitEthernet 0/0/0
[AR1-GigabitEthernet0/0/0]ip address 172.168.16.2 24
[AR1-GigabitEthernet0/0/0]quit
[AR1]quit
```

3）通信测试

路由器（IP:172.168.16.2）-ping- 物理主机（IP:172.168.16.1）- 通

```
<AR1>ping 172.168.16.1
  PING 172.168.16.1: 56  data bytes, press CTRL_C to break
  Reply from 172.168.16.1: bytes=56 Sequence=1 ttl=64 time=60 ms
  Reply from 172.168.16.1: bytes=56 Sequence=2 ttl=64 time=10 ms
      ……省略……
--- 172.168.16.1 ping statistics ---
  5 packet(s) transmitted
  5 packet(s) received
  0.00% packet loss
  round-trip min/avg/max = 1/18/60 ms
```

物理主机（172.168.16.1）-ping- 路由器（172.168.16.2）- 通

```
C:\Users\fsm>ping 172.168.16.2
正在 Ping 172.168.16.2 具有 32 字节的数据：
来自 172.168.16.2 的回复：字节 =32 时间 =16ms TTL=255
……省略……
172.168.16.2 的 Ping 统计信息：
    数据包：已发送 = 4，已接收 = 4，丢失 = 0（0% 丢失），
往返行程的估计时间 ( 以毫秒为单位 )：
    最短 = 5ms，最长 = 16ms，平均 = 10ms
C:\Users\fsm>
```

4）物理主机 telnet 访问路由器，见图 2-12-27。

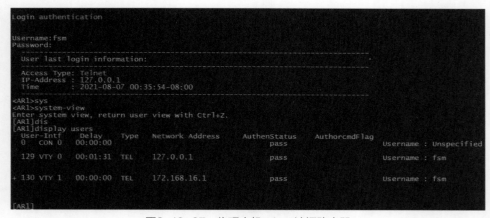

图2-12-27　物理主机telnet访问路由器

4. 华为 eNSP 虚拟网卡与物理主机通信

1）Cloud 配置见图 2-12-28。拓扑图见图 2-12-29。

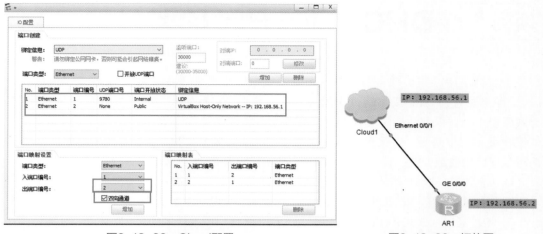

图2-12-28　Cloud配置　　　　　图2-12-29　拓扑图

2）路由器配置

[AR1-GigabitEthernet0/0/0]ip address 192.168.56.2 24

3）通信测试

路由器（IP：192.168.56.2）- ping - 物理主机（IP：192.168.56.1）- 通

```
<AR1>ping 192.168.56.1
  PING 192.168.56.1: 56  data bytes, press CTRL_C to break
  Reply from 192.168.56.1: bytes=56 Sequence=1 ttl=128 time=50 ms
……省略……
  Reply from 192.168.56.1: bytes=56 Sequence=5 ttl=128 time=20 ms
--- 192.168.56.1 ping statistics ---
  5 packet(s) transmitted
  5 packet(s) received
  0.00% packet loss
  round-trip min/avg/max = 10/22/50 ms
```

4）物理主机通过 telnet 访问 AR，见图 2-12-30，登录信息见图 2-12-31。路由器 AR telnet 协议配置略。

图2-12-30　物理主机通过telnet访问AR

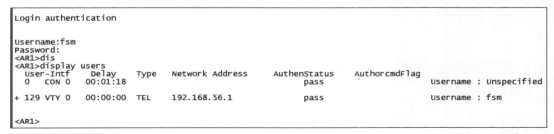

图2-12-31　登录信息

项目 13　DHCP 动态获取 IP 地址

企业网络中，主机或者网络设备联网需要配置相关 IP 参数，手工配置工作量比较大（属于静态配置），且容易出错，如配置重复或者用户自行修改，都可能引起 IP 地址冲突，导致不能上网。采取动态获取 IP 地址等网络参数，能避免 IP 地址冲突，并且减少运维人员的工作量。大型网络一般采用 Windows 或者 Linux 服务器搭建专用的 DHCP 服务，小型网络可以在路由器或者交换机上配置 DHCP，动态获取 IP 地址。

DHCP 动态获取 IP 地址的过程简要如下：

DHCP 通过发送（DHCP Discover）广播报文，发现 DHCP 服务器；DHCP 服务器分配一个 IP 地址通过（DHCP Offer）单播报文；DHCP 客户端发送（DHCP Request）广播报文，请求 IP 地址；DHCP 服务器收到请求报文后，服务器向 DHCP 客户端发送（DHCP ACK）单播确认报文，包含 IP 地址相关信息。

通过抓取报文，清楚看到 DHCP 工作过程：在 Filter 中输入"bootp"过滤"DHCP"报文。图 2-13-1 是 DHCP 客户端获取 IP 地址的报文，分别是红、蓝色框内报文。

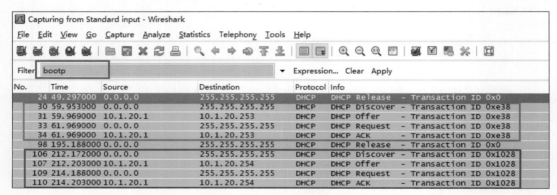

图2-13-1　"DHCP"报文

项目简介

DHCP 常见的部署方法有 interface 与 global 两种模式。PC 配置为"自动获取 IP 地址"，这种方式获取的 IP 地址是不固定的。对于特殊情况，终端电脑获取的 IP 地址需要固定不变的情况，可以在路由器中配置 IP 地址与 MAC 地址绑定来实现。

一、路由器接口配置 DHCP（interface 模式）

1. 项目拓扑图

如图 2-13-2 所示，实现功能：路由器配置 DHCP 服务，PC 设置自动获取 IP 地址，实现网络通信。

2. 项目配置步骤

步骤1　AR1 路由器基本配置

```
<Huawei>system-view
Enter system view, return user view with Ctrl+Z.
[Huawei]sysname DHCP-Server
# 使能 DHCP
[DHCP-Server]dhcp enable
Info: The operation may take a few seconds. Please wait for a moment.done.
# 接口配置 IP 地址（以下两条）
[DHCP-Server]interface GigabitEthernet 0/0/0
[DHCP-Server-GigabitEthernet0/0/0]ip address 10.1.20.1 24
# 接口下 DHCP 选择 interface
[DHCP-Server-GigabitEthernet0/0/0]
[DHCP-Server-GigabitEthernet0/0/0]dhcp select interface
# 配置 DNS 地址（主备）
[DHCP-Server-GigabitEthernet0/0/0]dhcp server dns-list 192.168.9.99
[DHCP-Server-GigabitEthernet0/0/0]dhcp server dns-list 192.168.9.98
[DHCP-Server-GigabitEthernet0/0/0]quit
[DHCP-Server]quit
<DHCP-Server>save
```

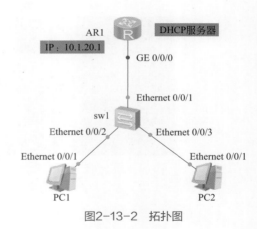

图2-13-2　拓扑图

步骤2　PC 配置

PC1：

PC1-IPv4 配置为 DHCP，如图 2-13-3 所示。

图2-13-3　PC1配置自动获取IP地址

在 PC1 的菜单"命令行"中输入 ipconfig，查看自动获取的 IP 地址相关信息，如图 2-13-4 所示。

145

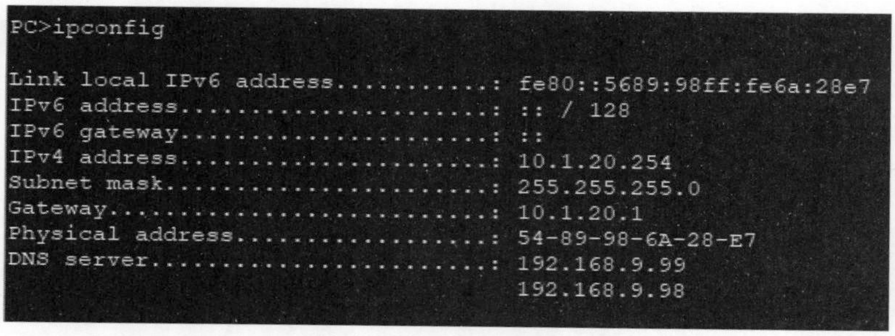

```
PC>ipconfig

Link local IPv6 address...........: fe80::5689:98ff:fe6a:28e7
IPv6 address......................: :: / 128
IPv6 gateway......................: ::
IPv4 address......................: 10.1.20.254
Subnet mask.......................: 255.255.255.0
Gateway...........................: 10.1.20.1
Physical address..................: 54-89-98-6A-28-E7
DNS server........................: 192.168.9.99
                                    192.168.9.98
```

图2-13-4　PC1自动获取的IP地址（IP：10.1.20.254）

PC2:

PC2-IPv4 配置为 DHCP，如图 2-13-5 所示。

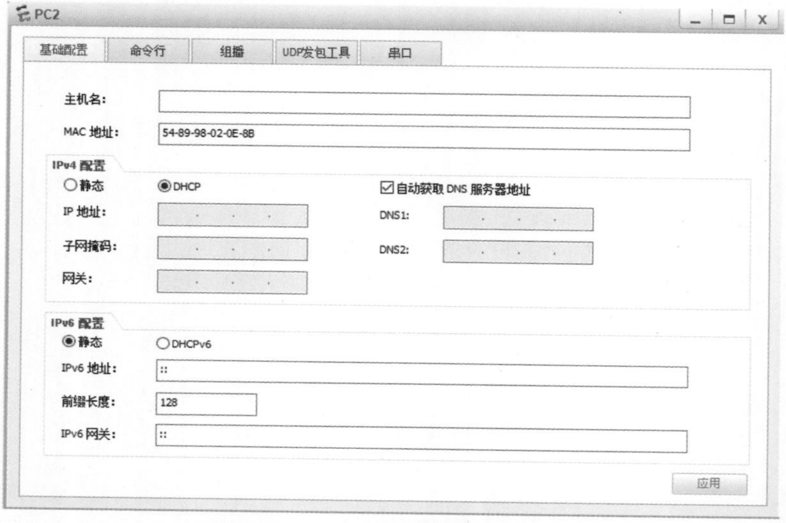

图2-13-5　PC2配置自动获取IP地址

在 PC2 的菜单“命令行”输入 ipconfig，查看自动获取的 IP 地址相关信息，如图 2-13-6 所示。

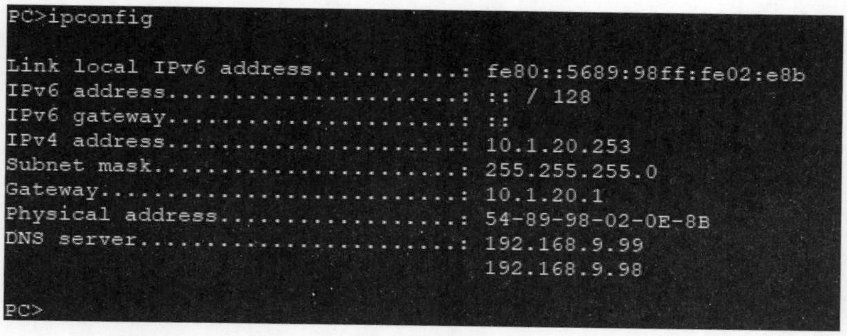

```
PC>ipconfig

Link local IPv6 address...........: fe80::5689:98ff:fe02:e8b
IPv6 address......................: :: / 128
IPv6 gateway......................: ::
IPv4 address......................: 10.1.20.253
Subnet mask.......................: 255.255.255.0
Gateway...........................: 10.1.20.1
Physical address..................: 54-89-98-02-0E-8B
DNS server........................: 192.168.9.99
                                    192.168.9.98

PC>
```

图2-13-6　PC2自动获取的IP地址（IP：10.1.20.253）

从以上可以看出 IP 地址分配是从地址大的往小的分配，PC1 的 IP 地址 -10.1.20.254，PC2 的 IP 地址 -10.1.20.253。

测试：PC1（IP：10.1.20.254）-ping-PC2（IP：10.1.20.253）- 通，见图 2-13-7。

```
PC>ping 10.1.20.253

Ping 10.1.20.253: 32 data bytes, Press Ctrl_C to break
From 10.1.20.253: bytes=32 seq=1 ttl=128 time=47 ms
From 10.1.20.253: bytes=32 seq=2 ttl=128 time=47 ms
From 10.1.20.253: bytes=32 seq=3 ttl=128 time=31 ms
From 10.1.20.253: bytes=32 seq=4 ttl=128 time=31 ms
From 10.1.20.253: bytes=32 seq=5 ttl=128 time=16 ms

--- 10.1.20.253 ping statistics ---
  5 packet(s) transmitted
  5 packet(s) received
  0.00% packet loss
  round-trip min/avg/max = 16/34/47 ms

PC>
```

图2-13-7　PC1（IP：10.1.20.254）-ping-PC2（IP：10.1.20.253）

二、全局配置 DHCP，地址池配置（global 模式）

1. 拓扑图（图2-13-2）
2. 项目配置步骤

步骤1　AR1 路由器配置

```
<Huawei>system-view
Enter system view, return user view with Ctrl+Z.
[Huawei]sysname DHCP-Server
[DHCP-Server]interface GigabitEthernet 0/0/0
[DHCP-Server-GigabitEthernet0/0/0]ip address 10.1.20.1 24
[DHCP-Server-GigabitEthernet0/0/0]quit
[DHCP-Server]dhcp enable
# 配置地址池名字为 fsm
[DHCP-Server]ip pool fsm
Info: It's successful to create an IP address pool.
[DHCP-Server-ip-pool-fsm]
# 配置 DHCP 网段以及子网掩码
[DHCP-Server-ip-pool-fsm]network 10.1.20.0 mask 255.255.255.0
# 配置网关
[DHCP-Server-ip-pool-fsm]gateway-list 10.1.20.1
# 配置 DNS（主备）
[DHCP-Server-ip-pool-fsm]dns-list 192.168.9.99
[DHCP-Server-ip-pool-fsm]dns-list 192.168.9.98
[DHCP-Server-ip-pool-fsm]quit
[DHCP-Server]interface GigabitEthernet 0/0/0
# 进入接口视图，DHCP 选择"global"
[DHCP-Server-GigabitEthernet0/0/0]dhcp select global
[DHCP-Server-GigabitEthernet0/0/0]quit
[DHCP-Server]quit
<DHCP-Server>save
```

步骤2　查看地址池信息

如图 2-13-8 所示。

步骤3　IP 地址释放命令

IP 地址释放命令：ipconfig /release，如图 2-13-9 所示。从信息中看到当输入释放 IP 地址命令后，IP 地址参数消失了。

```
[DHCP-Server]display ip pool
-----------------------------------------------------------------
Pool-name            : fsm
Pool-No              : 0
Position             : Local         Status           : Unlocked
Gateway-0            : 10.1.20.1
Mask                 : 255.255.255.0
VPN instance         : --

IP address Statistic
   Total             :253
   Used              :3             Idle             :250
   Expired           :0             Conflict         :0       Disable    :0
```

图2-13-8　地址池信息

```
PC>ipconfig /release

IP Configuration

Link local IPv6 address.............: fe80::5689:98ff:fe6a:28e7
IPv6 address........................: :: / 128
IPv6 gateway........................: ::
IPv4 address........................: 0.0.0.0
Subnet mask.........................: 0.0.0.0
Gateway.............................: 0.0.0.0
Physical address....................: 54-89-98-6A-28-E7
DNS server..........................:
```

图2-13-9　IP地址释放命令

步骤4　IP 地址重新获取

IP 地址重新获取命令：ipconfig /renew，如图 2-13-10 所示。

```
PC>ipconfig /renew

IP Configuration

Link local IPv6 address.............: fe80::5689:98ff:fe6a:28e7
IPv6 address........................: :: / 128
IPv6 gateway........................: ::
IPv4 address........................: 10.1.20.253
Subnet mask.........................: 255.255.255.0
Gateway.............................: 10.1.20.1
Physical address....................: 54-89-98-6A-28-E7
DNS server..........................: 192.168.9.99
                                      192.168.9.98
```

图2-13-10　IP地址重新获取

从以上看到，IP 地址重新获取相关参数。

步骤5　固定 IP 地址配置

在某些特殊场合，网络设备需要固定 IP 地址，配置办法如下：

比如 PC2 固定 IP：10.1.20.66

配置 IP 绑定 MAC 地址

[DHCP-Server-ip-pool-fsm]static-bind ip-address 10.1.20.66 mac-address 5489-9802-0E8B

提示错误信息，说明在地址池中已经存在地址

Error:The static-MAC is exist in this IP-pool.

释放 IP 地址，如图 2-13-11 所示。

```
PC>ipconfig /release

IP Configuration

Link local IPv6 address...........: fe80::5689:98ff:fe02:e8b
IPv6 address.....................: :: / 128
IPv6 gateway.....................: ::
IPv4 address.....................: 0.0.0.0
Subnet mask......................: 0.0.0.0
Gateway..........................: 0.0.0.0
Physical address.................: 54-89-98-02-0E-8B
DNS server.......................:
```

图2-13-11　释放IP地址

重新配置 IP 地址与 MAC 绑定：
[DHCP-Server-ip-pool-fsm]static-bind ip-address 10.1.20.66 mac-address 5489-9802-0E8B
重新租用新 IP 地址，如图 2-13-12 所示。

```
PC>ipconfig /renew

IP Configuration

Link local IPv6 address...........: fe80::5689:98ff:fe02:e8b
IPv6 address.....................: :: / 128
IPv6 gateway.....................: ::
IPv4 address.....................: 10.1.20.66
Subnet mask......................: 255.255.255.0
Gateway..........................: 10.1.20.1
Physical address.................: 54-89-98-02-0E-8B
DNS server.......................: 192.168.9.99
                                   192.168.9.98
```

图2-13-12　重新获取IP地址

查询 IP 地址使用详细情况，如图 2-13-13 所示。

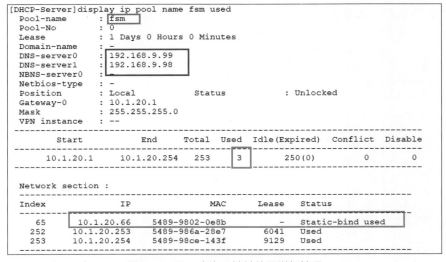

图2-13-13　查询IP地址使用详细情况

从图中可以看出，DHCP 名字为 fsm 地址池，一共使用了三个地址，其中 IP：10.1.20.66 为静态绑定，还有两个地址是 10.1.20.253/10.1.20.254。

三、通过中继配置 DHCP 服务（relay 模式）

如果 DHCP 客户端和 DHCP 服务器不在同一网段内，避免在每个网段都部署 DHCP 服务器，DHCP 中继负责 DHCP 服务器与 DHCP 客户端之间的 DHCP 报文转发。

1. 拓扑图

见图 2-13-14。

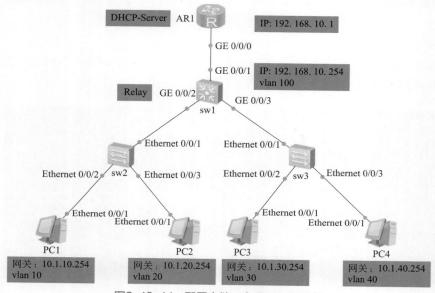

图2-13-14　配置中继，实现DHCP服务

2. 项目配置步骤

步骤1　交换机基本配置

① 交换机 sw2 配置
```
<Huawei>system-view
Enter system view, return user view with Ctrl+Z.
[Huawei]sysname sw2
[sw2]vlan batch 10 20
[sw2]interface Ethernet 0/0/1
[sw2-Ethernet0/0/1]port link-type trunk
[sw2-Ethernet0/0/1]port trunk allow-pass vlan 10 20
[sw2]interface Ethernet 0/0/2
[sw2-Ethernet0/0/2]port link-type access
[sw2-Ethernet0/0/2]port default vlan 10
[sw2-Ethernet0/0/2]quit
[sw2]interface Ethernet 0/0/3
[sw2-Ethernet0/0/3]port link-type access
[sw2-Ethernet0/0/3]port default vlan 20
[sw2-Ethernet0/0/3]quit
```

```
[sw2]quit
<sw2>save
```
② 交换机 sw3 配置
```
<Huawei>system-view
Enter system view, return user view with Ctrl+Z.
[Huawei]sysname sw3
[sw3]vlan batch 30 40
[sw3]interface Ethernet 0/0/1
[sw3-Ethernet0/0/1]port link-type trunk
[sw3-Ethernet0/0/1]port trunk allow-pass vlan 30 40
[sw3-Ethernet0/0/1]quit
[sw3]interface Ethernet 0/0/2
[sw3-Ethernet0/0/2]port link-type access
[sw3-Ethernet0/0/2]port default vlan 30
[sw3-Ethernet0/0/2]quit
[sw3]interface Ethernet 0/0/3
[sw3-Ethernet0/0/3]port link-type access
[sw3-Ethernet0/0/3]port default vlan 40
[sw3-Ethernet0/0/3]quit
[sw3]quit
<sw3>save
```
③ 交换机 sw1 配置
```
<Huawei>system-view
[Huawei]sysname sw1
[sw1]vlan batch 10 20 30 40 100
[sw1]interface GigabitEthernet 0/0/1
[sw1-GigabitEthernet0/0/1]port link-type access
[sw1-GigabitEthernet0/0/1]port default vlan 100
[sw1-GigabitEthernet0/0/1]quit
[sw1]interface GigabitEthernet 0/0/2
[sw1-GigabitEthernet0/0/2]port link-type trunk
[sw1-GigabitEthernet0/0/2]port trunk allow-pass vlan 10 20
[sw1-GigabitEthernet0/0/2]quit
[sw1]interface GigabitEthernet 0/0/3
[sw1-GigabitEthernet0/0/3]port link-type trunk
[sw1-GigabitEthernet0/0/3]port trunk allow-pass vlan 30 40
[sw1-GigabitEthernet0/0/3]quit
[sw1]interface Vlanif 100
[sw1-Vlanif100]ip address 192.168.10.254 24
[sw1-Vlanif100]quit
[sw1]interface Vlanif 10
[sw1-Vlanif10]ip address 10.1.10.254 24
[sw1-Vlanif10]quit
[sw1]interface Vlanif 20
[sw1-Vlanif20]ip address 10.1.20.254 24
[sw1-Vlanif20]quit
[sw1]interface Vlanif 30
[sw1-Vlanif30]ip address 10.1.30.254 24
[sw1-Vlanif30]quit
[sw1]interface Vlanif 40
[sw1-Vlanif40]ip address 10.1.40.254 24
```

```
[sw1-Vlanif40]quit
[sw1]quit
<sw1>save
```
④ 路由器 DHCP-Server 配置
```
<Huawei>system-view
[Huawei]sysname DHCP-Server
[DHCP-Server]interface GigabitEthernet 0/0/0
[DHCP-Server-GigabitEthernet0/0/0]ip address 192.168.10.1 24
[DHCP-Server-GigabitEthernet0/0/0]quit
# 配置默认路由，下一跳 IP 地址 192.168.10.254
[DHCP-Server]ip route-static 0.0.0.0 0 192.168.10.254
```

步骤 2　配置地址池
```
[DHCP-Server]dhcp enable
[DHCP-Server]ip pool fsm10
Info: It's successful to create an IP address pool.
[DHCP-Server-ip-pool-fsm10]network 10.1.10.0 mask 255.255.255.0
[DHCP-Server-ip-pool-fsm10]gateway-list 10.1.10.254
# 配置 DNS，可以用一条语句完成主备配置
[DHCP-Server-ip-pool-fsm10]dns-list 192.168.9.99 192.168.9.98
[DHCP-Server-ip-pool-fsm10]quit
[DHCP-Server]ip pool fsm20
Info: It's successful to create an IP address pool.
[DHCP-Server-ip-pool-fsm20]network 10.1.20.0 mask 24
[DHCP-Server-ip-pool-fsm20]gateway-list 10.1.20.254
[DHCP-Server-ip-pool-fsm20]dns-list 192.168.9.99 192.168.9.98
[DHCP-Server-ip-pool-fsm20]quit
[DHCP-Server]ip pool fsm30
Info: It's successful to create an IP address pool.
[DHCP-Server-ip-pool-fsm30]network 10.1.30.0 mask 24
[DHCP-Server-ip-pool-fsm30]gateway-list 10.1.30.254
[DHCP-Server-ip-pool-fsm30]dns-list 192.168.9.99 192.168.9.98
[DHCP-Server-ip-pool-fsm30]quit
[DHCP-Server]ip pool fsm40
Info: It's successful to create an IP address pool.
[DHCP-Server-ip-pool-fsm40]network 10.1.40.0 mask 24
[DHCP-Server-ip-pool-fsm40]gateway-list 10.1.40.254
[DHCP-Server-ip-pool-fsm40]dns-list 192.168.9.99 192.168.9.98
[DHCP-Server-ip-pool-fsm40]quit
[DHCP-Server]interface GigabitEthernet 0/0/0
[DHCP-Server-GigabitEthernet0/0/0]dhcp select global
[DHCP-Server-GigabitEthernet0/0/0]quit
[DHCP-Server]quit
<DHCP-Server>save
<DHCP-Server>save
```

步骤 3　中继配置
```
[sw1]dhcp enable
[sw1]interface Vlanif 10
#DHCP 选择 relay 中继模式
[sw1-Vlanif10]dhcp select relay
#DHCP 服务 IP 地址 192.168.10.1
```

```
[sw1-Vlanif10]dhcp relay server-ip 192.168.10.1
[sw1-Vlanif10]quit
[sw1]interface Vlanif 20
[sw1-Vlanif20]dhcp select relay
[sw1-Vlanif20]dhcp relay server-ip 192.168.10.1
[sw1-Vlanif20]quit
[sw1]interface Vlanif 30
[sw1-Vlanif30]dhcp select relay
[sw1-Vlanif30]dhcp relay server-ip 192.168.10.1
[sw1-Vlanif30]quit
[sw1]interface Vlanif 40
[sw1-Vlanif40]dhcp select relay
[sw1-Vlanif40]dhcp relay server-ip 192.168.10.1
[sw1-Vlanif40]quit
[sw1]quit
<sw1>save
```

步骤 4　测试通信情况

将全部 PC 设置为 DHCP 获取 IP 地址方式，并且命令行输入 "ipconfig"，查看获取的 IP 参数。见图 2-13-15。

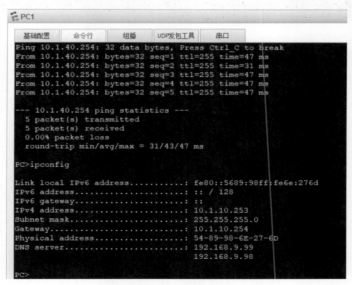

(a) PC1自动获取IP地址

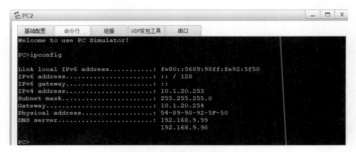

(b) PC2自动获取IP地址

图2-13-15

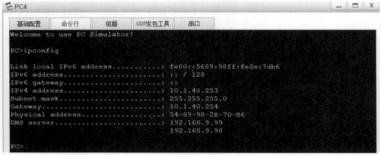

(c) PC3自动获取IP地址

(d) PC4自动获取IP地址

图2-13-15　获取IP地址

```
PC1(IP:10.1.10.253)-ping-PC2(IP:10.1.20.253)-通
PC>ping 10.1.20.253
Ping 10.1.20.253: 32 data bytes, Press Ctrl_C to break
From 10.1.20.253: bytes=32 seq=1 ttl=127 time=78 ms
……省略……
--- 10.1.20.253 ping statistics ---
  5 packet(s) transmitted
  5 packet(s) received
  0.00% packet loss
  round-trip min/avg/max = 78/81/94 ms
PC1(IP:10.1.10.253)-ping-PC3(IP:10.1.30.253) -通
PC>ping 10.1.30.253
Ping 10.1.30.253: 32 data bytes, Press Ctrl_C to break
From 10.1.30.253: bytes=32 seq=1 ttl=127 time=78 ms
……省略……
--- 10.1.30.253 ping statistics ---
  5 packet(s) transmitted
  5 packet(s) received
  0.00% packet loss
  round-trip min/avg/max = 63/78/94 ms
PC1(IP:10.1.10.253)-ping-PC4 (IP:10.1.40.253)-通
PC>ping 10.1.40.253
Ping 10.1.40.253: 32 data bytes, Press Ctrl_C to break
From 10.1.40.253: bytes=32 seq=1 ttl=127 time=78 ms
……省略……
--- 10.1.40.253 ping statistics ---
  5 packet(s) transmitted
```

```
5 packet(s) received
0.00% packet loss
round-trip min/avg/max = 62/81/94 ms
```

项目 14　telnet 远程访问的安全性分析

本项目介绍在 telnet 远程访问协议中，可以通过 Wireshark 抓取报文，就会看到用户名以及密码，说明在网络中传输有风险。而使用 ssh 远程协议访问，抓取报文时就会加密显示。所以在实际工作中，建议使用 ssh 远程协议访问网络设备。

一、eNSP 模拟器网络设备接口通过 Wireshark 抓取报文

1. 拓扑图

如图 2-14-1 所示。

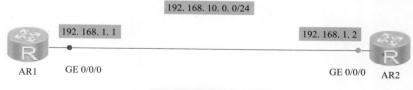

图2-14-1　拓扑图

IP 地址如表 2-14-1 所示。

表 2-14-1　IP 地址表

序号	设备	IP
1	AR1	192.168.1.1
2	AR2	192.168.1.2

2. 路由器基本配置

步骤 1　AR1 基本配置

```
[AR1]interface GigabitEthernet 0/0/0
[AR1-GigabitEthernet0/0/0]ip address 192.168.10.1 24
```

步骤 2 　 AR2 基本配置

```
[AR1]interface GigabitEthernet 0/0/0
[AR1-GigabitEthernet0/0/0]ip address 192.168.10.2 24
```

步骤 3 　 ping 测试

```
AR2(IP:192.168.10.2)-ping-AR1(IP: 192.168.10.1)-通
PING 192.168.10.1: 56  data bytes, press CTRL_C to break
Reply from 192.168.10.1: bytes=56 Sequence=1 ttl=255 time=1 ms
Reply from 192.168.10.1: bytes=56 Sequence=2 ttl=255 time=1 ms
```

3. telnet 远程配置

步骤 1 　 AR1 配置 telnet 服务

```
[Huawei]sysname AR1
[AR1]aaa
[AR1-aaa]local-user fsm password cipher a123456
[AR1-aaa]local-user fsm service-type telnet
[AR1-aaa]local-user fsm privilege level 15
[AR1-ui-vty0-4]authentication-mode aaa
[AR1-ui-vty0-4]user privilege level 15
```

步骤 2 　 AR2 配置 telnet 服务

配置与 AR1 类似。

4. telnet 远程登录

步骤 1 　 AR1-telnet-AR2

```
<AR1>telnet 192.168.10.2
Press CTRL_] to quit telnet mode
Trying 192.168.10.2 ...
Connected to 192.168.10.2 ...
Login authentication
# 输入用户名以及密码
Username:
Password:
------------------------------------------------------------------------------
User last login information:
------------------------------------------------------------------------------
Access Type: Telnet
IP-Address : 192.168.10.1
Time       : 2021-02-19 11:14:20-08:00
------------------------------------------------------------------------------
<AR2>
```

步骤 2 　 AR2-telnet-AR1

```
<AR2>telnet 192.168.10.1
Press CTRL_] to quit telnet mode
Trying 192.168.10.1 ...
Connected to 192.168.10.1 ...
Login authentication
# 输入用户名以及密码
Username:
```

```
Password:
--------------------------------------------------------------------------------
User last login information:
--------------------------------------------------------------------------------
Access Type: Telnet
IP-Address : 192.168.10.2
Time       : 2021-02-19 11:14:49-08:00
--------------------------------------------------------------------------------
<AR1>
```

5. Wireshark 抓取报文

步骤1　在 AR1- GigabitEthernet0/0/0 抓取报文，如图 2-14-2 所示。

步骤2　抓取报文，如图 2-14-3 所示。

步骤3　选择 Follow TCP Steam，如图 2-14-4 所示。

步骤4　用户名与密码一览无余，内容也是明文。如图 2-14-5 所示。

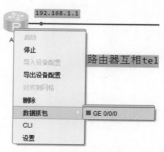

图2-14-2　Wireshark软件抓取报文

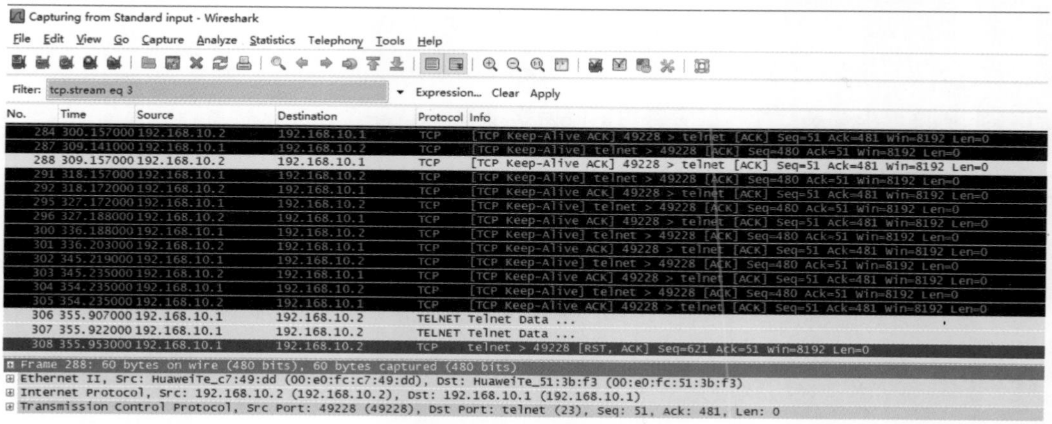

图2-14-3　抓取报文

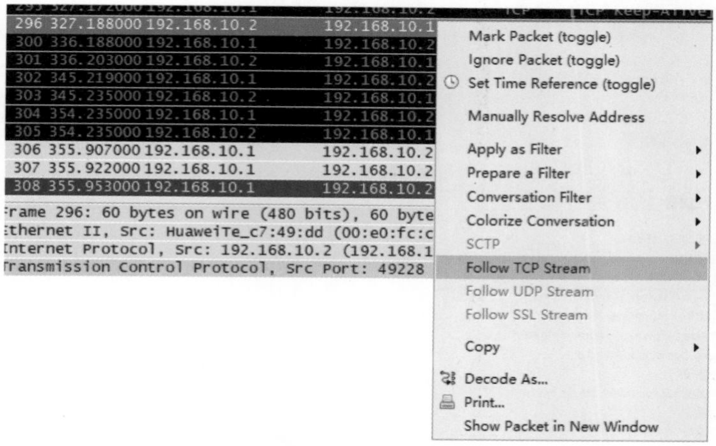

图2-14-4　选择Follow TCP Steam

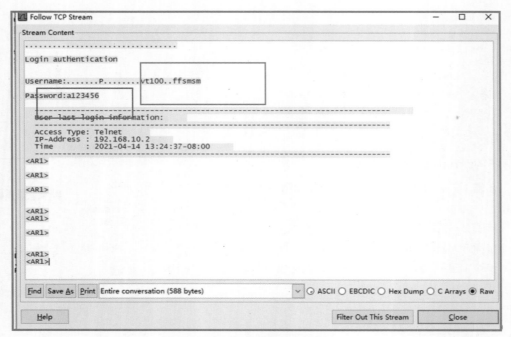

图2-14-5 显示用户名与密码

可以看出，用户名是 fsm，密码是 a123456。明文传输很不安全，在实际工作中，一般使用 ssh 协议远程登录交换机或者路由器等网络设备。

二、现网环境中物理主机 telnet 远程登录交换机

本项目中，本交换机无法配置 ssh，因此采用物理主机 telnet 远程登录交换机，查看信息明文传输情况。

步骤1 打开 Wireshark 软件，选择网卡（本机物理网卡）。如图 2-14-6 所示。

步骤2 在 SecureCRT 软件，telnet 登录现网中一台交换机（IP: 10.10.*.***），如图 2-14-7 所示。

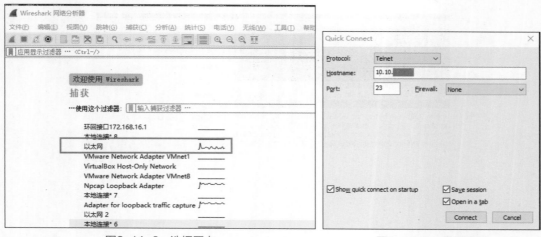

图2-14-6 选择网卡　　　　　图2-14-7 登录界面

步骤 3　输入用户名以及密码，如图 2-14-8 所示。

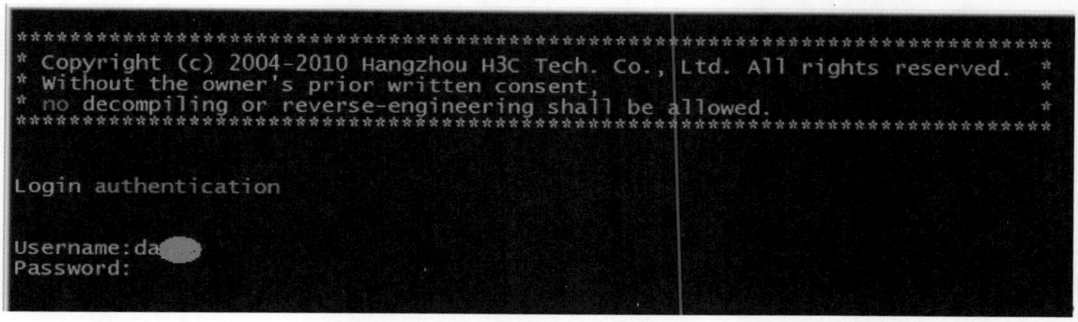

图2-14-8　输入名户名以及密码

步骤 4　在抓包软件 Wireshark 中过滤报文，如图 2-14-9 所示。

在 filter 处输入过滤规则，ip.addr==10.10.*.*** && tcp.port==23，过滤相应报文。

图2-14-9　过滤报文

选取 876 号报文，见图 2-14-10 所示。

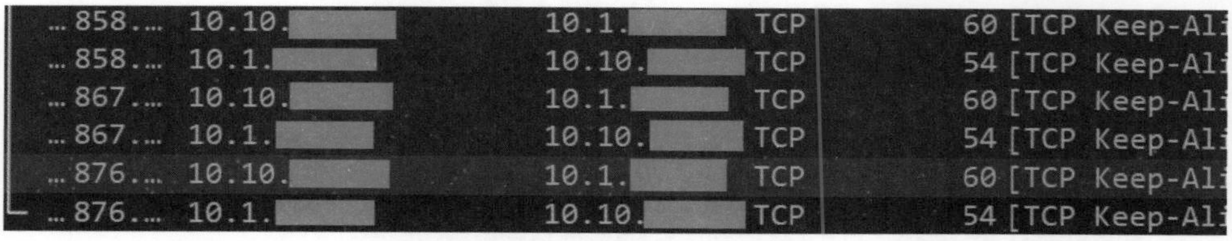

图2-14-10　显示报文

在 876 号报文上，右键选择"追踪流"，如图 2-14-11 所示。

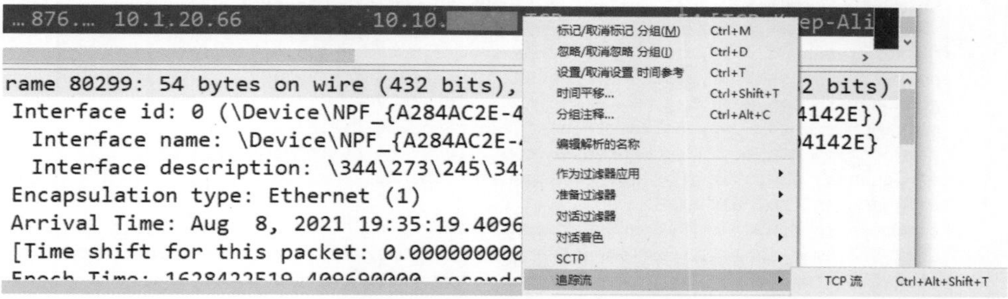

图2-14-11　选择"追踪流"

抓取报文结果见图 2-14-12，显示用户名以及密码。

步骤 5　用远程 ssh 协议登录交换机报文，如图 2-14-13 所示。

图 2-14-13 是报文以及追踪流显示，图 2-14-14 是无法正常显示用户名以及密码（加密传输的缘故）。

159

```
* no decompiling or reverse-engineering shall be allowed.
*
***************************************************************
*****
....vt100..........

Login authentication

Username:d██████
.
Password:d██████
.
<ISP_SW████>.ddiiss
.
                ^
% Incomplete command found at '^' position.
<ISP_SW████>.ddiiss   ccuurr
.
#
```

图2-14-12　显示用户名以及密码等信息

... 515....   10.1.20.66   10.10.4.13   TCP   54 62762 → 22 [

图2-14-13　显示报文

```
SSH-1.99-Comware-5.20
SSH-2.0-OpenSSH_for_Windows_7.7
...<       .wI.(...u..R...#....Ydiffie-hellman-group-exchange-sha1,diffie-
hellman-group14-sha1,diffie-hellman-group1-sha1....ssh-rsa....aes128-cbc,
3des-cbc,des-cbc....aes128-cbc,3des-cbc,des-cbc...+hmac-sha1,hmac-
sha1-96,hmac-md5,hmac-md5-96...+hmac-sha1,hmac-sha1-96,hmac-md5,hmac-
md5-96....none....none.........................$  ......4..f3....*....
0curve25519-sha256,curve25519-sha256@libssh.org,ecdh-sha2-nistp256,ecdh-
sha2-nistp384,ecdh-sha2-nistp521,diffie-hellman-group-exchange-
sha256,diffie-hellman-group16-sha512,diffie-hellman-group18-sha512,diffie-
hellman-group-exchange-sha1,diffie-hellman-group14-sha256,diffie-hellman-
group14-sha1,ext-info-c..."ecdsa-sha2-nistp256-cert-v01@openssh.com,ecdsa-
sha2-nistp384-cert-v01@openssh.com,ecdsa-sha2-nistp521-cert-
v01@openssh.com,ssh-ed25519-cert-v01@openssh.com,ssh-rsa-cert-
v01@openssh.com,ecdsa-sha2-nistp256,ecdsa-sha2-nistp384,ecdsa-sha2-
nistp521,ssh-ed25519,rsa-sha2-512,rsa-sha2-256,ssh-rsa...lchacha20-
poly1305@openssh.com,aes128-ctr,aes192-ctr,aes256-ctr,aes128-
gcm@openssh.com,aes256-gcm@openssh.com...lchacha20-
poly1305@openssh.com,aes128-ctr,aes192-ctr,aes256-ctr,aes128-
gcm@openssh.com,aes256-gcm@openssh.com....umac-64-etm@openssh.com,umac-128-
etm@openssh.com,hmac-sha2-256-etm@openssh.com,hmac-sha2-512-
etm@openssh.com,hmac-sha1-
etm@openssh.com,umac-64@openssh.com,umac-128@openssh.com,hmac-sha2-256,hmac-
sha2-512,hmac-sha1...umac-64-etm@openssh.com,umac-128-etm@openssh.com,hmac-
sha2-256-etm@openssh.com,hmac-sha2-512-etm@openssh.com,hmac-sha1-
etm@openssh.com,umac-64@openssh.com,umac-128@openssh.com,hmac-sha2-256,hmac-
sha2-512,hmac-sha1....none....none..........................<.........'The
connection is closed by SSH Server..... ......
```

图2-14-14　加密显示

项目 15 NAT 网络地址转换

网络转换技术 NAT 的主要功能是实现内网访问外网，实现 IP 地址转换。NAT 一般部署在出口防火墙或者路由器中，可以更加安全地访问 Internet，同时可以保护私有网络信息不直接暴露于公网，是一种主要解决 IP 地址资源短缺的技术。NAT 转换技术包括静态、动态以及地址端口转换 NAPT 三种方式。

NAT 转化主要解决 IPv4 地址资源枯竭制约网络发展的问题，IPv6 技术可以解决 IP 地址匮乏的问题，但是大多数网络设备以及服务应用还是基于 IPv4 技术。简单地讲，NAT 网络地址转换就是将私网地址转换为公网地址。

 项目简介

本项目通过 NAT 转换技术，实现私网地址能够访问互联网。当采用静态、动态两种 NAT 转换模式，可以实现私网地址访问互联网需求，但是不能节约公网地址，一般不采用。端口转换 NAPT（Easy IP），可以实现一个公网地址对应对个私网地址，因为进行了端口转换。

一、NAT 转换基本配置

1. 拓扑图

如图 2-15-1 所示。

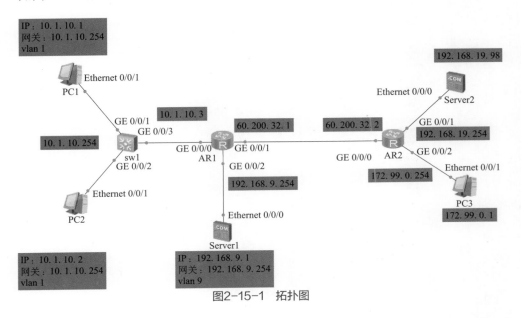

图2-15-1 拓扑图

PC 与 Server、sw、AR 等 IP 地址配置见表 2-15-1。

表 2-15-1　IP 地址配置表

设备	IP	网关	vlan
PC1	10.1.10.1	10.1.10.254	10
PC2	10.1.10.2	10.1.10.254	10
Server1	192.168.9.1	192.168.9.254	9
Server2	192.168.19.98	192.168.19.254	/
PC3	172.99.0.1	172.99.0.254	/
sw1	10.1.10.254	/	/
AR1	60.200.32.1	/	/
AR2	60.200.32.2	/	/

2. 项目配置

步骤 1　PC 以及 Server1 等 IP 地址配置如图 2-15-2 所示。

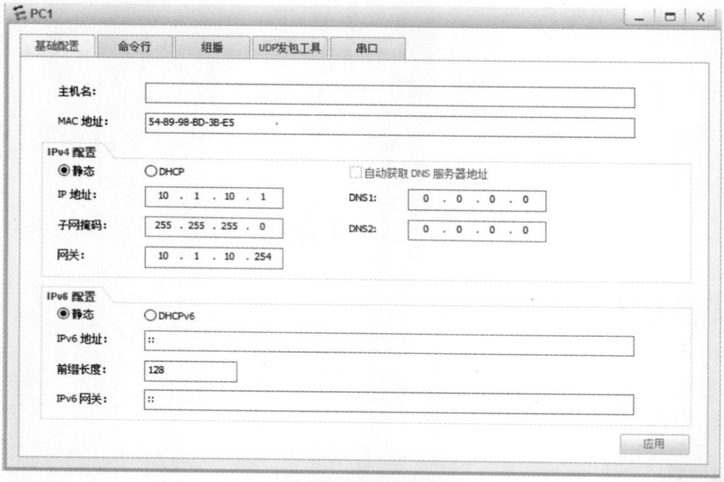

(a) PC1-IP 地址配置

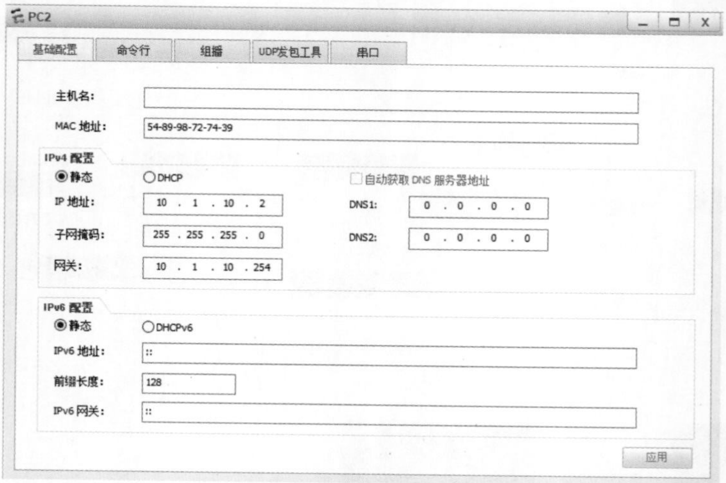

(b) PC2-IP 地址配置

(c) Server1-IP地址配置

(d) Server2-IP地址配置

(e) PC3-IP地址配置

图2-15-2　IP地址配置

步骤2　sw1 基本配置

```
<Huawei>system-view
Enter system view, return user view with Ctrl+Z.
[Huawei]sysname sw1
[sw1]interface Vlanif 1
# 配置 IP 地址
[sw1-Vlanif1]ip address 10.1.10.254 24
[sw1-Vlanif1]quit
```
测试 -sw1(IP: 10.1.10.254)-ping-PC1(IP: 10.1.10.1)- 通
```
[sw1]ping 10.1.10.1
PING 10.1.10.1: 56  data bytes, press CTRL_C to break
Reply from 10.1.10.1: bytes=56 Sequence=1 ttl=128 time=80 ms
```
……省略……
```
--- 10.1.10.1 ping statistics ---
  5 packet(s) transmitted
  5 packet(s) received
  0.00% packet loss
  round-trip min/avg/max = 10/36/80 ms
```

步骤3　AR1 基本配置

```
<Huawei>system-view
Enter system view, return user view with Ctrl+Z.
[Huawei]sysname AR1
[AR1]un in en
Info: Information center is disabled.
# 配置 IP 地址
[AR1]interface GigabitEthernet 0/0/0
[AR1-GigabitEthernet0/0/0]ip address 10.1.10.3 24
[AR1-GigabitEthernet0/0/0]quit
# 配置 IP 地址
[AR1]interface GigabitEthernet 0/0/1
[AR1-GigabitEthernet0/0/1]ip address 60.200.32.1 24
[AR1-GigabitEthernet0/0/1]quit
# 配置 IP 地址
[AR1]interface GigabitEthernet 0/0/2
[AR1-GigabitEthernet0/0/2]ip address 192.168.9.254 24
[AR1-GigabitEthernet0/0/2]quit
# 配置默认路由
[AR1]ip route-static 0.0.0.0 0 60.200.32.2
[AR1]quit
<AR1>save
```

sw1 配置两条静态路由：

```
[sw1]ip route-static 192.168.9.0 24 10.1.10.3
[sw1]ip route-static 60.200.32.0 24 10.1.10.3
```

PC1 的测试通信情况，如图 2-15-3 所示。

步骤4　AR2 基本配置

```
<Huawei>system-view
Enter system view, return user view with Ctrl+Z.
[Huawei]un in en
Info: Information center is disabled.
```

```
# 配置 IP 地址
[Huawei]sysname AR2
[AR2]interface GigabitEthernet 0/0/0
[AR2-GigabitEthernet0/0/0]ip address 60.200.32.2 24
[AR2-GigabitEthernet0/0/0]quit
# 配置 IP 地址
[AR2]interface GigabitEthernet 0/0/1
[AR2-GigabitEthernet0/0/1]ip address 192.168.19.254 24
[AR2-GigabitEthernet0/0/1]quit
# 配置 IP 地址
[AR2]interface GigabitEthernet 0/0/2
[AR2-GigabitEthernet0/0/2]ip address 172.99.0.254 24
[AR2-GigabitEthernet0/0/2]quit
[AR2]quit
<AR2>save
```

```
PC>ping 60.200.32.1

Ping 60.200.32.1: 32 data bytes, Press Ctrl_C to break
From 60.200.32.1: bytes=32 seq=1 ttl=255 time=62 ms
From 60.200.32.1: bytes=32 seq=2 ttl=255 time=47 ms
From 60.200.32.1: bytes=32 seq=3 ttl=255 time=31 ms
From 60.200.32.1: bytes=32 seq=4 ttl=255 time=47 ms
From 60.200.32.1: bytes=32 seq=5 ttl=255 time=47 ms

--- 60.200.32.1 ping statistics ---
  5 packet(s) transmitted
  5 packet(s) received
  0.00% packet loss
  round-trip min/avg/max = 31/46/62 ms

PC>ping 10.1.10.2

Ping 10.1.10.2: 32 data bytes, Press Ctrl_C to break
From 10.1.10.2: bytes=32 seq=1 ttl=128 time=31 ms
From 10.1.10.2: bytes=32 seq=2 ttl=128 time=47 ms
From 10.1.10.2: bytes=32 seq=3 ttl=128 time=47 ms
From 10.1.10.2: bytes=32 seq=4 ttl=128 time=31 ms
From 10.1.10.2: bytes=32 seq=5 ttl=128 time=31 ms

--- 10.1.10.2 ping statistics ---
  5 packet(s) transmitted
  5 packet(s) received
  0.00% packet loss
  round-trip min/avg/max = 31/37/47 ms

PC>ping 192.168.9.1

Ping 192.168.9.1: 32 data bytes, Press Ctrl_C to break
From 192.168.9.1: bytes=32 seq=1 ttl=254 time=31 ms
From 192.168.9.1: bytes=32 seq=2 ttl=254 time=31 ms
From 192.168.9.1: bytes=32 seq=3 ttl=254 time=47 ms
From 192.168.9.1: bytes=32 seq=4 ttl=254 time=31 ms
From 192.168.9.1: bytes=32 seq=5 ttl=254 time=47 ms

--- 192.168.9.1 ping statistics ---
  5 packet(s) transmitted
  5 packet(s) received
  0.00% packet loss
  round-trip min/avg/max = 31/37/47 ms
```

图2-15-3　通信情况

AR2-ping-Server2（IP:192.168.19.98）- 通，如图 2-15-4 所示。

```
<AR2>ping 192.168.19.98
  PING 192.168.19.98: 56  data bytes, press CTRL_C to break
    Reply from 192.168.19.98: bytes=56 Sequence=1 ttl=255 time=50 ms
    Reply from 192.168.19.98: bytes=56 Sequence=2 ttl=255 time=10 ms
    Reply from 192.168.19.98: bytes=56 Sequence=3 ttl=255 time=10 ms
    Reply from 192.168.19.98: bytes=56 Sequence=4 ttl=255 time=10 ms
    Reply from 192.168.19.98: bytes=56 Sequence=5 ttl=255 time=20 ms

  --- 192.168.19.98 ping statistics ---
    5 packet(s) transmitted
    5 packet(s) received
    0.00% packet loss
    round-trip min/avg/max = 10/20/50 ms

<AR2>ping 172.99.0.1
  PING 172.99.0.1: 56  data bytes, press CTRL_C to break
    Reply from 172.99.0.1: bytes=56 Sequence=1 ttl=128 time=30 ms
    Reply from 172.99.0.1: bytes=56 Sequence=2 ttl=128 time=10 ms
    Reply from 172.99.0.1: bytes=56 Sequence=3 ttl=128 time=20 ms
    Reply from 172.99.0.1: bytes=56 Sequence=4 ttl=128 time=20 ms
    Reply from 172.99.0.1: bytes=56 Sequence=5 ttl=128 time=20 ms

  --- 172.99.0.1 ping statistics ---
    5 packet(s) transmitted
    5 packet(s) received
    0.00% packet loss
    round-trip min/avg/max = 10/20/30 ms

<AR2>
```

图2-15-4　通信情况

测试网络见图 2-15-5。PC3（IP：172.99.0.1）-ping-Server2（IP:192.168.19.98）- 通

但是 PC1-ping-Server2- 不通，是因为公网中没有私网路由地址（使用 60.200.32.0/24 模拟公网地址），需要配置 NAT 转换，将私网地址转换为公网地址。

```
PC>ping 192.168.19.98
  ping 192.168.19.98
  Ping 192.168.19.98: 32 data
bytes, Press Ctrl_C to break
  Request timeout!
  Request timeout!
  ……省略……
  --- 192.168.19.98 ping sta-
tistics ---
    5 packet(s) transmitted
    0 packet(s) received
    100.00% packet loss
```

在交换机 sw1 上配置两条静态路由。

```
[sw1]ip route-static 192.168.19.0 24 10.1.10.3
[sw1]ip route-static 172.99.0.0 24 10.1.10.3
[sw1]quit
```

图2-15-5　PC3与Server2、PC3与AR2的测试情况

二、静态 NAT 转换

静态 NAT 转换，将内部私有地址与公网地址实现一一对应转换，该技术需要有批量的公网地址，由于公网地址价格费用较高，现网中一般不使用。

1. 拓扑图

如图 2-15-6 所示。

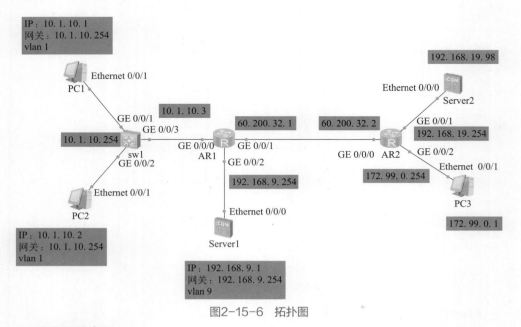

图2-15-6　拓扑图

2. 项目配置步骤

按照静态 NAT 转换的原理，设置地址如下：

私网地址：10.1.10.1；公网地址：60.200.32.10
基本命令如下：
nat static global 60.200.32.10 inside 10.1.10.1
在内网路由器 AR1 出口 GigabitEthernet0/0/1 进行地址转换。
将内部地址 10.1.10.1 转化为公网地址 60.200.32.10
[AR1-GigabitEthernet0/0/1]nat static global 60.200.32.10 inside 10.1.10.1
网络测试：
PC1(IP: 10.1.10.1)-ping-Server2（IP:192.168.19.98）- 通，如图 2-15-7 所示。

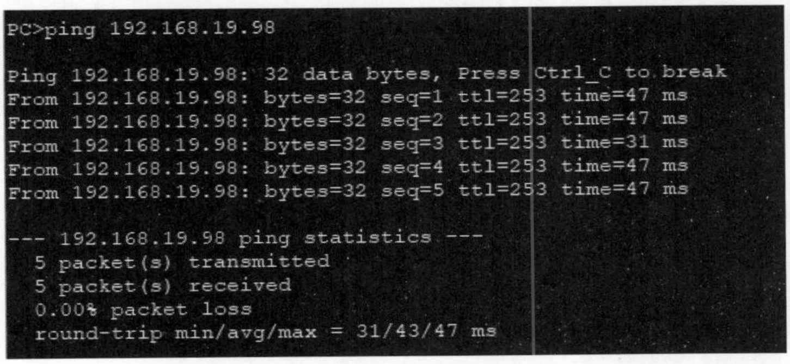

```
PC>ping 192.168.19.98

Ping 192.168.19.98: 32 data bytes, Press Ctrl_C to break
From 192.168.19.98: bytes=32 seq=1 ttl=253 time=47 ms
From 192.168.19.98: bytes=32 seq=2 ttl=253 time=47 ms
From 192.168.19.98: bytes=32 seq=3 ttl=253 time=31 ms
From 192.168.19.98: bytes=32 seq=4 ttl=253 time=47 ms
From 192.168.19.98: bytes=32 seq=5 ttl=253 time=47 ms

--- 192.168.19.98 ping statistics ---
  5 packet(s) transmitted
  5 packet(s) received
  0.00% packet loss
  round-trip min/avg/max = 31/43/47 ms
```

图2-15-7　测试情况

查看静态 NAT 地址转换信息（命令：display nat session all）
```
[AR1]display nat session all
  NAT Session Table Information:
    Protocol           : ICMP(1)
    SrcAddr    Vpn    : 10.1.10.1        # 私网地址
    DestAddr   Vpn    : 192.168.19.98    # 目的地址
    Type Code IcmpId  : 0   8   29681
    NAT-Info
      New SrcAddr     : 60.200.32.10     # 公网地址
      New DestAddr    : ----
      New IcmpId      : ----
    Protocol          : ICMP(1)
    SrcAddr    Vpn    : 10.1.10.1
    DestAddr   Vpn    : 192.168.19.98
    Type Code IcmpId  : 0   8   29680
    NAT-Info
      New SrcAddr     : 60.200.32.10
      New DestAddr    : ----
      New IcmpId      : ----
    Protocol          : ICMP(1)
    SrcAddr    Vpn    : 10.1.10.1
    DestAddr   Vpn    : 192.168.19.98
    Type Code IcmpId  : 0   8   29678
    NAT-Info
      New SrcAddr     : 60.200.32.10
      New DestAddr    : ----
      New IcmpId      : ----
  Total : 3
[AR1]
```

测试,PC1(IP:10.1.10.1)-ping-PC3(IP:172.99.0.1)- 通,见图2-15-8。

```
PC>ping 172.99.0.1

Ping 172.99.0.1: 32 data bytes, Press Ctrl_C to break
From 172.99.0.1: bytes=32 seq=1 ttl=126 time=46 ms
From 172.99.0.1: bytes=32 seq=2 ttl=126 time=63 ms
From 172.99.0.1: bytes=32 seq=3 ttl=126 time=47 ms
From 172.99.0.1: bytes=32 seq=4 ttl=126 time=47 ms
From 172.99.0.1: bytes=32 seq=5 ttl=126 time=47 ms

--- 172.99.0.1 ping statistics ---
  5 packet(s) transmitted
  5 packet(s) received
  0.00% packet loss
  round-trip min/avg/max = 46/50/63 ms
```

图2-15-8 测试网络

如果 PC2 也有需求连接外网,那同样也需要进行私网与公网地址的转换。

私网地址:10.1.10.2;公网地址:60.200.32.20

```
[AR1-GigabitEthernet0/0/1]nat static global 60.200.32.20 inside 10.1.10.2
```
查看静态 NAT 转换信息
```
[AR1]display nat session all
  NAT Session Table Information:
    Protocol        : ICMP(1)
    SrcAddr   Vpn   : 10.1.10.2
    DestAddr  Vpn   : 192.168.19.98
    Type Code IcmpId : 0   8   30347
    NAT-Info
      New SrcAddr   : 60.200.32.20
      New DestAddr  : ----
      New IcmpId    : ----
    Protocol        : ICMP(1)
    SrcAddr   Vpn   : 10.1.10.2
    DestAddr  Vpn   : 192.168.19.98
    Type Code IcmpId : 0   8   30343
    NAT-Info
      New SrcAddr   : 60.200.32.20
      New DestAddr  : ----
      New IcmpId    : ----
    Protocol        : ICMP(1)
    SrcAddr   Vpn   : 10.1.10.2
    DestAddr  Vpn   : 192.168.19.98
    Type Code IcmpId : 0   8   30340
    NAT-Info
      New SrcAddr   : 60.200.32.20
      New DestAddr  : ----
      New IcmpId    : ----
    Protocol        : ICMP(1)
    SrcAddr   Vpn   : 10.1.10.1
    DestAddr  Vpn   : 192.168.19.98
    Type Code IcmpId : 0   8   30344
    NAT-Info
```

```
            New SrcAddr         : 60.200.32.10
            New DestAddr        : ----
            New IcmpId          : ----
```

通过查询 AR1 GigabitEthernet 0/0/1 信息，可以明显看出两个内网地址访问外网就需要两个公网地址，在现网中很少使用。

查询 AR1- GigabitEthernet0/0/1 下 NAT 转换配置信息如下：

```
[AR1]int GigabitEthernet 0/0/1
[AR1-GigabitEthernet0/0/1]dis this
[V200R003C00]
#
interface GigabitEthernet0/0/1
  ip address 60.200.32.1 255.255.255.0
  nat static global 60.200.32.10 inside 10.1.10.1 netmask 255.255.255.255
  nat static global 60.200.32.20 inside 10.1.10.2 netmask 255.255.255.255
```

三、动态 NAT 转换

动态 NAT 也是将内部地址与公网地址对应转换，只是建立一个组，内部地址在转换的时候选择一个空闲的公网地址，如果地址池中公网地址使用完毕，其他内网地址就无法连接外部网络。动态 NAT 转换一般也不使用，因为同样不能节省公网地址。

1. 拓扑图

见图 2-15-6。

2. 基本配置步骤

步骤1　删除静态 NAT 配置，共计两条。

```
[AR1-GigabitEthernet0/0/1]undo nat static global 60.200.32.10 inside 10.1.10.1
netmask 255.255.255.255
[AR1-GigabitEthernet0/0/1]undo nat static global 60.200.32.20 inside 10.1.10.2 netmask 255.255.255.255
```

步骤2　动态 NAT 转换

```
# 配置基本策略
[AR1] acl 2000
# 配置规则，允许 IP 地址段。
[AR1-acl-basic-2000] rule 5 permit source 10.1.10.0 0.0.0.255
# 创建公网地址组 60.100.200.100-60.100.200.110（共计 11 个地址）
[AR1]nat address-group 1 60.200.32.100 60.200.32.110
# 路由器出口调用 ACL，不做端口转换。
[AR1-GigabitEthernet0/0/1]nat outbound 2000 address-group 1 no-pat
```

测试 PC1（IP:10.1.10.1）-ping-Server2（IP:192.168.19.98）- 通，PC1（IP:10.1.10.1）-ping-PC3（IP: 172.99.0.1）- 通，见图 2-15-9。

查询动态 NAT 转换会话信息，见图 2-15-10。

图2-15-9 网络测试

图2-15-10 动态NAT会话信息

四、NAPT 转换

NAPT（Easy IP）转换也是一种动态转换，是将内部所有地址转换为一个公网地址，用不同的端口号进行区分。目前普遍采用该技术，只需使用一个公网 IP 地址，就可将多个内部 IP 用户连接到因特网。其核心之处就在于采用端口号实现公网和私网的转换。使用接口公网地址（就是路由器出口配置地址）进行转换，本项目中用 AR1 GigabitEthernet0/0/1 IP 地址 60.200.32.1 进行转换。

本项目实现效果：10.1.10.0/24 与 192.168.9.0/24 都能与外网通信。

1. 拓扑图

见图 2-15-6。

2. 基本配置步骤

步骤1 删除配置

```
[AR1-GigabitEthernet0/0/1]undo nat outbound 2000 address-group 1 no-pat
   [AR1]undo nat address-group 1
```

步骤2 NAPT（Easy IP）转换配置

由于动态转换项目已经在 ACL 2000 中，对应 rule 5 允许 10.1.10.0 段的网络地址，仅需要配置允许 192.168.9.0 网段。

```
[AR1-acl-basic-2000]rule 10 permit source 192.168.9.0 0.0.0.255
# 路由器接口调用 ACL，进行端口转换。
[AR1-GigabitEthernet0/0/1]nat outbound 2000
```

3. 网络测试

PC1（IP:10.1.10.1）-ping-Server2（IP: 192.168.19.98）- 通，见图 2-15-11。

```
PC>ping 192.168.19.98
Ping 192.168.19.98: 32 data bytes, Press Ctrl_C to break
From 192.168.19.98: bytes=32 seq=1 ttl=253 time=47 ms
From 192.168.19.98: bytes=32 seq=2 ttl=253 time=47 ms
From 192.168.19.98: bytes=32 seq=3 ttl=253 time=63 ms
From 192.168.19.98: bytes=32 seq=4 ttl=253 time=31 ms
From 192.168.19.98: bytes=32 seq=5 ttl=253 time=47 ms

--- 192.168.19.98 ping statistics ---
  5 packet(s) transmitted
  5 packet(s) received
  0.00% packet loss
  round-trip min/avg/max = 31/47/63 ms
```

图 2-15-11 网络测试

PC2（IP:10.1.10.2）-ping-Server2（IP: 192.168.19.98）- 通，见图 2-15-12。

```
PC>ping 192.168.19.98
Ping 192.168.19.98: 32 data bytes, Press Ctrl_C to break
From 192.168.19.98: bytes=32 seq=1 ttl=253 time=46 ms
From 192.168.19.98: bytes=32 seq=2 ttl=253 time=32 ms
From 192.168.19.98: bytes=32 seq=3 ttl=253 time=47 ms
From 192.168.19.98: bytes=32 seq=4 ttl=253 time=31 ms
From 192.168.19.98: bytes=32 seq=5 ttl=253 time=62 ms

--- 192.168.19.98 ping statistics ---
  5 packet(s) transmitted
  5 packet(s) received
  0.00% packet loss
  round-trip min/avg/max = 31/43/62 ms
```

图 2-15-12 网络测试

查询会话信息如下：

```
<AR1>display nat session all
 NAT Session Table Information:
   Protocol            : ICMP(1)
   SrcAddr    Vpn      : 10.1.10.1
   DestAddr   Vpn      : 192.168.19.98
   Type Code IcmpId    : 0   8   3097
   NAT-Info
     New SrcAddr       : 60.200.32.1
     New DestAddr      : ----
     New IcmpId        : 10429
   Protocol            : ICMP(1)
   SrcAddr    Vpn      : 10.1.10.1
   DestAddr   Vpn      : 192.168.19.98
   Type Code IcmpId    : 0   8   3098
   NAT-Info
     New SrcAddr       : 60.200.32.1
     New DestAddr      : ----
     New IcmpId        : 10430
   Protocol            : ICMP(1)
   SrcAddr    Vpn      : 10.1.10.1
   DestAddr   Vpn      : 192.168.19.98
   Type Code IcmpId    : 0   8   3100
   NAT-Info
     New SrcAddr       : 60.200.32.1
     New DestAddr      : ----
     New IcmpId        : 10432
   Protocol            : ICMP(1)
```

```
  SrcAddr    Vpn          : 10.1.10.1
  DestAddr   Vpn          : 192.168.19.98
  Type Code IcmpId        : 0    8    3099
  NAT-Info
    New SrcAddr           : 60.200.32.1
    New DestAddr          : ----
    New IcmpId            : 10431
  Protocol                : ICMP(1)
  SrcAddr    Vpn          : 10.1.10.1
  DestAddr   Vpn          : 192.168.19.98
  Type Code IcmpId        : 0    8    3096
  NAT-Info
    New SrcAddr           : 60.200.32.1
    New DestAddr          : ----
    New IcmpId            : 10428
Total : 5
```

PC1 与 Server 2 共通信了 5 个包，从表 2-15-2 中可以看出，公网地址通过 5 个不同的端口与外网通信。

表 2-15-2　PC1 与 Server 2 内部地址与公网地址之间端口对应关系

PC1	Server 2	公网地址	端口
10.1.10.1	192.168.19.98	60.200.32.1	10429
10.1.10.1	192.168.19.98	60.200.32.1	10430
10.1.10.1	192.168.19.98	60.200.32.1	10432
10.1.10.1	192.168.19.98	60.200.32.1	10431
10.1.10.1	192.168.19.98	60.200.32.1	10428

Server1（IP:192.168.9.1）-ping-Server2（IP: 192.168.19.98）- 通，见图 2-15-13。

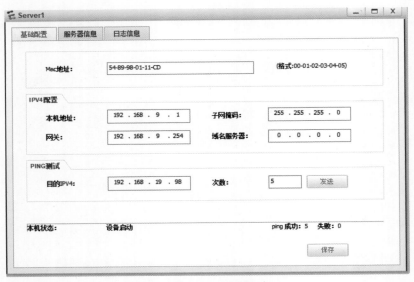

图2-15-13　测试通信

五、配置 DNS 服务器与测试网络

在 Server 2 服务器中配置 DNS 域名解析，就是将域名解析为 IP 地址，本例中当 PC2-ping-www.fsmdiy.com 时，Server2 服务器将域名解析 IP 地址为 172.99.0.1。

步骤1 在服务器 Server2 中设置 DNS 解析服务器，Server 服务器 DNS 配置见图 2-15-14，选择"服务器信息"-DNSServer，在右侧配置中填写主机域名（www.fsmdiy.com）以及 IP 地址（IP：172.99.0.1），然后点击"增加"，最后点击"启动"。

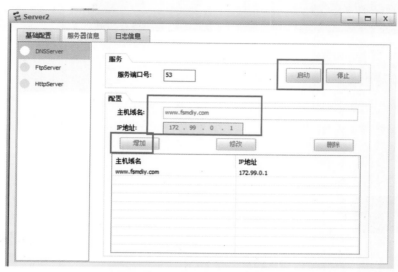

图2-15-14 配置DNS服务器

步骤2 在 PC2 中，网络参数中配置 DNS 地址，见图 2-15-15。

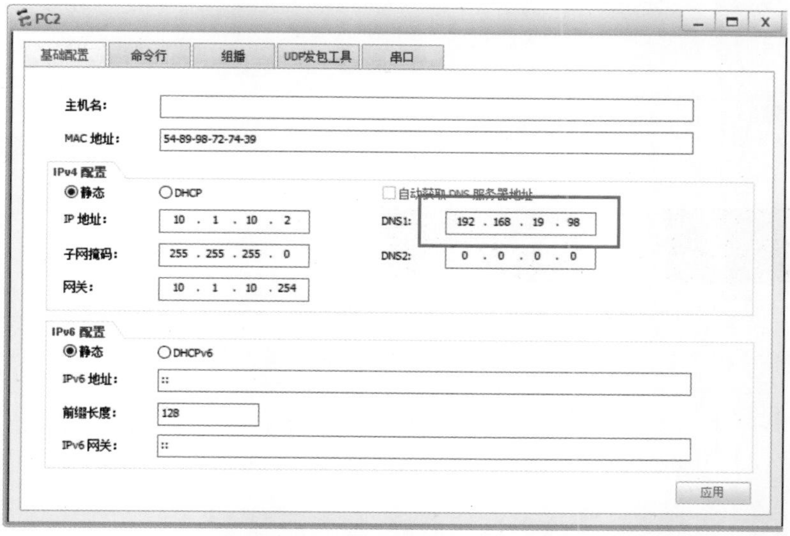

图2-15-15 PC配置DNS服务器地址

步骤3

测试：PC2（IP：10.1.10.2）- ping – www.fsmdiy.com（域名）- 通，如图 2-15-16 所示。

```
PC>ping www.fsmdiy.com
Ping www.fsmdiy.com [172.99.0.1]: 32 data bytes, Press Ctrl_C to break
From 172.99.0.1: bytes=32 seq=1 ttl=126 time=63 ms
From 172.99.0.1: bytes=32 seq=2 ttl=126 time=62 ms
From 172.99.0.1: bytes=32 seq=3 ttl=126 time=47 ms
From 172.99.0.1: bytes=32 seq=4 ttl=126 time=47 ms
From 172.99.0.1: bytes=32 seq=5 ttl=126 time=31 ms

--- 172.99.0.1 ping statistics ---
  5 packet(s) transmitted
  5 packet(s) received
  0.00% packet loss
  round-trip min/avg/max = 31/50/63 ms
```

图2-15-16 网络测试

六、隐藏内网服务器 IP 地址

在互联网中访问企业中的服务器时，不是直接访问内部地址，而是访问到公网地址，达到隐藏企业服务器真实 IP 地址的目的。

1. 拓扑图

如图 2-15-17 所示。

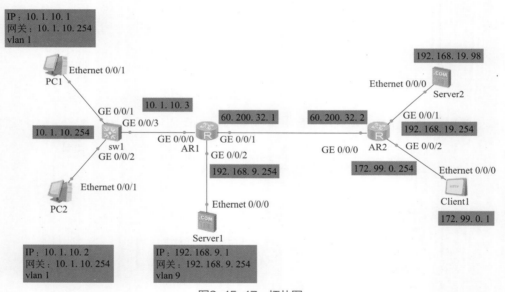

图2-15-17 拓扑图

将拓扑图中 PC3 更换为 Client 1，并给予 IP 参数配置，见图 2-15-18。

2. 网络配置

[AR1-GigabitEthernet0/0/1]nat static global 60.200.32.10 inside 192.168.9.1

查询会话信息，见图 2-15-19。

3. 网络测试

测试 Client 1（IP：172.99.0.1）-ping- 公网地址（IP：60.200.32.10）- 通，Client 1（IP：172.99.0.1）ping IP（60.200.32.10），实际上是 ping（IP:192.168.9.1），见图 2-15-20。

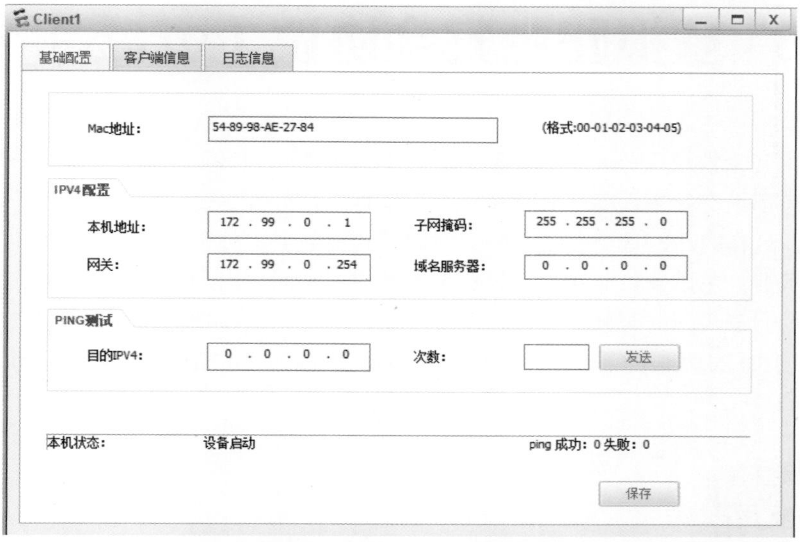

图2-15-18　Client 1参数配置

```
[AR1]display nat session all
  NAT Session Table Information:

     Protocol          : ICMP(1)
       SrcAddr   Vpn     :  192.168.9.1
       DestAddr  Vpn     :  192.168.19.254
       Type Code IcmpId  :  0    8    1
       NAT-Info
         New SrcAddr     :  60.200.32.10
         New DestAddr    :  ----
         New IcmpId      :  ----

    Total  :  1
```

图2-15-19　会话信息

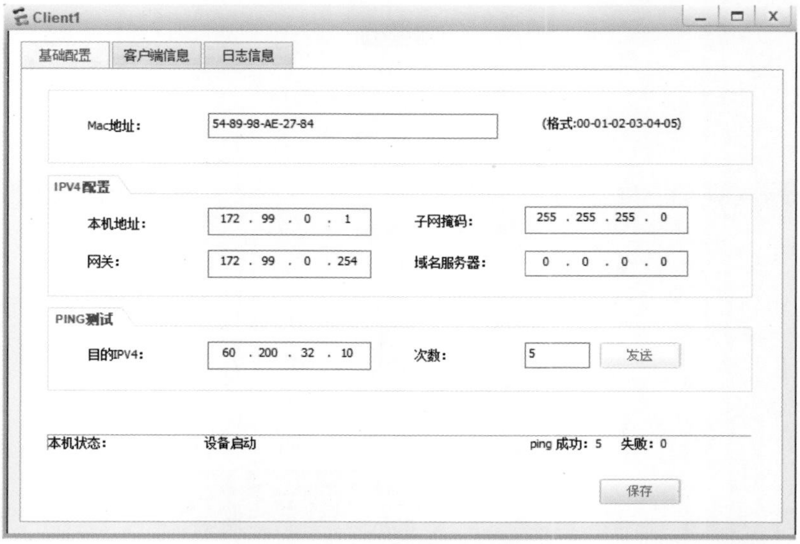

图2-15-20　网络测试

项目 16　虚拟路由冗余协议（vrrp）

虚拟路由冗余协议（vrrp）主要解决网络中网关的单点故障。在 vrrp 默认配置情况下，所有的数据包都是交给主设备（Master）来处理的，只有当主设备出现故障时，备用设备才会处理数据，这样的话，备用设备大多数时间不转发数据，处于闲置状态，同时由于数据全部由主设备承担，主设备负荷变大，导致数据处理不及时，出现延时现象。那怎么能做到负载分担呢？我们在实施的时候，可以在网络中使用两个网段，10.1.10.0/24、10.1.20.0/24，在配置时，10.1.10.0/24 网段的 Master 为 sw1，Backup 为 sw2；10.1.20.0/24 网段的 Master 为 sw2，Backup 为 sw2。

 项目简介

vrrp 是企业中提供网络高可靠性的主要协议，主要解决网关单点故障引起的网络中断事故，通过热备份路由器的方式，当网络中路由设备故障时，能自动在线切换，保障通信的连续与稳定。

一、vrrp 虚拟路由网关基本配置

1. 拓扑图

如图 2-16-1 所示。

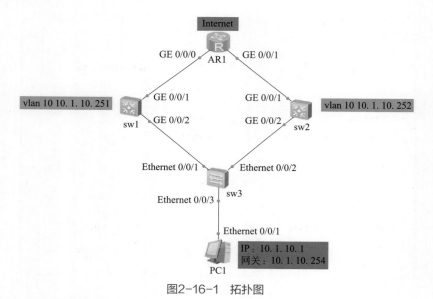

图 2-16-1　拓扑图

vlan 以及 IP 地址、虚拟网关见表 2-16-1。

表 2-16-1　vlan 以及 IP 地址、虚拟网关

vlan	IP	虚拟网关
10	10.1.10.0/24	10.1.10.254

2. 基本配置

步骤 1　PC1 配置

如图 2-16-2 所示。

IP：10.1.10.1，子网掩码：255.255.255.0 网关：10.1.10.254。

图2-16-2　PC1配置

步骤 2　sw3 交换机配置

```
<Huawei>
<Huawei>system-view
[Huawei]sysname sw3
# 创建 vlan 10
[sw3]vlan 10
# Ethernet 0/0/1 配置为 trunk，并且允许所有 vlan 通过
[sw3]interface  Ethernet 0/0/1
[sw3-Ethernet0/0/1]port link-type trunk
[sw3-Ethernet0/0/1]port trunk allow-pass vlan all
[sw3-Ethernet0/0/1]quit
# Ethernet 0/0/2 配置为 trunk，并且允许所有 vlan 通过
[sw3]interface  Ethernet 0/0/2
[sw3-Ethernet0/0/2]port link-type trunk
[sw3-Ethernet0/0/2]port trunk allow-pass vlan all
[sw3]quit
<sw3>save
# Ethernet 0/0/3 配置为 access，并且允许 vlan 10 通过
[sw3]interface Ethernet 0/0/3
[sw3-Ethernet0/0/3]port link-type access
[sw3-Ethernet0/0/3]port default vlan 10
```

步骤 3　sw1 基本配置

```
<Huawei>system-view
[Huawei]sysname sw1
```

```
[sw1]vlan 10
[sw1-vlan10]quit
# GigabitEthernet 0/0/2 配置为 trunk，并且允许所有 vlan 通过
[sw1]interface GigabitEthernet 0/0/2
[sw1-GigabitEthernet0/0/2]port link-type trunk
[sw1-GigabitEthernet0/0/2]port trunk allow-pass vlan all
[sw1-GigabitEthernet0/0/2]quit
# 配置地址网关地址 10.1.10.251
[sw1]interface Vlanif 10
[sw1-Vlanif10]ip address 10.1.10.251 24
[sw1-Vlanif10]quit
[sw1]quit
<sw1>save
```

步骤4　sw2 基本配置

```
<Huawei>system-view
[Huawei]sysname sw2
[sw2]un in en
[sw2]vlan 10
[sw2-vlan10]quit
[sw2]interface  GigabitEthernet 0/0/2
[sw2-GigabitEthernet0/0/2]port link-type trunk
[sw2-GigabitEthernet0/0/2]port trunk allow-pass vlan all
[sw2-GigabitEthernet0/0/2]quit
[sw2]interface Vlanif 10
[sw2-Vlanif10]ip address 10.1.10.252 24
[sw2-Vlanif10]quit
[sw2]quit
<sw2>save
```

步骤5　关于 vrrp 虚拟网关配置

① sw1 交换机

```
# 创建 vrrp 工作组为 10（数字自定），虚拟 IP 地址为 10.1.10.254
[sw1-Vlanif10]vrrp vrid 10 virtual-ip 10.1.10.254
# 配置优先级为 120，目的是让 sw1 成为 Master。
[sw1-Vlanif10]vrrp vrid 10 priority 120
[sw1-Vlanif10]quit
[sw1]quit
<sw1>save
```

② sw2 交换机

```
# 创建 vrrp 工作组为 10，虚拟 IP 地址为 10.1.10.254，优先级默认是 100。
[sw2-Vlanif10]vrrp vrid 10 virtual-ip 10.1.10.254
[sw2-Vlanif10]quit
[sw2]quit
<sw2>save
```

步骤6　查看 vrrp 配置

① sw1 交换机

查看 vrrp 配置信息

```
[sw1]display vrrp
Vlanif10 | Virtual Router 10
State : Master # sw1 为 Master 角色
```

```
    Virtual IP : 10.1.10.254    #虚拟网关地址
    Master IP : 10.1.10.251     #物理网关地址
    PriorityRun : 120           #运行优先级
    PriorityConfig : 120        #配置优先级
    MasterPriority : 120        #主设备优先级
    Preempt : YES   Delay Time : 0 s
    TimerRun : 1 s
    TimerConfig : 1 s
    Auth type : NONE
    Virtual MAC : 0000-5e00-010a   #虚拟 MAC
    Check TTL : YES
    Config type : normal-vrrp
    Create time : 2021-08-27 12:29:47 UTC-08:00
    Last change time : 2021-08-27 12:30:03 UTC-08:00
```

查看 vrrp 配置简要信息

```
[sw1]display vrrp brief
VRID    State        Interface           Type         Virtual IP
----------------------------------------------------------------
10      Master       Vlanif10            Normal       10.1.10.254
----------------------------------------------------------------
Total:1    Master:1    Backup:0    Non-active:0
```

以上信息可以看出 sw1 State= Master；Interface= Vlanif10；Type= Normal Virtual IP= 10.1.10.254

② sw2 交换机

查看 vrrp 配置信息

```
[sw2]display vrrp
  Vlanif10 | Virtual Router 10
    State : Backup
    Virtual IP : 10.1.10.254
    Master IP : 10.1.10.251
    PriorityRun : 100           #运行优先级 100
    PriorityConfig : 100        #配置优先级 100
    MasterPriority : 120        #主设备优先级 120
    Preempt : YES   Delay Time : 0 s
    TimerRun : 1 s
    TimerConfig : 1 s
    Auth type : NONE
    Virtual MAC : 0000-5e00-010a
    Check TTL : YES
    Config type : normal-vrrp
    Create time : 2021-08-27 12:29:48 UTC-08:00
Last change time : 2021-08-27 12:30:01 UTC-08:00
```

查看 vrrp 配置简要信息

```
[sw2]display vrrp brief
VRID    State        Interface           Type         Virtual IP
----------------------------------------------------------------
10      Backup       Vlanif10            Normal       10.1.10.254
----------------------------------------------------------------
Total:1    Master:0    Backup:1    Non-active:0
[sw2]
```

步骤7 测试通信情况

PC1(IP:10.1.10.1)-ping- 网关（IP：10.1.10.254）- 通，见图 2-16-3。

查看 PC 的 arp 信息，见图 2-16-4。

图2-16-3 网络测试　　　　　　　　　　　图2-16-4 PC-arp信息

查看交换机 sw3 的 mac-address，目的是查看数据是从交换机的哪个接口出去，可以看到数据从 sw3-Eth0/0/1 出去。

```
[sw3]display mac-address
MAC address table of slot 0:

MAC Address      VLAN/       PEVLAN CEVLAN Port          Type        LSP/LSR-ID
                 VSI/SI                                              MAC-Tunnel

4c1f-cc4b-441b   10          -      -      Eth0/0/1      dynamic     0/-
5489-98cf-0de8   10          -      -      Eth0/0/3      dynamic     0/-
0000-5e00-010a   10          -      -      Eth0/0/1      dynamic     0/-

Total matching items on slot 0 displayed = 3
```

将 sw1 交换机关机，见图 2-16-5。测试见图 2-16-6。

PC1（IP：10.1.10.1）-ping- 网关（IP：10.1.10.254）- 通

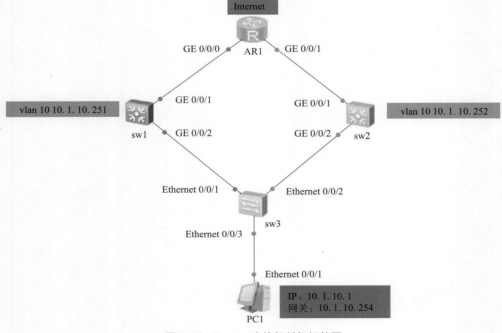

图2-16-5 sw1交换机关机拓扑图

查看 PC 的 arp 信息，见图 2-16-7。

图2-16-6　网络测试　　　　　　　　　图2-16-7　PC-arp

查看交换机 sw3 的 mac-address，目的是查看数据是从交换机的哪个接口出去，可以看到数据从 sw3-Eth0/0/2 出去。

```
[sw3]display mac-address
MAC address table of slot 0:
-------------------------------------------------------------------------------
MAC Address      VLAN/       PEVLAN  CEVLAN  Port        Type       LSP/LSR-ID
                 VSI/SI                                              MAC-Tunnel
-------------------------------------------------------------------------------
4c1f-cc1d-2c46   10          -       -       Eth0/0/2    dynamic    0/-
5489-98cf-0de8   10          -       -       Eth0/0/3    dynamic    0/-
0000-5e00-010a   10          -       -       Eth0/0/2    dynamic    0/-
-------------------------------------------------------------------------------
Total matching items on slot 0 displayed = 3
```

 sw1 交换机设备正常工作，但是当上行口 GigabitEthernet 0/0/1 线路故障时，即便数据发到 sw1，也是不能与外界通信，如何解决呢？通过配置链路检测功能来解决该故障。

1）查询目前配置

```
[sw1]interface  Vlanif 10
[sw1-Vlanif10]dis this
interface Vlanif10
 ip address 10.1.10.251 255.255.255.0
 vrrp vrid 10 virtual-ip 10.1.10.254
 vrrp vrid 10 priority 120
Return
# 删除优先级
[sw1-Vlanif10]undo  vrrp vrid 10 priority
# 配置当上行线路故障是后，优先级降低 30
[sw1-Vlanif10]vrrp vrid 10 track interface GigabitEthernet 0/0/1 reduced 30
# 配置 GigabitEthernet 0/0/1 用来做链路跟踪。
[sw1-Vlanif10]vrrp vrid 10 track interface GigabitEthernet 0/0/1
# 配置优先级为 120
[sw1-Vlanif10]vrrp vrid 10 priority 120
[sw1-Vlanif10]quit
[sw1]quit
```

```
<sw1>save
```

2）模拟上行链路断开

执行 PC1（IP:10.1.10.1）-ping - 网关（IP：10.1.10.254），然后在 sw3 交换机通过 mac-address 查询，数据从交换机 sw3 的 Eth0/0/1 接口出去。查询 sw3 交换机的 mac-address，见图 2-16-8。

```
<sw3>dis mac-address
MAC address table of slot 0:
------------------------------------------------------------------------
MAC Address      VLAN/         PEVLAN CEVLAN Port            Type        LSP/LSR-ID
                 VSI/SI                                                  MAC-Tunnel
------------------------------------------------------------------------
5489-98cf-0de8   10            -      -      Eth0/0/3        dynamic     0/-
4c1f-cc4b-441b   10            -      -      Eth0/0/1        dynamic     0/-
0000-5e00-010a   10            -      -      Eth0/0/1        dynamic     0/-
------------------------------------------------------------------------
Total matching items on slot 0 displayed = 3
```

图2-16-8　sw3交换机mac-address

3）手动关闭 sw1-GigabitEthernet 0/0/1，模拟链路故障。

```
[sw1-GigabitEthernet0/0/1]shutdown
```

在 sw1 交换机查询 vrrp。

```
[sw1]display vrrp 10
  Vlanif10 | Virtual Router 10
    State : Backup #
    Virtual IP : 10.1.10.254
    Master IP : 10.1.10.252
    PriorityRun : 90 # 优先级由原来120 降低到 90
    PriorityConfig : 120# 配置优先级是 120
    MasterPriority : 100# 主设备优先级是 100
    Preempt : YES    Delay Time : 0 s
    TimerRun : 1 s
    TimerConfig : 1 s
    Auth type : NONE
    Virtual MAC : 0000-5e00-010a
    Check TTL : YES
    Config type : normal-vrrp
    Track IF : GigabitEthernet0/0/1   Priority reduced : 30
    IF state : DOWN
    Create time : 2021-08-27 14:45:55 UTC-08:00
    Last change time : 2021-08-27 15:00:56 UTC-08:00
```

查询 sw1 交换机 vrrp 简要信息，发现转态已经变为 Backup。

```
[sw1]display vrrp brief
VRID   State      Interface               Type        Virtual IP
--------------------------------------------------------------------
10     Backup     Vlanif10                Normal      10.1.10.254
--------------------------------------------------------------------
Total:1    Master:0     Backup:1     Non-active:0
[sw1]
```

查询 sw2 交换机 vrrp 信息

```
[sw2]display vrrp 10
  Vlanif10 | Virtual Router 10
    State : Master
```

```
    Virtual IP : 10.1.10.254
    Master IP : 10.1.10.252
    PriorityRun : 100  # 运行优先级
    PriorityConfig : 100# 配置优先级
    MasterPriority : 100# 主设备优先级
    Preempt : YES    Delay Time : 0 s
    TimerRun : 1 s
    TimerConfig : 1 s
    Auth type : NONE
    Virtual MAC : 0000-5e00-010a
    Check TTL : YES
    Config type : normal-vrrp
    Create time : 2021-08-27 12:29:48 UTC-08:00
    Last change time : 2021-08-27 15:00:56 UTC-08:00
```

查询 sw2 交换机简要信息，已经变为 Master。

```
[sw2]display vrrp brief
VRID   State      Interface            Type      Virtual IP
-------------------------------------------------------------
10     Master     Vlanif10             Normal    10.1.10.254
-------------------------------------------------------------
Total:1    Master:1    Backup:0    Non-active:0
[sw2]
```

测试通信情况 PC1(IP:10.1.10.1)-ping- 网关（IP: 10.1.10.254）- 通，见图 2-16-9。

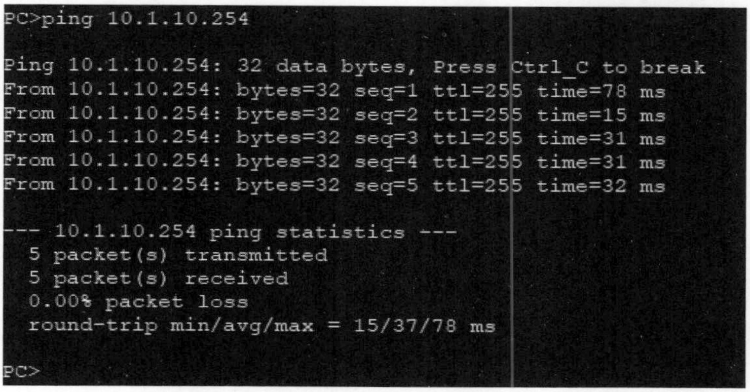

```
PC>ping 10.1.10.254

Ping 10.1.10.254: 32 data bytes, Press Ctrl_C to break
From 10.1.10.254: bytes=32 seq=1 ttl=255 time=78 ms
From 10.1.10.254: bytes=32 seq=2 ttl=255 time=15 ms
From 10.1.10.254: bytes=32 seq=3 ttl=255 time=31 ms
From 10.1.10.254: bytes=32 seq=4 ttl=255 time=31 ms
From 10.1.10.254: bytes=32 seq=5 ttl=255 time=32 ms

--- 10.1.10.254 ping statistics ---
  5 packet(s) transmitted
  5 packet(s) received
  0.00% packet loss
  round-trip min/avg/max = 15/37/78 ms
PC>
```

图2-16-9 网络测试

查询交换机 sw3 的 mac-address，发现 PC1 访问网关交换机接口已经变为 Eth0/0/2。

```
[sw3]display mac-address
MAC address table of slot 0:
-------------------------------------------------------------------------------
MAC Address    VLAN/PEVLAN CEVLAN Port    Type    LSP/LSR-ID
VSI/SI    MAC-Tunnel
-------------------------------------------------------------------------------
5489-98cf-0de8  10      -        -       Eth0/0/3     dynamic     0/-
4c1f-cc1d-2c46  10      -        -       Eth0/0/2     dynamic     0/-
0000-5e00-010a  10      -        -       Eth0/0/2     dynamic     0/-
-------------------------------------------------------------------------------
Total matching items on slot 0 displayed = 3
```

二、负载均衡模式

vrrp 优先级默认为 100（范围 0～255），数值越大，越有可能成为 Master，优先级最高设为 255，如果配置的 vrrp 虚拟 IP 地址与接口的 IP 地址相同，那么优先级自动变为 255。

1. 拓扑图

如图 2-16-10 所示。

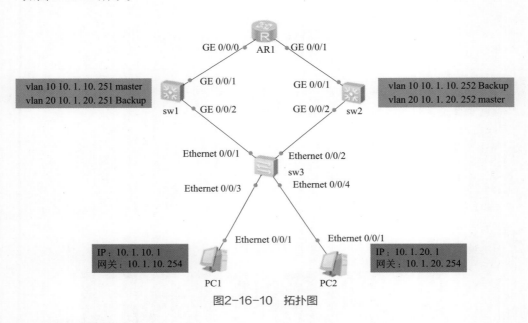

图2-16-10　拓扑图

2. 基本配置

1）PC2 网络参数配置，见图 2-16-11。

图2-16-11　PC2参数配置

2）sw3 交换机增加配置

```
<sw3>sys
[sw3]vlan 20
[sw3-vlan20]quit
[sw3]interface Ethernet 0/0/4
[sw3-Ethernet0/0/4]port link-type access
[sw3-Ethernet0/0/4]port default vlan 20
[sw3-Ethernet0/0/4]quit
[sw3]quit
<sw3>save
```

3）sw1 交换机增加配置

```
<sw1>sys
[sw1]vlan 20
[sw1]interface Vlanif 20
[sw1-Vlanif20]ip address 10.1.20.251 24
# 配置 vrrp 工作组为 20，虚拟 IP 地址：10.1.20.254
[sw1-Vlanif20]vrrp vrid 20 virtual-ip 10.1.20.254
[sw1-Vlanif20]quit
[sw1]quit
<sw1>save
```

4）sw2 交换机增加配置

```
<sw2>sys
Enter system view, return user view with Ctrl+Z.
[sw2]vlan 20
[sw2]interface Vlanif 20
[sw2-Vlanif20]ip address 10.1.20.252 24
# 配置 vrrp 工作组为 20，虚拟 IP 地址：10.1.20.254
[sw2-Vlanif20]vrrp vrid 20 virtual-ip 10.1.20.254
# 配置优先级为 105，为 Backup
[sw2-Vlanif20]vrrp vrid 20 priority 105
```

3. 查看 vrrp 信息

sw1

```
[sw1]display vrrp
  Vlanif10 | Virtual Router 10
    State : Master  #sw1 在工作组 10 中为 Master
    Virtual IP : 10.1.10.254
    Master IP : 10.1.10.251
    PriorityRun : 120  # 运行优先级
    PriorityConfig : 120  # 配置优先级
    MasterPriority : 120  # 主设备的优先级
    Preempt : YES   Delay Time : 0 s
    TimerRun : 1 s
    TimerConfig : 1 s
    Auth type : NONE
    Virtual MAC : 0000-5e00-010a
    Check TTL : YES
    Config type : normal-vrrp
    Track IF : GigabitEthernet0/0/1   Priority reduced : 30
```

```
    IF state : up
    Create time : 2021-08-27 18:08:27 UTC-08:00
    Last change time : 2021-08-27 18:36:31 UTC-08:00
  Vlanif20 | Virtual Router 20
    State : Backup #sw1 在工作组 20 中为 Backup
    Virtual IP : 10.1.20.254
    Master IP : 10.1.20.252
    PriorityRun : 100 # 运行优先级 100
    PriorityConfig : 100 # 配置优先级 100
    MasterPriority : 105 # 主设备优先级 105
    Preempt : YES    Delay Time : 0 s
    TimerRun : 1 s
    TimerConfig : 1 s
    Auth type : NONE
    Virtual MAC : 0000-5e00-0114
    Check TTL : YES
    Config type : normal-vrrp
    Create time : 2021-08-27 18:26:46 UTC-08:00
    Last change time : 2021-08-27 21:53:43 UTC-08:00
sw2
[sw2]display vrrp
  Vlanif10 | Virtual Router 10
    State : Backup #sw2 在工作组 10 中为 Backup
    Virtual IP : 10.1.10.254
    Master IP : 10.1.10.251
    PriorityRun : 100
    PriorityConfig : 100
    MasterPriority : 120
    Preempt : YES    Delay Time : 0 s
    TimerRun : 1s
    TimerConfig : 1s
    Auth type : NONE
    Virtual MAC : 0000-5e00-010a
    Check TTL : YES
    Config type : normal-vrrp
    Create time : 2021-08-27 18:08:33 UTC-08:00
    Last change time : 2021-08-27 21:51:09 UTC-08:00
  Vlanif20 | Virtual Router 20
    State : Master #sw2 在工作组 20 中为 Master
    Virtual IP : 10.1.20.254
    Master IP : 10.1.20.252
    PriorityRun : 105
    PriorityConfig : 105
    MasterPriority : 105
    Preempt : YES    Delay Time : 0 s
    TimerRun : 1 s
    TimerConfig : 1 s
    Auth type : NONE
    Virtual MAC : 0000-5e00-0114
    Check TTL : YES
    Config type : normal-vrrp
```

```
        Create time : 2021-08-27 18:30:38 UTC-08:00
        Last change time : 2021-08-27 21:53:43 UTC-08:00
```

查看 vrrp 简要信息

```
sw1
[sw1]display vrrp brief
VRID   State         Interface              Type        Virtual IP
----------------------------------------------------------------------
10     Master        Vlanif10               Normal      10.1.10.254
20     Backup        Vlanif20               Normal      10.1.20.254
----------------------------------------------------------------------
Total:2     Master:1       Backup:1     Non-active:0
sw2
[sw2]display vrrp brief
VRID   State         Interface              Type        Virtual IP
----------------------------------------------------------------------
10     Backup        Vlanif10               Normal      10.1.10.254
20     Master        Vlanif20               Normal      10.1.20.254
----------------------------------------------------------------------
Total:2     Master:1       Backup:1     Non-active:0
[sw2]
```

4. 测试通信情况

PC1（IP:10.1.10.1）-ping-PC2（IP:10.1.20.1）- 通，见图 2-16-12。

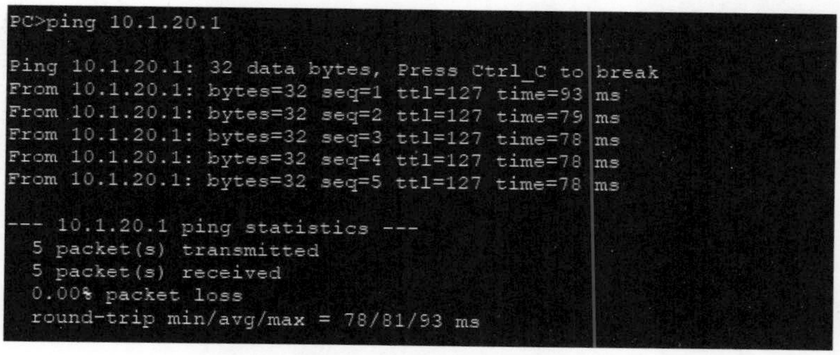

```
PC>ping 10.1.20.1

Ping 10.1.20.1: 32 data bytes, Press Ctrl_C to break
From 10.1.20.1: bytes=32 seq=1 ttl=127 time=93 ms
From 10.1.20.1: bytes=32 seq=2 ttl=127 time=79 ms
From 10.1.20.1: bytes=32 seq=3 ttl=127 time=78 ms
From 10.1.20.1: bytes=32 seq=4 ttl=127 time=78 ms
From 10.1.20.1: bytes=32 seq=5 ttl=127 time=78 ms

--- 10.1.20.1 ping statistics ---
  5 packet(s) transmitted
  5 packet(s) received
  0.00% packet loss
  round-trip min/avg/max = 78/81/93 ms
```

图2-16-12　网络测试

三、IP 地址拥有者

vrrp 接口地址与虚拟 IP 地址相同的路由称之为 IP 地址拥有者，优先级为 255，并且无法修改。将 10.1.20.252 作为 10.1.20.0 段的网关。进行配置工作，观察优先级。

1. 基本配置

sw1 配置进行修改

```
# 删除 vrrp vrid 20
[sw1-Vlanif20]undo vrrp vrid 20 virtual-ip 10.1.20.254
# 配置 vrid 20 虚拟网关是 10.1.20.252
[sw1-Vlanif20]vrrp vrid 20 virtual-ip 10.1.20.252
[sw1-Vlanif20]quit
[sw1]quit
<sw1>save
```

sw2 配置进行修改

```
[sw2-Vlanif20]undo vrrp vrid 20 virtual-ip 10.1.20.254
[sw2-Vlanif20]vrrp vrid 20 virtual-ip 10.1.20.252
      Warning: The priority of this VRRP backup group has changed to 255 and will not
change.
[sw2-Vlanif20]quit
[sw2]quit
<sw2>save
```

PC2 修改网关地址（IP：10.1.20.252），见图 2-16-13。

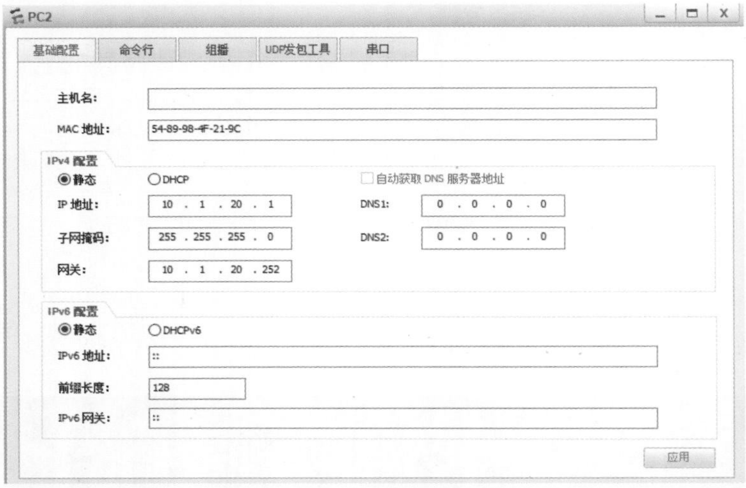

图2-16-13　PC2配置

2. 测试通信

PC1（IP:10.1.10.1）-ping-PC2（IP:10.1.20.1）- 通，见图 2-16-14。

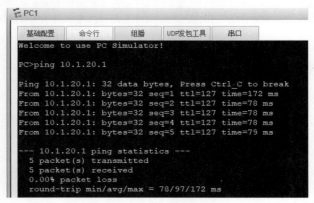

图2-16-14　网络测试

3. 查询 vrrp 配置信息

sw1

```
[sw1]display vrrp
```

```
Vlanif10 | Virtual Router 10
  State : Master
  Virtual IP : 10.1.10.254
  Master IP : 10.1.10.251
  PriorityRun : 120
  PriorityConfig : 120
  MasterPriority : 120
  Preempt : YES   Delay Time : 0 s
  TimerRun : 1 s
  TimerConfig : 1 s
  Auth type : NONE
  Virtual MAC : 0000-5e00-010a
  Check TTL : YES
  Config type : normal-vrrp
  Track IF : GigabitEthernet0/0/1   Priority reduced : 30
  IF state : up
  Create time : 2021-08-28 13:57:00 UTC-08:00
  Last change time : 2021-08-28 13:57:44 UTC-08:00
Vlanif20 | Virtual Router 20
  State : Backup
  Virtual IP : 10.1.20.252
  Master IP : 10.1.20.252
  PriorityRun : 100   #运行优先级
  PriorityConfig : 100   #配置优先级
  MasterPriority : 255   #当虚拟网关地址与物理地址相同时，优先级变为最大 255
  Preempt : YES   Delay Time : 0 s
  TimerRun : 1 s
  TimerConfig : 1 s
  Auth type : NONE
  Virtual MAC : 0000-5e00-0114
  Check TTL : YES
  Config type : normal-vrrp
  Create time : 2021-08-28 15:08:19 UTC-08:00
  Last change time : 2021-08-28 15:10:10 UTC-08:00
sw2
[sw2]display vrrp
Vlanif10 | Virtual Router 10
  State : Backup
  Virtual IP : 10.1.10.254
  Master IP : 10.1.10.251
  PriorityRun : 100
  PriorityConfig : 100
  MasterPriority : 120
  Preempt : YES   Delay Time : 0 s
  TimerRun : 1 s
  TimerConfig : 1 s
  Auth type : NONE
  Virtual MAC : 0000-5e00-010a
  Check TTL : YES
  Config type : normal-vrrp
  Create time : 2021-08-28 14:15:11 UTC-08:00
```

```
Last change time : 2021-08-28 14:15:22 UTC-08:00
Vlanif20 | Virtual Router 20
State : Master
Virtual IP : 10.1.20.252
Master IP : 10.1.20.252
PriorityRun : 255  #运行优先级
PriorityConfig : 100  #配置优先级
MasterPriority : 255  #主设备优先级
Preempt : YES   Delay Time : 0 s
TimerRun : 1 s
TimerConfig : 1 s
Auth type : NONE
Virtual MAC : 0000-5e00-0114
Check TTL : YES
Config type : normal-vrrp
Create time : 2021-08-28 15:10:10 UTC-08:00
Last change time : 2021-08-28 15:10:10 UTC-08:00
[sw2]
```

项目 17　动态路由 ospf

在中大型企业中一般使用 ospf 动态路由协议，ospf 是典型的链路状态路由协议，使用最短路径算法计算与选择路由。ospf 协议的优点：无环路、收敛速度快、扩展性好、支持认证。ospf 报文封装在 IP 报文中，协议号是 89。

项目简介

通过配置 ospf 动态路由协议，实现全网互通。本节重点讲解单区域与多区域配置办法。

一、单区域 ospf

把所有的路由器划分到区域，如是单区域时，一般建议采用 area0。
见图 2-17-1，实现功能：通过配置 ospf 动态路由协议，实现全网网络互通。

1. 项目配置步骤

步骤 1　AR1 路由器接口 IP 地址配置
```
<Huawei>system-view
Enter system view, return user view with Ctrl+Z.
```

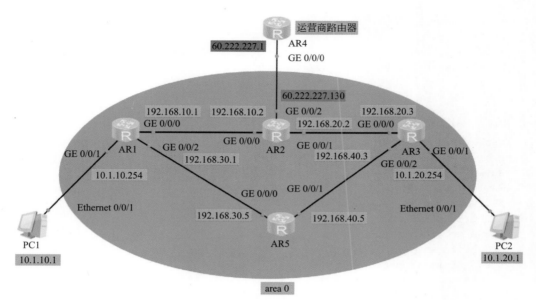

图2-17-1 拓扑图

```
[Huawei]sysname AR1
[AR1]un in en
Info: Information center is disabled.
# AR1-GigabitEthernet0/0/0 配置 IP 地址 192.168.10.1
[AR1]interface GigabitEthernet 0/0/0
[AR1-GigabitEthernet0/0/0]ip address 192.168.10.1 24
[AR1-GigabitEthernet0/0/0]quit
# AR1-GigabitEthernet0/0/1 配置 IP 地址 10.1.10.254
[AR1]interface GigabitEthernet 0/0/1
[AR1-GigabitEthernet0/0/1]ip address 10.1.10.254 24
[AR1-GigabitEthernet0/0/1]quit
# AR1-GigabitEthernet0/0/2 配置 IP 地址 192.168.30.1
[AR1]interface  GigabitEthernet 0/0/2
[AR1-GigabitEthernet0/0/2]ip address 192.168.30.1 24
[AR1-GigabitEthernet0/0/2]quit
[AR1]quit
<AR1>save
```

步骤2 AR2 路由器接口 IP 地址配置

```
<Huawei>system-view
[Huawei]sysname AR2
# AR2-GigabitEthernet0/0/0 配置 IP 地址 192.168.10.2
[AR2]interface GigabitEthernet 0/0/0
[AR2-GigabitEthernet0/0/0]ip address 192.168.10.2 24
[AR2-GigabitEthernet0/0/0]quit
[AR2]un in en
Info: Information center is disabled.
# AR2-GigabitEthernet0/0/1 配置 IP 地址 192.168.20.2
[AR2]interface  GigabitEthernet 0/0/1
[AR2-GigabitEthernet0/0/1]ip address 192.168.20.2 24
[AR2-GigabitEthernet0/0/1]quit
```

```
# AR2-GigabitEthernet0/0/2 配置IP地址 60.222.227.130
[AR2]interface  GigabitEthernet 0/0/2
[AR2-GigabitEthernet0/0/2]ip address 60.222.227.130 24
[AR2-GigabitEthernet0/0/2]quit
[AR2]quit
<AR2>save
```

步骤3　AR3路由器接口IP地址配置

```
<Huawei>system-view
Enter system view, return user view with Ctrl+Z.
[Huawei]sysname AR3
# AR3-GigabitEthernet0/0/0 配置IP地址 192.168.20.3
[AR3]interface GigabitEthernet 0/0/0
[AR3-GigabitEthernet0/0/0]ip address 192.168.20.3 24
[AR3-GigabitEthernet0/0/0]quit
[AR3]un in en
Info: Information center is disabled.
# AR3-GigabitEthernet0/0/1 配置IP地址 10.1.20.254
[AR3]interface  GigabitEthernet 0/0/1
[AR3-GigabitEthernet0/0/1]ip address 10.1.20.254 24
[AR3-GigabitEthernet0/0/1]quit
# AR3-GigabitEthernet0/0/2 配置IP地址 192.168.40.3
[AR3]interface  GigabitEthernet 0/0/2
[AR3-GigabitEthernet0/0/2]ip address 192.168.40.3 24
[AR3-GigabitEthernet0/0/2]quit
[AR3]quit
<AR3>save
```

步骤4　AR4路由器接口IP地址配置

```
<Huawei>system-view
Enter system view, return user view with Ctrl+Z.
[Huawei]sysname AR4
# AR4-GigabitEthernet0/0/0 配置IP地址 60.222.227.1
[AR4]interface GigabitEthernet 0/0/0
[AR4-GigabitEthernet0/0/0]ip address 60.222.227.1 24
[AR4-GigabitEthernet0/0/0]quit
[AR4]quit
<AR4>save
```

步骤5　AR5路由器接口IP地址配置

```
<Huawei>sys
[Huawei]sysname AR5
# AR5-GigabitEthernet0/0/0 配置IP地址 192.168.30.5
[AR5]interface GigabitEthernet 0/0/0
[AR5-GigabitEthernet0/0/0]ip address 192.168.30.5 24
[AR5-GigabitEthernet0/0/0]quit
# AR5-GigabitEthernet0/0/1 配置IP地址 192.168.40.5
[AR5]interface GigabitEthernet 0/0/1
[AR5-GigabitEthernet0/0/1]ip address 192.168.40.5 24
[AR5-GigabitEthernet0/0/1]quit
[AR5]quit
<AR5>save
```

路由器 IP 地址配置见表 2-17-1。

表 2-17-1　路由器 IP 地址表

AR1	G 0/0/0	192.168.10.1	G 0/0/1	10.1.10.254	G 0/0/2	192.168.30.1
AR2	G 0/0/0	192.168.10.2	G 0/0/1	192.168.20.2	G 0/0/2	60.222.227.130
AR3	G 0/0/0	192.168.20.3	G 0/0/1	10.1.20.254	G 0/0/2	192.168.40.3
AR4	G 0/0/0	60.222.227.1	G 0/0/1	/	G 0/0/2	/
AR5	G 0/0/0	192.168.30.5	G 0/0/1	192.168.40.5	G 0/0/2	/

2. 检查路由器配置接口是否正常

检查各个路由器配置 IP 地址的接口，Physical（物理）与 Protocol（协议）是否为 up 状态。

步骤 1　AR1 配置

```
[AR1]dis ip interface brie
*down: administratively down
down: standby
(l): loopback
(s): spoofing
The number of interface that is up in Physical is 4
The number of interface that is down in Physical is 0
The number of interface that is up in Protocol is 4
The number of interface that is down in Protocol is 0
Interface                 IP Address/Mask      Physical   Protocol
GigabitEthernet0/0/0      192.168.10.1/24      up         up
GigabitEthernet0/0/1      10.1.10.254/24       up         up
GigabitEthernet0/0/2      192.168.30.1/24      up         up
NULL0                     unassigned           up         up(s)
[AR1]
```

步骤 2　AR2 配置

```
[AR2]display ip interface brief
*down: administratively down
^down: standby
(l): loopback
(s): spoofing
The number of interface that is up in Physical is 4
The number of interface that is down in Physical is 0
The number of interface that is up in Protocol is 4
The number of interface that is down in Protocol is 0
Interface                 IP Address/Mask      Physical   Protocol
GigabitEthernet0/0/0      192.168.10.2/24      up         up
GigabitEthernet0/0/1      192.168.20.2/24      up         up
GigabitEthernet0/0/2      60.222.227.130/24    up         up
NULL0                     unassigned           up         up(s)
[AR2]
```

步骤 3　AR3 配置

```
[AR3]display ip interface brief
*down: administratively down
^down: standby
```

```
(l): loopback
(s): spoofing
The number of interface that is up in Physical is 4
The number of interface that is down in Physical is 0
The number of interface that is up in Protocol is 4
The number of interface that is down in Protocol is 0
Interface                 IP Address/Mask        Physical    Protocol
GigabitEthernet0/0/0      192.168.20.3/24        up          up
GigabitEthernet0/0/1      10.1.20.254/24         up          up
GigabitEthernet0/0/2      192.168.40.3/24        up          up
NULL0                     unassigned             up          up(s)
[AR3]
```

步骤 4 AR4 配置

```
[AR4]display ip interface brief
*down: administratively down
^down: standby
(l): loopback
(s): spoofing
The number of interface that is up in Physical is 2
The number of interface that is down in Physical is 2
The number of interface that is up in Protocol is 2
The number of interface that is down in Protocol is 2
Interface                 IP Address/Mask        Physical    Protocol
GigabitEthernet0/0/0      60.222.227.1/24        up          up
GigabitEthernet0/0/1      unassigned             down        down
GigabitEthernet0/0/2      unassigned             down        down
NULL0
```

步骤 5 AR5 配置

```
[AR5]display ip interface brief
*down: administratively down
^down: standby
(l): loopback
(s): spoofing
The number of interface that is up in Physical is 3
The number of interface that is down in Physical is 1
The number of interface that is up in Protocol is 2
The number of interface that is down in Protocol is 2
Interface                 IP Address/Mask        Physical    Protocol
GigabitEthernet0/0/0      192.168.30.5/24        up          up
GigabitEthernet0/0/1      192.168.40.5/24        up          up
GigabitEthernet0/0/2      unassigned             down        down
NULL0                     unassigned             up          up(s)
[AR5]
......................
```

3. 路由器配置环回接口地址

步骤 1 AR1 配置

```
[AR1]interface LoopBack 0
# 配置环回接口地址 LoopBack 0 IP: 1.1.1.1
```

```
[AR1-LoopBack0]ip address 1.1.1.1 32
[AR1-LoopBack0]quit
[AR1]quit
<AR1>save
```

步骤 2　AR2 配置

```
[AR2]interface  LoopBack 0
# 配置环回接口地址 LoopBack 0 IP：2.2.2.2
[AR2-LoopBack0]ip address 2.2.2.2 32
[AR2-LoopBack0]quit
[AR2]quit
<AR2>save
```

步骤 3　AR3 配置

```
[AR3]interface LoopBack 0
# 配置环回接口地址 LoopBack 0 IP：3.3.3.3
[AR3-LoopBack0]ip address 3.3.3.3 32
[AR3-LoopBack0]quit
[AR3]quit
<AR3>save
```

步骤 4　AR4 配置

```
[AR4]interface LoopBack 0
# 配置环回接口地址 LoopBack 0 IP：4.4.4.4
[AR4-LoopBack0]ip address 4.4.4.4 32
[AR4-LoopBack0]quit
[AR4]quit
<AR4>save
```

步骤 5　AR5 配置

```
[AR5]interface LoopBack 0
# 配置环回接口地址 LoopBack 0 IP：5.5.5.5
[AR5-LoopBack0]ip address 5.5.5.5 32
[AR5-LoopBack0]quit
[AR5]quit
<AR5>save
```

4. 配置 ospf 动态路由

基础知识

配置 ospf 时，用 Router ID 标识路由器身份，Router ID 与 IP 地址格式类似，它作为路由器标识符，相当于身份证。一般建议手动配置 Router ID，如没有手动配置，默认选举路由器中环回 IP 地址大的作为 Router ID，当路由器没有配置环回的时候，路由器选择物理接口 IP 地址大的作为 Router ID。

通配符，一般用 255.255.255.255 减去网段掩码，比如 192.168.10.0/24，在 ospf 中宣告的时候：network　192.168.10.0　0.0.0.255，通配符中的 0 表示精确值，而 255 表示任意值。

步骤1 AR1 配置

配置 ospf 1 的 router-id，作为本路由器的标识。
[AR1]ospf 1 router-id 1.1.1.1
进入区域 area 0
[AR1-ospf-1]area 0
宣告 192.168.10.0/24 网段
[AR1-ospf-1-area-0.0.0.0]network 192.168.10.0 0.0.0.255
宣告 192.168.30.0 /24 网段
[AR1-ospf-1-area-0.0.0.0]network 192.168.30.0 0.0.0.255
宣告 10.1.10.0 /24 网段
[AR1-ospf-1-area-0.0.0.0]network 10.1.10.0 0.0.0.255
宣告 1.1.1.1 /32 网段
[AR1-ospf-1-area-0.0.0.0]network 1.1.1.1 0.0.0.0
[AR1-ospf-1]quit
[AR1]quit
<AR1>save

步骤2 AR2 配置

[AR2]ospf 1 router-id 2.2.2.2
[AR2-ospf-1]area 0
[AR2-ospf-1-area-0.0.0.0]network 192.168.10.0 0.0.0.255
[AR2-ospf-1-area-0.0.0.0]network 192.168.20.0 0.0.0.255
[AR2-ospf-1-area-0.0.0.0]network 2.2.2.2 0.0.0.0
[AR2-ospf-1-area-0.0.0.0]quit
[AR2-ospf-1]quit
[AR2]quit
<AR2>save

步骤3 AR3 配置

[AR3]ospf 1 router-id 3.3.3.3
[AR3-ospf-1]area 0
[AR3-ospf-1-area-0.0.0.0]network 192.168.20.0 0.0.0.255
[AR3-ospf-1-area-0.0.0.0]network 10.1.20.0 0.0.0.255
[AR3-ospf-1-area-0.0.0.0]network 3.3.3.3 0.0.0.0
[AR3-ospf-1-area-0.0.0.0]network 192.168.40.0 0.0.0.255
[AR3-ospf-1-area-0.0.0.0]quit
[AR3-ospf-1]quit
[AR3]quit
<AR3>save

步骤4 AR5 配置

[AR5]ospf 1 router-id 5.5.5.5
[AR5-ospf-1]area 0
[AR5-ospf-1-area-0.0.0.0]network 192.168.30.0 0.0.0.255
[AR5-ospf-1-area-0.0.0.0]network 192.168.40.0 0.0.0.255
[AR5-ospf-1-area-0.0.0.0]network 5.5.5.5 0.0.0.0
[AR5-ospf-1-area-0.0.0.0]quit
[AR5-ospf-1]quit
[AR5]quit
<AR5>sa

5. 查询路由器邻居关系

通过命令 display ospf peer brief 查看邻居关系，确保邻居关系正常（full 代表正常）。

步骤 1　AR2 邻居关系，见图 2-17-2。

```
[AR2]display ospf peer brief

        OSPF Process 1 with Router ID 2.2.2.2
                Peer Statistic Information
----------------------------------------------------------------------------
Area Id          Interface                       Neighbor id      State
0.0.0.0          GigabitEthernet0/0/0            1.1.1.1          Full
0.0.0.0          GigabitEthernet0/0/1            3.3.3.3          Full

----------------------------------------------------------------------------
[AR2]
```

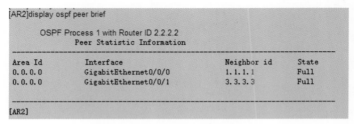

图2-17-2　AR2邻居关系

AR2 邻居关系字段含义如表 2-17-2。

表 2-17-2　AR2 邻居关系

Area Id	Interface	Neighbor id	State
Ospf 区域	本路由器接口	对端路由器 router-id	最终建立状态（full）

步骤 2　AR5 邻居关系，见图 2-17-3。

```
[AR5]display ospf peer brief

        OSPF Process 1 with Router ID 5.5.5.5
                Peer Statistic Information
----------------------------------------------------------------------------
Area Id          Interface                       Neighbor id      State
0.0.0.0          GigabitEthernet0/0/1            3.3.3.3          Full
0.0.0.0          GigabitEthernet0/0/0            1.1.1.1          Full
----------------------------------------------------------------------------
```

图2-17-3　AR5邻居关系

6. PC 配置以及测试

步骤 1　PC1 配置，见图 2-17-4。

IP：10.1.10.1，子网掩码：255.255.255.0，网关：10.1.10.254。

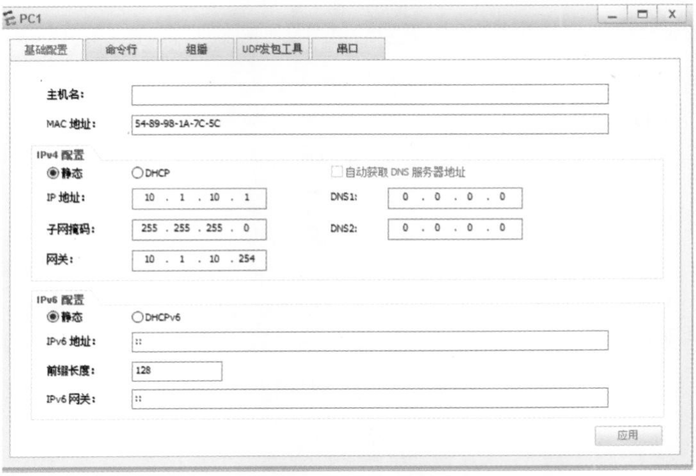

图2-17-4　PC1参数配置

步骤2　PC2 配置，见图 2-17-5。

IP：10.1.20.1，子网掩码：255.255.255.0，网关：10.1.20.254。

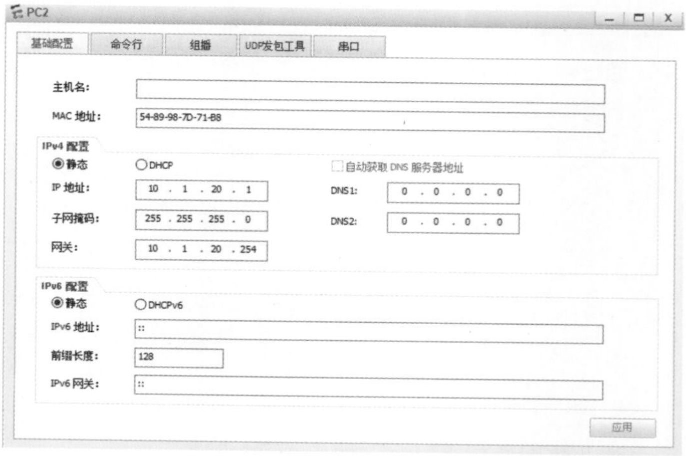

图2-17-5　PC2参数配置

7. 通信测试

见图 2-17-6。

测试 PC1(IP:10.1.10.1)-ping-PC2(10.1.20.1)-通

图2-17-6　通信测试

8. 查看路由表

命令：display ip rou-ting-table protocol ospf（查看 ospf 路由信息）

步骤1　AR1 路由表，见图 2-17-7。

通过查看路由表发现去往 3.3.3.3 以及 10.1.20.0 网段有两条路由。

图2-17-7　AR1路由表

在 PC1 tracert 去往 PC2(IP:10.1.20.1) 的路径见图 2-17-8，可以看出：PC1（10.10.10.1）-AR1（10.1.10.254）-AR2(192.168.10.2)-AR3(192.168.20.3)-PC2(10.1.20.1)

以上去往目标网段具备两条路径，开销值（图 2-17-7 中 cost 值）等参数相同，实现功能就是负载分担功能。PC1 去往 PC2 的路径共计两条（PC1-AR1-AR2-AR3-PC2、PC1-AR1-AR5-AR3-PC2），通信数据负载自动选择路径。

步骤 2　通过修改开销值 cost，改变数据流量走向。

通过修改带宽或者在路由器出接口直接修改 cost 值（默认 1）都可以改变流量走向切换，一般采取后者（修改开销值）。修改 AR1 GigabitEthernet0/0/0 接口 cost 为 100，开销值越小，路径越优先。命令如下

```
[AR1-GigabitEthernet0/0/0]ospf cost 100
```

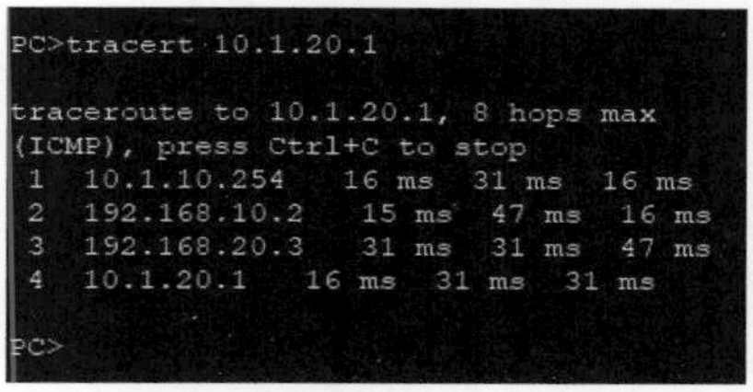

```
PC>tracert 10.1.20.1

traceroute to 10.1.20.1, 8 hops max
(ICMP), press Ctrl+C to stop
 1  10.1.10.254   16 ms  31 ms  16 ms
 2  192.168.10.2   15 ms  47 ms  16 ms
 3  192.168.20.3   31 ms  31 ms  47 ms
 4  10.1.20.1   16 ms  31 ms  31 ms

PC>
```

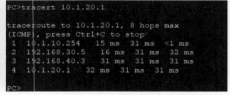

图2-17-8　PC1 tracert去往PC2(IP:10.1.20.1)的路径

再次测试

PC1(IP:10.1.10.1)-ping-PC2(10.1.20.1)- 通，见图 2-17-9。

在 PC1 tracert 去往 PC2(IP:10.1.20.1) 的路径，见图 2-17-10。

路径 PC1(10.10.10.1)-AR1(10.1.10.254)-AR5(192.168.30.5)-AR3(192.168.40.3)-PC2(10.1.20.1)

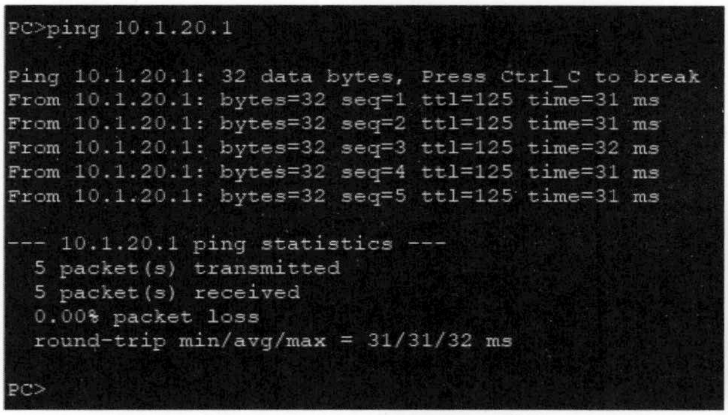

```
PC>ping 10.1.20.1

Ping 10.1.20.1: 32 data bytes, Press Ctrl_C to break
From 10.1.20.1: bytes=32 seq=1 ttl=125 time=31 ms
From 10.1.20.1: bytes=32 seq=2 ttl=125 time=31 ms
From 10.1.20.1: bytes=32 seq=3 ttl=125 time=32 ms
From 10.1.20.1: bytes=32 seq=4 ttl=125 time=31 ms
From 10.1.20.1: bytes=32 seq=5 ttl=125 time=31 ms

--- 10.1.20.1 ping statistics ---
  5 packet(s) transmitted
  5 packet(s) received
  0.00% packet loss
  round-trip min/avg/max = 31/31/32 ms

PC>
```

图2-17-9　通信测试

```
PC>tracert 10.1.20.1

traceroute to 10.1.20.1, 8 hops max
(ICMP), press Ctrl+C to stop
 1  10.1.10.254   15 ms  31 ms  <1 ms
 2  192.168.30.5   16 ms  31 ms  32 ms
 3  192.168.40.3   31 ms  31 ms  31 ms
 4  10.1.20.1   32 ms  31 ms  31 ms

PC>
```

图2-17-10　PC1 tracert去往PC2(IP:10.1.20.1)的路径

9. 关于 ospf 动态路由协议三张表，如表 2-17-3 所示。

表 2-17-3　ospf 动态路由协议三张表清单

邻居表	display ospf peer brief
路由表	display ip routing-table protocol ospf
LSDB 数据库表	display ospf lsdb

AR2 为例，介绍 ospf 动态路由协议三张表。

步骤 1　邻居表见图 2-17-11。

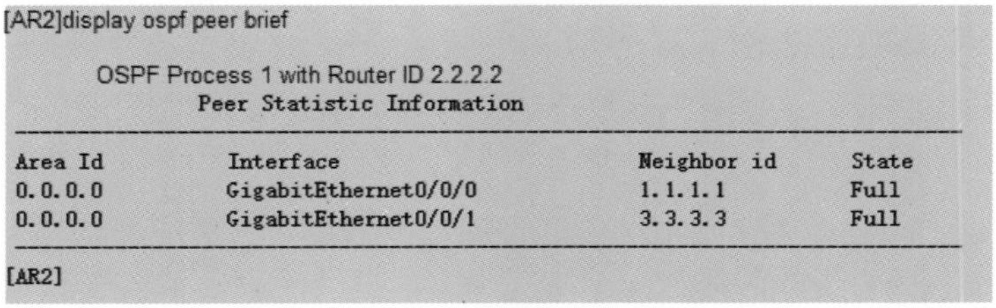

```
[AR2]display ospf peer brief

     OSPF Process 1 with Router ID 2.2.2.2
              Peer Statistic Information
 ----------------------------------------------------------------------------
 Area Id          Interface                          Neighbor id      State
 0.0.0.0          GigabitEthernet0/0/0               1.1.1.1          Full
 0.0.0.0          GigabitEthernet0/0/1               3.3.3.3          Full
 ----------------------------------------------------------------------------
[AR2]
```

图2-17-11　邻居表

> **知识链接**
>
> 如修改 Router ID，需要执行以下命令进行生效。
>
> 在路由器用户视图下：
>
> ```
> <AR1>reset ospf process
> Warning: The OSPF process will be reset. Continue? [Y/N]:y
> ```

步骤 2　路由表见图 2-17-12。

```
[AR2]display ip routing-table protocol ospf
Route Flags: R - relay, D - download to fib

Public routing table : OSPF
         Destinations : 7        Routes : 8

OSPF routing table status : <Active>
         Destinations : 7        Routes : 8

Destination/Mask    Proto    Pre   Cost       Flags  NextHop         Interface
      1.1.1.1/32    OSPF     10    1             D   192.168.10.1    GigabitEthernet
0/0/0
      3.3.3.3/32    OSPF     10    1             D   192.168.20.3    GigabitEthernet
0/0/1
      5.5.5.5/32    OSPF     10    2             D   192.168.10.1    GigabitEthernet
0/0/0
                    OSPF     10    2             D   192.168.20.3    GigabitEthernet
0/0/1
     10.1.10.0/24   OSPF     10    2             D   192.168.10.1    GigabitEthernet
0/0/0
     10.1.20.0/24   OSPF     10    2             D   192.168.20.3    GigabitEthernet
0/0/1
    192.168.30.0/24 OSPF     10    2             D   192.168.10.1    GigabitEthernet
0/0/0
    192.168.40.0/24 OSPF     10    2             D   192.168.20.3    GigabitEthernet
0/0/1

OSPF routing table status : <Inactive>
         Destinations : 0        Routes : 0

[AR2]
```

图 2-17-12　路由表

> **知识链接**
>
> **查询 DR 与 BDR 的方法**
>
> 在 ospf 广播类型（Broadcast）网络中需要在路由器中选举 DR（Designated Router，指定路由器）和 BDR（Backup Designated Router，备份指定路由器）。DROTHER 路由器需要与 DR/BDR 建立邻居关系，并将自己的 LAS 信息发送给 DRDR/BDR，犹如班长角色，而 BDR 属于副班长角色，ospf 网络中，DR 和 BDR 的 LSDB 包含有整个网络的完整拓扑。
>
> 当路由器端口优先级设为 0 时，这台路由器将不能参与 DR/BDR 选举，只能为 DROTHER。
>
> 接口优先级（默认端口上的优先级都为 1，优先级范围 1～255）：优先级最高的为 DR，其次为 BDR。在优先级相同的情况下比较 Router ID，最高者为 DR，

其次为 BDR。

① 查看 DR 与 BDR 路由器

以 AR1 为例，通过"display ospf interface GigabitEthernet 0/0/0"查询，见图 2-17-13。

```
<AR1>display ospf interface GigabitEthernet 0/0/0
        OSPF Process 1 with Router ID 1.1.1.1
           Interfaces

 Interface: 192.168.10.1 (GigabitEthernet0/0/0)
 Cost: 100    State: BDR    Type: Broadcast    MTU: 1500
 Priority: 1
 Designated Router: 192.168.10.2
 Backup Designated Router: 192.168.10.1
 Timers: Hello 10 , Dead 40 , Poll 120 , Retransmit 5 , Transmit Delay 1
<AR1>
```

图2-17-13　查询DR与BDR路由器

以 AR2 为例，通过"display ospf interface GigabitEthernet 0/0/0"查询，见图 2-17-14。

```
<AR2>display ospf interface GigabitEthernet 0/0/0
        OSPF Process 1 with Router ID 2.2.2.2
           Interfaces

 Interface: 192.168.10.2 (GigabitEthernet0/0/0)
 Cost: 1     State: DR     Type: Broadcast    MTU: 1500
 Priority: 1
 Designated Router: 192.168.10.2
 Backup Designated Router: 192.168.10.1
 Timers: Hello 10 , Dead 40 , Poll 120 , Retransmit 5 , Transmit Delay 1
<AR2>
```

图2-17-14　查询DR与BDR路由器

在图 2-7-14 中，可以看出 DR 设备的 IP 地址：192.168.10.2；BRD 设备的 IP 地址：192.168.10.1。

② 修改 AR1 GigabitEthernet0/0/0 接口优先级，改变 DR 与 BDR 角色。

通过"ospf dr-priority"进行接口修改优先级。

[AR1-GigabitEthernet0/0/0]ospf dr-priority 100

再次查看 DR 与 BDR 变化情况

在路由器 AR1 中查看，见图 2-17-15。

```
[AR1]display ospf interface GigabitEthernet 0/0/0
        OSPF Process 1 with Router ID 1.1.1.1
           Interfaces

 Interface: 192.168.10.1 (GigabitEthernet0/0/0)
 Cost: 1     State: DR     Type: Broadcast    MTU: 1500
 Priority: 100
 Designated Router: 192.168.10.1
 Backup Designated Router: 192.168.10.2
 Timers: Hello 10 , Dead 40 , Poll 120 , Retransmit 5 , Transmit Delay 1
[AR1]
```

图2-17-15　查询DR与BDR路由器

在路由器 AR2 中查看，见图 2-17-16。

```
[AR2]display ospf interface GigabitEthernet 0/0/0
       OSPF Process 1 with Router ID 2.2.2.2
           Interfaces

 Interface: 192.168.10.2 (GigabitEthernet0/0/0)
 Cost: 1        State: BDR        Type: Broadcast    MTU: 1500
 Priority: 1
 Designated Router: 192.168.10.1
 Backup Designated Router: 192.168.10.2
 Timers: Hello 10 , Dead 40 , Poll 120 , Retransmit 5 , Transmit Delay 1
[AR2]
```

图2-17-16 查询DR与BDR路由器

步骤3 LSDB 数据库表

通过路由器间的路由信息交换，LSDB 链路状态数据库内部可以达到信息同步。以 AR2 为例，参见图 2-17-17。

```
[AR2]display ospf lsdb
       OSPF Process 1 with Router ID 2.2.2.2
              Link State Database

                   Area: 0.0.0.0
 Type       LinkState ID    AdvRouter       Age    Len    Sequence    Metric
 Router     2.2.2.2         2.2.2.2         872    60     80000096    1
 Router     1.1.1.1         1.1.1.1         847    72     8000009B    100
 Router     5.5.5.5         5.5.5.5         874    60     80000095    1
 Router     3.3.3.3         3.3.3.3         643    72     80000094    1
 Network    192.168.20.2    2.2.2.2         795    32     8000008B    0
 Network    192.168.10.2    2.2.2.2         872    32     80000002    0
 Network    192.168.30.5    5.5.5.5         874    32     80000002    0
 Network    192.168.40.3    3.3.3.3         643    32     8000008B    0

[AR2]
```

图2-17-17 LSDB数据库表

LSDB 数据库表字段含义

Type:LSA 类型。

Link State ID:LSA 链路状态 ID，一般用 Router ID 表示。

AdvRouter: 通告链路状态信息的路由器的 ID 号。

Age:LSA 的老化时间。

Len:LSA 的长度。

Sequence: LSA 的序列号。

Metric: 度量值。

10. 静默路由

AR1 的路由器 GigabitEthernet0/0/0 连接的是 PC，而 PC 并不会发送 hello 报文，更不会建立邻居关系，可以将该端口配置为静默路由，该端口不收发 hello 包，从而节省设备的性能。

图 2-17-18 所示是没有配置静默路由的抓包截图，通过 224.0.0.5 IP 地址发送 hello 报文。

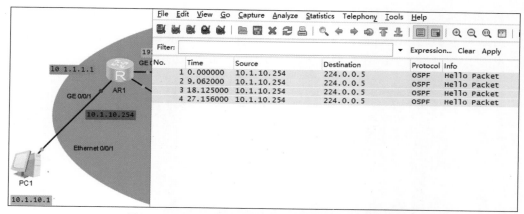

图2-17-18　AR1 g 0/0/1没有配置静默路由抓包数据

配置静默接口：

```
[AR1-ospf-1]silent-interface GigabitEthernet 0/0/1
```

配置完毕，再次 GigabitEthernet 0/0/1 抓包，已经看不到报文信息，见图 2-17-19。

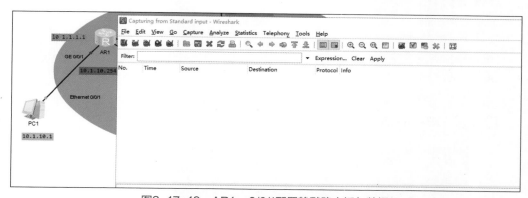

图2-17-19　AR1 g 0/0/1配置静默路由抓包数据

网络测试：

测试 PC1(10.10.10.1)-ping-PC2(10.10.20.1)- 通，见图 2-17-20。

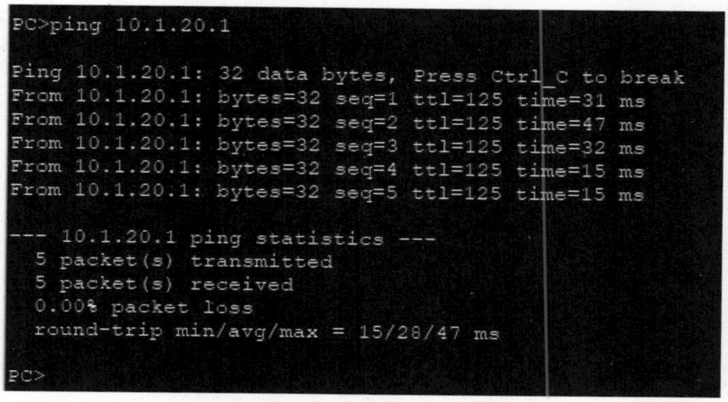

图2-17-20　网络测试

同理也可以将 AR3 的 GigabitEthernet 0/0/1 也可以设置为静默接口。

11. 路由下发

路由下发命令为"default-route-advertise always",测试 AR3-ping-IP:60.222.227.130- 不通,见图 2-17-21,原因是 AR3 路由器没有 60.222.227.0 网段的路由。

```
[AR3]ping 60.222.227.130
  PING 60.222.227.130: 56  data bytes, press CTRL_C to break
    Request time out
    Request time out
    Request time out
    Request time out
    Request time out

  --- 60.222.227.130 ping statistics ---
    5 packet(s) transmitted
    0 packet(s) received
    100.00% packet loss

[AR3]
```

图2-17-21 网络测试

具体配置如下:

AR2 配置默认路由

```
[AR2]ip  route-static 0.0.0.0 0 60.222.227.1
```

再配置路由下发命令,AR1、AR3、AR5 路由器在路由表中都会产生一条默认路由。

```
[AR2-ospf-1]default-route-advertise always
```

配置完毕,AR1、AR3、AR5 路由表如图 2-17-22 所示。

再次测试 AR3-ping -AR2 的 60.222.227.130- 通,如图 2-17-23 所示。

```
[AR1]display ip routing-table
Route Flags: R - relay, D - download to fib
------------------------------------------------------------
Routing Tables: Public
         Destinations : 21      Routes : 23

Destination/Mask    Proto   Pre  Cost     Flags NextHop         Interface
     0.0.0.0/0      O_ASE   150  1          D   192.168.10.2    GigabitEthernet
0/0/0
     1.1.1.1/32     Direct  0    0          D   127.0.0.1       LoopBack0
     2.2.2.2/32     OSPF    10   1          D   192.168.10.2    GigabitEthernet
0/0/0
     3.3.3.3/32     OSPF    10   1          D   192.168.30.5    GigabitEthernet
0/0/2
```

(a) AR1路由表

```
[AR3]display ip routing-table
Route Flags: R - relay, D - download to fib
------------------------------------------------------------
Routing Tables: Public
         Destinations : 23      Routes : 25

Destination/Mask    Proto   Pre  Cost     Flags NextHop         Interface
     0.0.0.0/0      O_ASE   150  1          D   192.168.20.2    GigabitEthernet
0/0/0
     1.1.1.1/32     OSPF    10   2          D   192.168.40.5    GigabitEthernet
0/0/2
```

(b) AR3路由表

```
[AR5]display ip routing-table
Route Flags: R - relay, D - download to fib
─────────────────────────────────────────────
Routing Tables: Public
        Destinations : 19    Routes : 21

Destination/Mask    Proto   Pre  Cost    Flags  NextHop        Interface
    0.0.0.0/0       O_ASE   150  1         D    192.168.40.3   GigabitEthernet
0/0/1
                    O_ASE   150  1         D    192.168.30.1   GigabitEthernet
0/0/0
    1.1.1.1/32      OSPF    10   1         D    192.168.30.1   GigabitEthernet
0/0/0
```

(c) AR5路由表

图2-17-22　路由表

```
[AR3]ping 60.222.227.130
PING 60.222.227.130: 56  data bytes, press CTRL_C to break
  Reply from 60.222.227.130: bytes=56 Sequence=1 ttl=255 time=20 ms
  Reply from 60.222.227.130: bytes=56 Sequence=2 ttl=255 time=40 ms
  Reply from 60.222.227.130: bytes=56 Sequence=3 ttl=255 time=30 ms
  Reply from 60.222.227.130: bytes=56 Sequence=4 ttl=255 time=30 ms
  Reply from 60.222.227.130: bytes=56 Sequence=5 ttl=255 time=30 ms

 --- 60.222.227.130 ping statistics ---
   5 packet(s) transmitted
   5 packet(s) received
   0.00% packet loss
   round-trip min/avg/max = 20/30/40 ms
```

图2-17-23　网络测试

12. ospf 认证

为了网络安全，建议 ospf 之间配置互相认证信息，分为接口认证以及区域认证，主要是防范非法流量侵入网络系统。

认证之前，先查看邻居关系。邻居关系建立，状态为 Full，见图 2-17-24。

图2-17-24　AR1邻居关系

接口认证配置：

[AR1]interface GigabitEthernet 0/0/0
密码 a123456　采用 Md5 认证
[AR1-GigabitEthernet0/0/0]ospf authentication-mode md5 1 plain a123456

再次在 AR1 路由器上查看邻居关系，已经无法建立。因为 AR2 也需要认证，见图 2-17-25。

[AR2]interface GigabitEthernet 0/0/0
密码 a123456　Md5 认证
[AR2-GigabitEthernet0/0/0]ospf authentication-mode md5 1 plain a123456

再次查看邻居关系，已经建立，见图 2-17-26。

ospf 区域的划分是以路由器接口进行的，与网段无关。如果是单区域的话，一般建议采用 area 0。

```
[AR1]display ospf peer brief
         OSPF Process 1 with Router ID 1.1.1.1
                 Peer Statistic Information
 -------------------------------------------------------------
 Area Id         Interface              Neighbor id      State
 0.0.0.0         GigabitEthernet0/0/2   5.5.5.5          Full
 -------------------------------------------------------------
[AR1]
```

图2-17-25　AR1邻居关系已经无法建立

```
[AR1]
[AR1]display ospf peer brief
         OSPF Process 1 with Router ID 1.1.1.1
                 Peer Statistic Information
 -------------------------------------------------------------
 Area Id         Interface              Neighbor id      State
 0.0.0.0         GigabitEthernet0/0/0   2.2.2.2          Init
 0.0.0.0         GigabitEthernet0/0/2   5.5.5.5          Full
 -------------------------------------------------------------
[AR1]
```

图2-17-26　ospf 接口认证之后邻居关系

二、多区域 ospf 配置

在多区域 ospf 配置中，非核心区域一定要与核心区域 area 0（骨干区域）连接，核心区域相当于一个国家的首都。

1. 拓扑图

如图 2-17-27 所示，实现功能：多个区域 ospf 之间建立邻居关系，实现网络全互通。

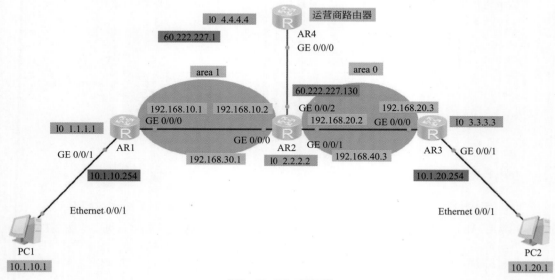

图2-17-27　拓扑图

2. 项目配置步骤

步骤1　删除单区域中关于 ospf 的配置

```
[AR1]undo ospf 1
Warning: The OSPF process will be deleted. Continue? [Y/N]:y
[AR2]undo ospf 1
Warning: The OSPF process will be deleted. Continue? [Y/N]:y
[AR3]undo ospf 1
Warning: The OSPF process will be deleted. Continue? [Y/N]:y
```

步骤2 AR1 配置

```
[AR1]ospf 1 router-id 1.1.1.1
# 进入 ospf 区域 1
[AR1-ospf-1]area 1
[AR1-ospf-1-area-0.0.0.1]network 1.1.1.1 0.0.0.0
[AR1-ospf-1-area-0.0.0.1]network 192.168.10.0 0.0.0.255
[AR1-ospf-1-area-0.0.0.1]network 10.1.10.0 0.0.0.255
[AR1-ospf-1-area-0.0.0.1]quit
[AR1-ospf-1]quit
[AR1]quit
<AR1>save
```

步骤3 AR2 配置

```
[AR2]ospf 1 router-id 2.2.2.2
[AR2-ospf-1]area 1
[AR2-ospf-1-area-0.0.0.1]network 192.168.10.0 0.0.0.255
[AR2-ospf-1-area-0.0.0.1]quit
[AR2-ospf-1]area 0
[AR2-ospf-1-area-0.0.0.0]network 2.2.2.2 0.0.0.0
[AR2-ospf-1-area-0.0.0.0]network 192.168.20.0 0.0.0.255
[AR2-ospf-1-area-0.0.0.0]quit
[AR2]quit
<AR2>save
```

步骤4 AR3 配置

```
[AR3]ospf 1 router-id 3.3.3.3
[AR3-ospf-1]area 0
[AR3-ospf-1-area-0.0.0.0]network 3.3.3.3 0.0.0.0
[AR3-ospf-1-area-0.0.0.0]network 192.168.20.0 0.0.0.255
[AR3-ospf-1-area-0.0.0.0]network 10.1.20.0 0.0.0.255
[AR3-ospf-1-area-0.0.0.0]quit
[AR3-ospf-1]quit
[AR3]quit
<AR3>save
```

步骤5 查看邻居关系

AR1、AR2、AR3 之间邻居关系见图 2-17-28。

步骤6 测试 PC1 与 PC2 通信情况

PC1(IP: 10.1.10.1)-ping-PC2(IP:10.1.20.1)- 通，见图 2-17-29。

PC1(IP: 10.1.10.1)- tracert -PC2(IP:10.1.20.1) 路由信息如下：

```
PC>tracert 10.1.20.1
traceroute to 10.1.20.1, 8 hops max
(ICMP), press Ctrl+C to stop
1  10.1.10.254    16 ms   15 ms   <1 ms
2  192.168.10.2   31 ms   <1 ms   16 ms
```

3   192.168.20.3   31 ms   16 ms   16 ms
4   10.1.20.1   31 ms   15 ms   16 ms

```
[AR1]display ospf peer brief
      OSPF Process 1 with Router ID 1.1.1.1
            Peer Statistic Information
 ----------------------------------------------------------------
 Area Id          Interface                    Neighbor id      State
 0.0.0.1          GigabitEthernet0/0/0         2.2.2.2          Full
 ----------------------------------------------------------------
[AR1]
```

(a) AR1邻居关系

```
[AR2]display ospf peer brief
      OSPF Process 1 with Router ID 2.2.2.2
            Peer Statistic Information
 ----------------------------------------------------------------
 Area Id          Interface                    Neighbor id      State
 0.0.0.0          GigabitEthernet0/0/1         3.3.3.3          Full
 0.0.0.1          GigabitEthernet0/0/0         1.1.1.1          Full
 ----------------------------------------------------------------
[AR2]
```

(b) AR2邻居关系

```
[AR3]display ospf peer brief
      OSPF Process 1 with Router ID 3.3.3.3
            Peer Statistic Information
 ----------------------------------------------------------------
 Area Id          Interface                    Neighbor id      State
 0.0.0.0          GigabitEthernet0/0/0         2.2.2.2          Full
 ----------------------------------------------------------------
[AR3]
```

(c) AR3邻居关系

图2-17-28 邻居关系

```
PC>ping 10.1.20.1

Ping 10.1.20.1: 32 data bytes, Press Ctrl_C to break
From 10.1.20.1: bytes=32 seq=1 ttl=125 time=15 ms
From 10.1.20.1: bytes=32 seq=2 ttl=125 time=16 ms
From 10.1.20.1: bytes=32 seq=3 ttl=125 time=31 ms
From 10.1.20.1: bytes=32 seq=4 ttl=125 time=16 ms
From 10.1.20.1: bytes=32 seq=5 ttl=125 time=15 ms

--- 10.1.20.1 ping statistics ---
  5 packet(s) transmitted
  5 packet(s) received
  0.00% packet loss
  round-trip min/avg/max = 15/18/31 ms

PC>tracert 10.1.20.1

traceroute to 10.1.20.1, 8 hops max
(ICMP), press Ctrl+C to stop
 1  10.1.10.254    16 ms   15 ms   <1 ms
 2  192.168.10.2   31 ms   <1 ms   16 ms
 3  192.168.20.3   31 ms   16 ms   16 ms
 4  10.1.20.1      31 ms   15 ms   16 ms
```

图2-17-29 网络测试

步骤7 区域认证

采用混合配置，在一个网络中交换机接口与区域认证混合使用。

AR1 区域认证配置

[AR1-ospf-1-area-0.0.0.1]authentication-mode md5 2 plain a123456

AR2 区域认证配置

[AR2-ospf-1-area-0.0.0.1]authentication-mode md5 2 plain a123456

```
[AR2-ospf-1-area-0.0.0.1]quit
[AR2-ospf-1]area 0
[AR2-ospf-1-area-0.0.0.0]authentication-mode md5 2 plain a123456
```
AR3 接口认证
```
[AR3-GigabitEthernet0/0/0]ospf authentication-mode md5 2 plain a123456
[AR3-GigabitEthernet0/0/0]quit
```

3. 测试网络情况

PC1(IP：10.1.10.1)-ping-PC2(IP:10.1.20.1)- 通，见图 2-17-30。

```
PC>ping 10.1.20.1

Ping 10.1.20.1: 32 data bytes, Press Ctrl_C to break
From 10.1.20.1: bytes=32 seq=1 ttl=125 time=15 ms
From 10.1.20.1: bytes=32 seq=2 ttl=125 time=16 ms
From 10.1.20.1: bytes=32 seq=3 ttl=125 time=31 ms
From 10.1.20.1: bytes=32 seq=4 ttl=125 time=15 ms
From 10.1.20.1: bytes=32 seq=5 ttl=125 time=16 ms

--- 10.1.20.1 ping statistics ---
  5 packet(s) transmitted
  5 packet(s) received
  0.00% packet loss
  round-trip min/avg/max = 15/18/31 ms

PC>tracert 10.1.20.1

traceroute to 10.1.20.1, 8 hops max
(ICMP), press Ctrl+C to stop
 1 10.1.10.254   <1 ms  16 ms  16 ms
 2 192.168.10.2  15 ms  16 ms  31 ms
 3 192.168.20.3  16 ms  15 ms  16 ms
 4 10.1.20.1     31 ms  16 ms  15 ms
```

图2-17-30　网络测试

PC1(IP：10.1.10.1)- tracert -PC2(IP:10.1.20.1) 路由信息如下：

```
PC>tracert 10.1.20.1
traceroute to 10.1.20.1, 8 hops max
(ICMP), press Ctrl+C to stop
 1 10.1.10.254   <1 ms  16 ms  16 ms
 2 192.168.10.2  15 ms  16 ms  31 ms
 3 192.168.20.3  16 ms  15 ms  16 ms
 4 10.1.20.1     31 ms  16 ms  15 ms
```

区域认证以及接口认证混合配置后，各个路由器之间建立关系见图 2-17-31。

```
<AR1>display ospf peer brief

     OSPF Process 1 with Router ID 1.1.1.1
           Peer Statistic Information
----------------------------------------------------------
Area Id        Interface              Neighbor id      State
0.0.0.1        GigabitEthernet0/0/0   2.2.2.2          Full
----------------------------------------------------------
<AR1>
```

(a) AR1邻居关系

```
<AR2>display ospf peer brief

     OSPF Process 1 with Router ID 2.2.2.2
           Peer Statistic Information
----------------------------------------------------------
Area Id        Interface              Neighbor id      State
0.0.0.0        GigabitEthernet0/0/1   3.3.3.3          Full
0.0.0.1        GigabitEthernet0/0/0   1.1.1.1          Full
----------------------------------------------------------
<AR2>
```

(b) AR2邻居关系

图2-17-31

```
<AR3>display ospf peer brief
    OSPF Process 1 with Router ID 3.3.3.3
         Peer Statistic Information
 ----------------------------------------------------------------
 Area Id            Interface                  Neighbor id       State
 0.0.0.0            GigabitEthernet0/0/0       2.2.2.2           Full
 ----------------------------------------------------------------
<AR3>
```

(c) AR3 邻居关系

图2-17-31　邻居关系

项目 18　生成树 stp 配置

交换机之间通过多条链路连接，可以提高网络的"健壮"性，但是由于环路问题，导致网络不通畅，解决办法是通过配置生成树（stp）协议进行"破环"，达到网络畅通的目的。stp 主要解决二层网络中环路问题，如网络中有环路将产生网络风暴，导致网络瘫痪，无法运行。

STP 工作简介：刚开机，交换机之间互相发送 BPDU，通过比较发现环路，进行相应接口逻辑阻塞，最终将变为无环路的树形网络结构，同时当活动的链路发生故障后，自动激活备份链路，恢复网络稳定运行。

 项目简介

为了提高网络可靠性，采取链路冗余，但是会导致网络中形成环路。本项目重点讲解"stp"协议如何破环。stp 是二层网络中非常重要的协议，通过层层分解，逐渐掌握 stp 工作原理。目前 stp 协议有三种模式，将分别给予介绍。

一、为何增加链路反而通信异常

1. 拓扑图

如图 2-18-1 所示，实现功能：演示环路引起的网络风暴，导致交换机接口数据剧增，无法转发数据，网络瘫痪。

交换机 sw1 与 sw2 之间增加线路 2，达到线路冗余，为什么反而无法通信呢？谜底即将揭晓。

2. 项目配置步骤

步骤1　PC 参数设置见图 2-18-2。

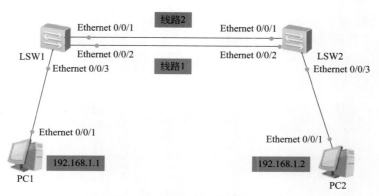

图2-18-1 拓扑图

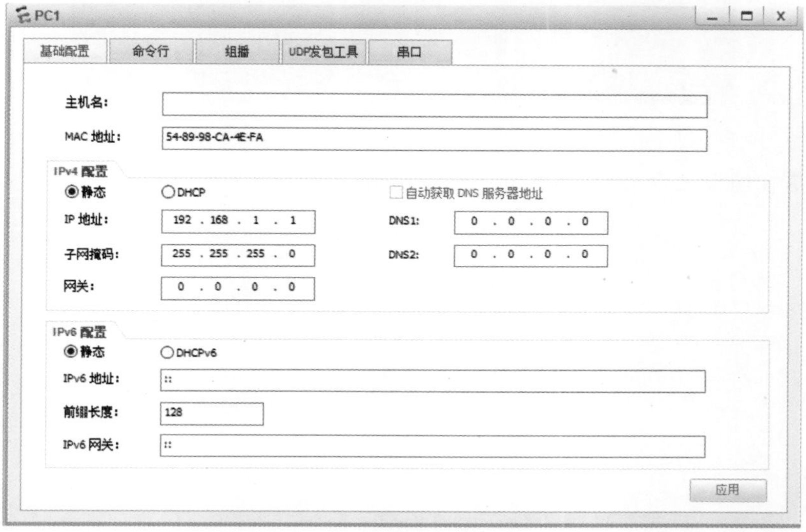

(a) PC1参数设置

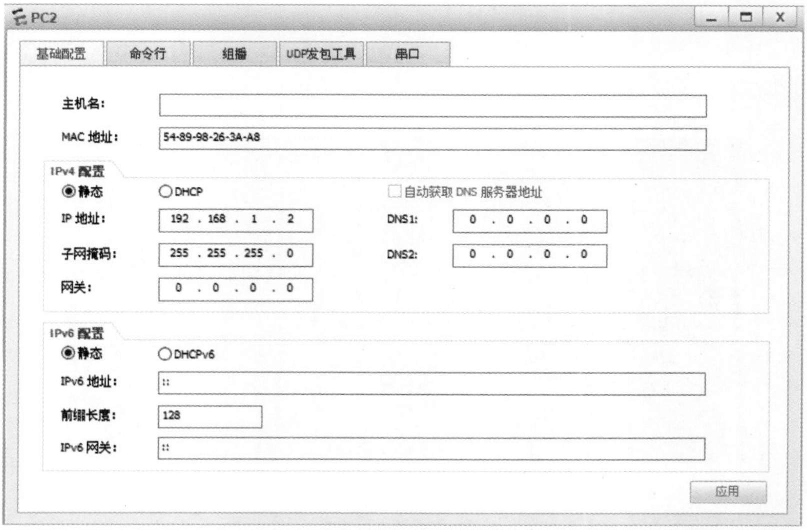

(b) PC2参数设置

图2-18-2 PC参数设置

步骤2 关闭stp

由于华为交换机默认开启stp协议，为便于实验，临时关闭stp，观察通信情况以及接口数据流量。图2-18-3所示是没有关闭stp协议时，PC1（IP：192.168.1.1）-ping-PC2（IP:192.168.1.2）- 通。

```
PC>ping 192.168.1.2

Ping 192.168.1.2: 32 data bytes, Press Ctrl_C to break
From 192.168.1.2: bytes=32 seq=1 ttl=128 time=78 ms
From 192.168.1.2: bytes=32 seq=2 ttl=128 time=47 ms
From 192.168.1.2: bytes=32 seq=3 ttl=128 time=62 ms
From 192.168.1.2: bytes=32 seq=4 ttl=128 time=47 ms
From 192.168.1.2: bytes=32 seq=5 ttl=128 time=47 ms

--- 192.168.1.2 ping statistics ---
  5 packet(s) transmitted
  5 packet(s) received
  0.00% packet loss
  round-trip min/avg/max = 47/56/78 ms

PC>
```

图2-18-3 测试通信

关闭生成树协议命令（全局模式：stp distable）
sw1
```
<Huawei>sys
[Huawei]sys sw1
# 关闭交换机生成树协议
[sw1]stp disable
Warning: The global STP state will be changed. Continue? [Y/N]y
Info: This operation may take a few seconds. Please wait for a moment...done.
<sw1>save
```
sw2
```
<Huawei>system-view
[Huawei]sysname sw2
[sw2]stp disable
Warning: The global STP state will be changed. Continue? [Y/N]y
Info: This operation may take a few seconds. Please wait for a moment...done.
```

关闭生成树stp，观察通信情况，已经不通。见图2-18-4。

```
PC>ping 192.168.1.2

Ping 192.168.1.2: 32 data bytes, Press Ctrl_C to break
Request timeout!
Request timeout!
Request timeout!
Request timeout!
Request timeout!

--- 192.168.1.2 ping statistics ---
  5 packet(s) transmitted
  0 packet(s) received
  100.00% packet loss

PC>
```

图2-18-4 通信情况

为什么为了网络稳定，设计两条链路，网络反而不通呢？观察交换机sw1 Ethernet 0/0/1接口数据流量，见图2-18-5。

```
[sw1]display interface Ethernet 0/0/1
Ethernet0/0/1 current state : UP
Line protocol current state : UP
Description:
Switch Port, PVID :        1, TPID : 8100(Hex), The Maximum Frame Length is 9216
IP Sending Frames' Format is PKTFMT_ETHNT_2, Hardware address is 4c1f-cc55-80f8
Last physical up time   : 2021-09-09 14:25:44 UTC-08:00
Last physical down time : 2021-09-09 14:23:39 UTC-08:00
Current system time: 2021-09-09 14:28:34-08:00
Hardware address is 4c1f-cc55-80f8
    Last 300 seconds input rate 0 bytes/sec, 0 packets/sec
    Last 300 seconds output rate 0 bytes/sec, 0 packets/sec
    Input:  44416484 bytes, 737605 packets
    Output: 58279917 bytes, 968546 packets
    Input:
      Unicast: 732499 packets, Multicast: 22 packets
      Broadcast: 5084 packets
    Output:
      Unicast: 963020 packets, Multicast: 5526 packets
      Broadcast: 0 packets
    Input bandwidth utilization  :     0%
    Output bandwidth utilization :     0%
```

图2-18-5　交换机sw1 Ethernet 0/0/1接口流量

可以看到，sw1 接口上数据量激增，此时如果在交换机上输入命令，你会发现卡顿延迟，通过在交换机接口数据抓包，出现大量的 arp 报文，见图 2-18-6。

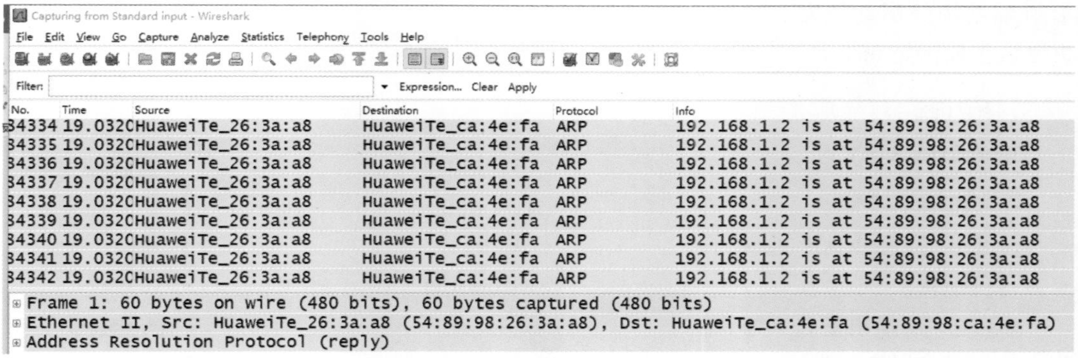

图2-18-6　交换机接口出现大量的arp报文

在实验中，当环路产生时候，电脑的 CPU 负荷也会飙升，见图 2-18-7。

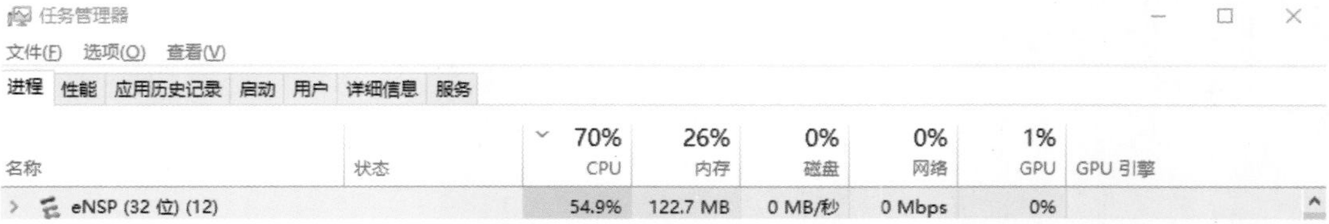

图2-18-7　电脑CPU飙升

如果要实现链路冗余，达到网络可靠性，可以通过聚合技术或者生成树协议来解决。

二、stp 协议（spanning tree protocol）

生成树协议是二层网络中非常重要的一个协议，主要解决二层网络中环路问题。如网络中存在环路，将引起网络风暴，导致网络瘫痪，一般建议开启 stp。生成树包括三种模式 stp、rstp、mstp。

在 stp 生成树端口角色见表 2-18-1。

表 2-18-1　stp 生成树端口角色

端口角色	描述
Root Port (根端口)	非桥交换机上距离根桥最近的接口，转发数据。非根设备上有且只有一个
Designated Port(指定端口)	向非根交换机转发配置 BPDU 的接口，转发数据
Alternate Port（阻塞端口）	阻塞端口，可以接收 BPDU，但是不转发数据

端口状态如表 2-18-2。

表 2-18-2　端口状态

端口状态	描述
Disabled	禁用状态，也就是关闭 stp
Listening	侦听状态，可以接收与转发 BPDU，不转发数据帧
Learning	学习状态，不转发数据帧，构建 MAC 地址表
Forwarding	转发状态，数据帧以及 BPDU 都可以接收与转发
Blocking	阻塞状态，接收 BPDU，不转发 BPDU，不转发数据帧

为了便于介绍生成树 stp，重新规划拓扑图。华为交换机默认生成树是 mstp。

1. 拓扑图

见图 2-18-8。

2. 项目配置步骤

步骤 1　设置 PC -IP 地址自行配置。

步骤 2　交换机配置（修改生成树模式以及优先级）。

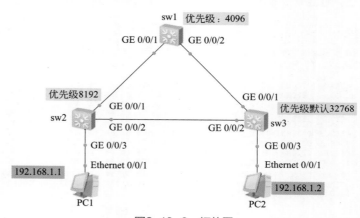

图2-18-8　拓扑图

 交换机 stp 优先级必须是 4096 的倍数，默认是 32768。

```
sw1:
<Huawei>system-view
[Huawei]sysname sw1
# 修改生成树模式为 stp
[sw1]stp mode stp
# 修改交换机优先级为 4096
[sw1]stp priority 4096
[sw1]save
sw2:
<Huawei>system-view
[Huawei]sysname sw2
```

```
[sw2]stp mode stp
[sw2]stp priority 8192
[sw2]save
```

sw3：默认优先级 32768。

```
<Huawei>system-view
[Huawei]sysname sw3
[sw3]stp mode stp
[sw3]save
```

步骤 3　观察交换机接口角色情况

sw1

```
[sw1]display  stp  brief
MSTID  Port                      Role  STP State    Protection
   0   GigabitEthernet0/0/1      DESI  FORWARDING   NONE
   0   GigabitEthernet0/0/2      DESI  FORWARDING   NONE
```
sw2
```
[sw2]display stp brief
MSTID  Port                      Role  STP State    Protection
   0   GigabitEthernet0/0/1      ROOT  FORWARDING   NONE
   0   GigabitEthernet0/0/2      DESI  FORWARDING   NONE
   0   GigabitEthernet0/0/3      DESI  FORWARDING   NONE
```
sw3
```
[sw3]display stp brief
MSTID  Port                      Role  STP State    Protection
   0   GigabitEthernet0/0/1      ROOT  FORWARDING   NONE
   0   GigabitEthernet0/0/2      ALTE  DISCARDING   NONE
   0   GigabitEthernet0/0/3      DESI  FORWARDING   NONE
```

步骤 4　标识接口角色

根据以上查询结果，在拓扑图中标识接口角色。见图 2-18-9。

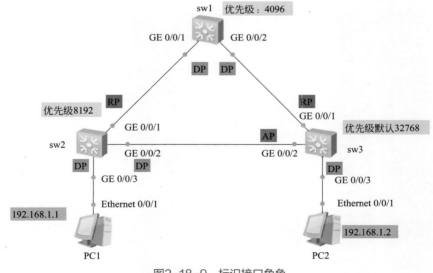

图 2-18-9　标识接口角色

知识链接

stp 选举的过程

① 网络中选举一个根桥（Root Bridge）。比较设备优先级与 MAC 地址，越小越优先，也就是比较设备的 BID（Bridge ID），它是交换机 STP 标识符，一共 8 个字节，由 2 个字节的优先级以及 6 个字节的 MAC 地址构成。为了便于观察，一般直接设置设备优先级，确定网络中根桥设备。

根桥设备上的端口都是指定端口（Designated Port）。

② 非根桥设备上选举一个根端口（Root Port），有且只有一个。根据开销值进行选举，也就是到达根桥设备路径（开销值 cost）最优的一个端口，可以接收与转发 BPDU。

③ 每个网段选举一个指定端口（Designated Port），可以接收与转发 BPDU。

④ 阻塞端口。不能转发 BPDU，但是可以接收 BPDU，主要作用就是检测链路状态，检测到链路中断后，逻辑阻塞的端口将进行数据转发。

PID（在选举指定端口以及根端口的时候，需要用到 PID）的端口 ID 由端口优先级（1 个字节）加端口编号（1 个字节）组成。端口优先级默认是 128，在设置的时候选取 16 的倍数。演示如下：

[sw2-GigabitEthernet0/0/1]stp port priority ?

INTEGER<0-240> Port priority, in steps of 16

步骤 5　查看端口 PID

图 2-18-10 和图 2-18-11 所示为查询 sw2 g 0/0/1、g 0/0/2 接口的 PID。

```
[sw2]display stp interface GigabitEthernet 0/0/1
-------[CIST Global Info][Mode STP]-------
CIST Bridge          :8192 .4c1f-ccad-4c7e
Config Times         :Hello 2s MaxAge 20s FwDly 15s MaxHop 20
Active Times         :Hello 2s MaxAge 20s FwDly 15s MaxHop 20
CIST Root/ERPC       :4096 .4c1f-ccea-1590 / 20000
CIST RegRoot/IRPC    :8192 .4c1f-ccad-4c7e / 0
CIST RootPortId      :128.1
BPDU-Protection      :Disabled
TC or TCN received   :379
TC count per hello   :0
STP Converge Mode    :Normal
Time since last TC   :0 days 0h:31m:39s
Number of TC         :45
Last TC occurred     :GigabitEthernet0/0/1
----[Port1(GigabitEthernet0/0/1)][FORWARDING]----
 Port Protocol        :Enabled
 Port Role            :Root Port
 Port Priority        :128
 Port Cost(Dot1T )    :Config=auto / Active=20000
 Designated Bridge/Port   :4096.4c1f-ccea-1590 / 128.1
 Port Edged           :Config=default / Active=disabled
 Point-to-point       :Config=auto / Active=true
 Transit Limit        :147 packets/hello-time
 Protection Type      :None
 Port STP Mode        :STP
 Port Protocol Type   :Config=auto / Active=dot1s
 BPDU Encapsulation   :Config=stp / Active=stp
 PortTimes            :Hello 2s MaxAge 20s FwDly 15s RemHop 0
 TC or TCN send       :4
 TC or TCN received   :72
 BPDU Sent            :5
          TCN: 4, Config: 1, RST: 0, MST: 0
 BPDU Received        :1173
          TCN: 0, Config: 1173, RST: 0, MST: 0
```

图2-18-10　sw2 g 0/0/1接口的PID

```
[sw2]display stp interface GigabitEthernet 0/0/2
-------[CIST Global Info][Mode STP]-------
CIST Bridge          :8192 .4c1f-ccad-4c7e
Config Times         :Hello 2s MaxAge 20s FwDly 15s MaxHop 20
Active Times         :Hello 2s MaxAge 20s FwDly 15s MaxHop 20
CIST Root/ERPC       :4096 .4c1f-ccea-1590 / 20000
CIST RegRoot/IRPC    :8192 .4c1f-ccad-4c7e / 0
CIST RootPortId      :128.1
BPDU-Protection      :Disabled
TC or TCN received   :379
TC count per hello   :0
STP Converge Mode    :Normal
Time since last TC   :0 days 0h:33m:21s
Number of TC         :45
Last TC occurred     :GigabitEthernet0/0/1
----[Port2(GigabitEthernet0/0/2)][FORWARDING]----
 Port Protocol        :Enabled
 Port Role            :Designated Port
 Port Priority        :128
 Port Cost(Dot1T )    :Config=auto / Active=20000
 Designated Bridge/Port   :8192.4c1f-ccad-4c7e / 128.2
 Port Edged           :Config=default / Active=disabled
 Point-to-point       :Config=auto / Active=true
 Transit Limit        :147 packets/hello-time
 Protection Type      :None
 Port STP Mode        :STP
 Port Protocol Type   :Config=auto / Active=dot1s
 BPDU Encapsulation   :Config=stp / Active=stp
 PortTimes            :Hello 2s MaxAge 20s FwDly 15s RemHop 20
 TC or TCN send       :193
 TC or TCN received   :111
 BPDU Sent            :20686
          TCN: 13, Config: 20673, RST: 0, MST: 0
 BPDU Received        :3292
          TCN: 2, Config: 3290, RST: 0, MST: 0
```

图2-18-11　sw2 g 0/0/2接口的PID

步骤 6　stp 协议的不足之处

1）图 2-18-12 所示链路发生故障，网络收敛时间约 30s（适应情景：交换机有 AP 端口，RP 连接的线路故障情况）。

时间计算：listening 到 learning 需要 15s，learning 到 forwarding 需要 15s，一共需要 30s 才能变成 forwarding，两者加起来约 30s。但是端口角色变化时间几乎为 0。

将 sw3 的 GigabitEthernet 0/0/1 关闭，模拟链路故障。

```
[sw3]interface GigabitEthernet 0/0/1
[sw3-GigabitEthernet0/0/1]shutdown
```

关闭后 PC1-ping-PC2- 不通，需要经过 30s 网络才能通。见图 2-18-13。

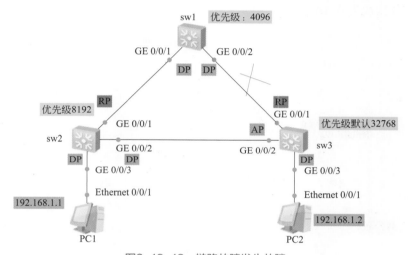

图2-18-12　链路故障发生故障

```
From 192.168.1.2: bytes=32 seq=41 ttl=128 time=78 ms
From 192.168.1.2: bytes=32 seq=42 ttl=128 time=79 ms
Request timeout!
Request timeout!
Request timeout!
Request timeout!
```

图2-18-13　测试网络

30s 左右后，网络通顺。见图 2-18-14。

```
PC>ping 192.168.1.2 -t

Ping 192.168.1.2: 32 data bytes, Press Ctrl_C to break
From 192.168.1.2: bytes=32 seq=1 ttl=128 time=47 ms
From 192.168.1.2: bytes=32 seq=2 ttl=128 time=62 ms
From 192.168.1.2: bytes=32 seq=3 ttl=128 time=63 ms
From 192.168.1.2: bytes=32 seq=4 ttl=128 time=62 ms
From 192.168.1.2: bytes=32 seq=5 ttl=128 time=63 ms
From 192.168.1.2: bytes=32 seq=6 ttl=128 time=78 ms
```

图2-18-14　测试网络

当线路没有故障前，PC1-ping-PC2，路径：PC1-sw2-sw1-sw3-PC2。

线路故障后，PC1-ping-PC2，路径：PC1-sw2-sw3-PC2，此时，我们可以查看 sw3 交换机端口角色变化情况。在交换机 sw3 查看如下：

```
[sw3]display stp brief
MSTID    Port                    Role  STP State   Protection
  0      GigabitEthernet0/0/2    ROOT  FORWARDING  NONE
  0      GigabitEthernet0/0/3    DESI  FORWARDING  NONE
```

sw3 交换机 GigabitEthernet0/0/2 接口由原来 AP 变化为 RP，进行数据转发状态。

2）如图 2-18-15 所示链路故障中，网络收敛时间约 50s（适应情景：交换机就没有 AP 角色，当 RP 端口与根桥的链路故障后情景）。

在线路正常时，sw3- AP 端口收到的 BPDU 比较优，当 sw1-sw2 之间的链路故障后，sw3-AP 端口收到次优 BPDU。

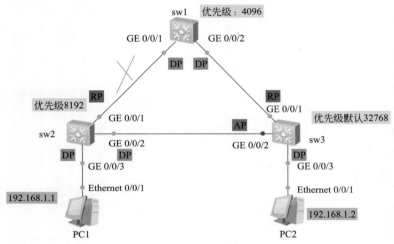

图 2-18-15 拓扑图

将交换机 sw2 的 GigabitEthernet 0/0/1 端口关闭。

```
[sw2]interface  GigabitEthernet 0/0/1
[sw2-GigabitEthernet0/0/1]shutdown
```

查看 sw2 交换机端口角色变化。

```
[sw2]display stp brief
MSTID    Port                    Role  STP State   Protection
  0      GigabitEthernet0/0/2    ROOT  FORWARDING  NONE
  0      GigabitEthernet0/0/3    DESI  FORWARDING  NONE
```

交换机 sw2 GigabitEthernet0/0/2 由原来的 DP 变为 RP。

查看 sw3 交换机端口角色变化。

```
[sw3]display stp brief
MSTID    Port                    Role  STP State   Protection
  0      GigabitEthernet0/0/1    ROOT  FORWARDING  NONE
  0      GigabitEthernet0/0/2    DESI  FORWARDING  NONE
  0      GigabitEthernet0/0/3    DESI  FORWARDING  NONE
```

交换机 sw3 GigabitEthernet0/0/2 由原来的 AP 变为 DP。

配置边缘端口：

当将 sw2 接口的 GigabitEthernet0/0/3 关闭，再打开，发现并不能立刻畅通，见图 2-18-16。

图2-18-16　测试网络

原因是 sw2 GigabitEthernet0/0/3 接口经过 listening 到 learning 需要 15s，learning 到 forwarding 需要 15s，约 30s 时间。由于 GigabitEthernet0/0/3 接的终端并非是交换机，它并不发送 BPDU，所以进行一般终端是 PC，交换机连接的接口设置为边缘端口，网络快速收敛。

```
[sw2-GigabitEthernet0/0/3]stp edged-port enable
[sw2- GigabitEthernet0/0/3]shutdown
[sw2-GigabitEthernet0/0/3]undo shutdown
```

交换机配置边缘接口后，模拟将 GigabitEthernet0/0/3 关闭再打开，网络迅速恢复。见图 2-18-17。

图2-18-17　网络测试

三、rstp

1. rstp 实现快速收敛，弥补 stp 不足

由于 stp 本身不足，收敛时间慢，rstp 协议应用而出。rstp 与 stp 相比增加了端口角色，减少了端口状态，达到快速收敛。主要通过以下几个方面加速收敛。

交换机接的是终端电脑，手工将端口修改为边缘端口；当新的根端口能收到指定端口发送的 BUDU 时，这个新端口迅速进入转发状态；PA 机制使用范围：点到点链路，全双工模式，指定端口（DP）与根端口（BP）之间发生 PA 机制，也就是 Proposal（提议）与 Agreement（同意）机制，运行机制的目的主要是让指定端口尽快进入转发状态。

Bachup 端口为 DP（指定端口）的备份端口，Alternate 端口为 RP（根端口）的备份端口。

219

2. stp 与 rstp 端口以及端口状态对比（表 2-18-3）

表 2-18-3 stp 与 rstp 端口以及端口状态对比表

stp 端口状态	rstp 端口状态	流量转发情况
Disabled	Discarding	不转发流量，也不学习 MAC 地址
Blocking		
Listening		
Learning	Learning	不转发流量，但是学习 MAC 地址
Forwarding	Forwarding	转发流量，学习 MAC 地址

3. stp 与 rstp 角色对比（表 2-18-4）

表 2-18-4 stp 与 rstp 角色对比

stp	rstp
Root Port（根端口）	Root Port（根端口）
Designated Port（指定端口）	Designated Port（指定端口）
Alternate Port（阻塞端口）	Alternate Port（阻塞端口）
/	Back Port（备份端口）

4. 项目配置步骤

首先，如果交换机连接终端是电脑，在交换机接口配置，开启边缘端口。

```
[sw3-GigabitEthernet0/0/3]stp edged-port enable
[sw1-GigabitEthernet0/0/3]stp edged-port enable
```

交换机 stp 模式修改为 rstp。

```
[sw1]stp mode rstp
[sw2]stp mode rstp
[sw3]stp mode rstp
```

1）当 sw1 GigabitEthernet0/0/1 的接口由于某种原因 shutdown 时，不需要等待 50s。拓扑图见图 2-18-18。

```
[sw1-GigabitEthernet0/0/1]shutdown
[sw1-GigabitEthernet0/0/1]undo shutdown
[sw1-GigabitEthernet0/0/1]quit
```

sw1 的 g 0/0/1 接口 shutdown，PC1-ping-PC2 通信几乎不受影响，见图 2-18-19。

2）当 sw1 的接口 g0/0/2 shut down 后，由于 sw3 一直能收到根桥发送的 BPDU，接口瞬间转发，不需要等待 30s。拓扑图见图 2-18-20。

```
[sw1-GigabitEthernet0/0/2]shutdown
# GigabitEthernet0/0/2 迅速进入转发状态
<sw3>dis stp brief
MSTID   Port                       Role   STP State    Protection
 0      GigabitEthernet0/0/2       ROOT   FORWARDING   NONE
 0      GigabitEthernet0/0/3       DESI   DISCARDING   NONE
```

sw1 的 g 0/0/2 接口 shutdown，PC1-ping-PC2 通信几乎不受影响，见图 2-18-21。

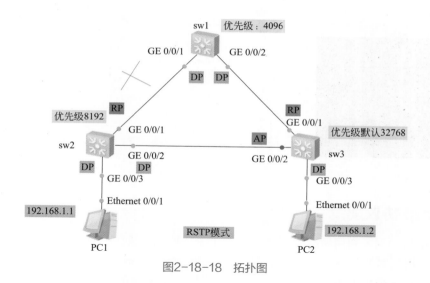

图2-18-18 拓扑图

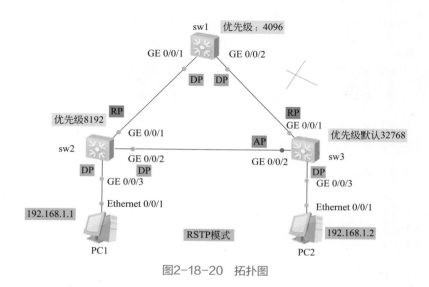

图2-18-19 网络测试

图2-18-20 拓扑图

221

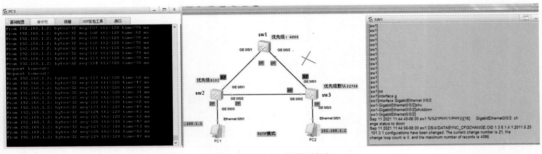

图2-18-21 网络测试

四、Mstp 协议配置

stp/rstp 同一局域网内所有的 vlan 共享一个生成树，没法在不同 vlan 间实现数据负载均衡，链路使用效率低，当某条链路阻塞后，这条链路将不承担任何流量，导致带宽浪费，同时导致部分 vlan 数据无法通信，正是由于以上不足，Mstp 应运而生，它兼容 stp/rstp。

1. 拓扑图

见图 2-18-22。

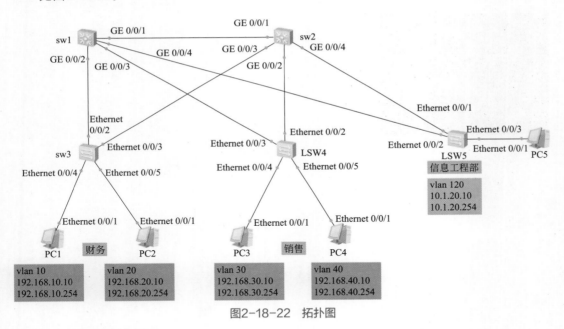

图2-18-22 拓扑图

2. 项目配置步骤

步骤1 基本配置

sw1:

```
<Huawei>sys
[Huawei]sysname sw1
# 批量创建 vlan
[sw1]vlan batch 10 20 30 40 120
[sw1]interface GigabitEthernet 0/0/1
```

222

```
[sw1-GigabitEthernet0/0/1]port link-type trunk
[sw1-GigabitEthernet0/0/1] port trunk allow-pass vlan 10 20 30 40 120
[sw1-GigabitEthernet0/0/1]quit
[sw1]interface GigabitEthernet 0/0/2
[sw1-GigabitEthernet0/0/2]port link-type trunk
[sw1-GigabitEthernet0/0/2] port trunk allow-pass vlan 10 20 30 40 120
[sw1-GigabitEthernet0/0/2]quit
[sw1]interface GigabitEthernet 0/0/3
[sw1-GigabitEthernet0/0/3]port link-type trunk
[sw1-GigabitEthernet0/0/3] port trunk allow-pass vlan 10 20 30 40 120
[sw1-GigabitEthernet0/0/3]quit
[sw1]interface GigabitEthernet 0/0/4
[sw1-GigabitEthernet0/0/4] port link-type trunk
[sw1-GigabitEthernet0/0/4] port trunk allow-pass vlan 10 20 30 40 120
<sw1>sa
```

sw2:

```
<Huawei>system-view
[Huawei]undo info-center enable
Info: Information center is disabled.
[Huawei]sysname sw2
[sw2]vlan batch 10 20 30 40 120
Info: This operation may take a few seconds. Please wait for a moment...done.
[sw2]interface  GigabitEthernet 0/0/1
[sw2-GigabitEthernet0/0/1]port link-type trunk
[sw2-GigabitEthernet0/0/1] port trunk allow-pass vlan 10 20 30 40 120
[sw2-GigabitEthernet0/0/1]quit
[sw2]interface GigabitEthernet 0/0/2
[sw2-GigabitEthernet0/0/2]port link-type trunk
[sw2-GigabitEthernet0/0/2] port trunk allow-pass vlan 10 20 30 40 120
[sw2-GigabitEthernet0/0/2]quit
[sw2]interface GigabitEthernet 0/0/3
[sw2-GigabitEthernet0/0/3]port link-type trunk
[sw2-GigabitEthernet0/0/3] port trunk allow-pass vlan 10 20 30 40 120
[sw2-GigabitEthernet0/0/3]quit
[sw2]interface GigabitEthernet 0/0/4
[sw2-GigabitEthernet0/0/4] port link-type trunk
[sw2-GigabitEthernet0/0/4] port trunk allow-pass vlan 10 20 30 40 120
[sw2]quit
<sw2>sa
<sw2>save
```

sw3:

```
<Huawei>system-view
[Huawei]sysname sw3
[sw3]vlan batch 10 20 30 40 120
[sw3-Ethernet0/0/2] port link-type trunk
[sw3-Ethernet0/0/2] port trunk allow-pass vlan 10 20 30 40 120
[sw3-Ethernet0/0/2]quit
[sw3]interface Ethernet 0/0/3
[sw3-Ethernet0/0/3] port link-type trunk
[sw3-Ethernet0/0/3] port trunk allow-pass vlan 10 20 30 40 120
[sw3]interface Ethernet 0/0/4
[sw3-Ethernet0/0/4]port link-type access
[sw3-Ethernet0/0/4]port default vlan 10
```

```
[sw3-Ethernet0/0/4]quit
[sw3]interface Ethernet 0/0/5
[sw3-Ethernet0/0/5]port link-type access
[sw3-Ethernet0/0/5]port default vlan 20
[sw3-Ethernet0/0/5]quit
[sw3]quit
<sw3>save
```

sw4:

```
<Huawei>sys
[Huawei]sysname sw4
[sw4]vlan batch 10 20 30 40 120
[sw4]interface Ethernet 0/0/2
[sw4-Ethernet0/0/2]port link-type trunk
[sw4-Ethernet0/0/2]port trunk allow-pass vlan 10 20 30 40 120
[sw4-Ethernet0/0/2]quit
[sw4]interface Ethernet 0/0/3
[sw4-Ethernet0/0/3]port link-type trunk
[sw4-Ethernet0/0/3]port trunk allow-pass vlan 10 20 30 40 120
[sw4-Ethernet0/0/3]quit
[sw4]interface Ethernet 0/0/4
[sw4-Ethernet0/0/4]port link-type access
[sw4-Ethernet0/0/4]port default vlan 30
[sw4-Ethernet0/0/4]quit
[sw4]interface Ethernet 0/0/5
[sw4-Ethernet0/0/5]port link-type access
[sw4-Ethernet0/0/5]port default vlan 40
[sw4-Ethernet0/0/5]quit
[sw4]quit
<sw4>save
```

sw5:

```
<Huawei>sys
[Huawei]sysname sw5
[sw5]vlan batch 10 20 30 40 120
[sw5]interface Ethernet 0/0/1
[sw5-Ethernet0/0/1port link-type trunk
[sw5-Ethernet0/0/1port trunk allow-pass vlan 10 20 30 40 120
[sw5-Ethernet0/0/1]quit
[sw5]interface Ethernet 0/0/2
[sw5-Ethernet0/0/2]port link-type trunk
[sw5-Ethernet0/0/2]port trunk allow-pass vlan 10 20 30 40 120
[sw5-Ethernet0/0/2]quit
[sw5]interface Ethernet 0/0/3
[sw5-Ethernet0/0/3]port link-type access
[sw5-Ethernet0/0/3]port default vlan 120
[sw4-Ethernet0/0/3]quit
```

步骤 2　PC 配置如表 2-18-5 所示，配置如图 2-18-23 所示。

表 2-18-5　PC 配置表

PC1	PC2	PC3	PC4	PC5
192.168.10.10	192.168.20.10	192.168.30.10	192.168.40.10	10.1.20.10

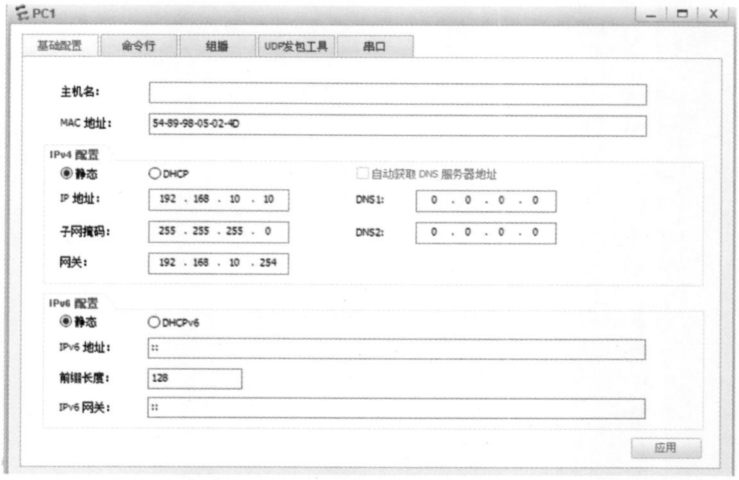

(a) PC1配置

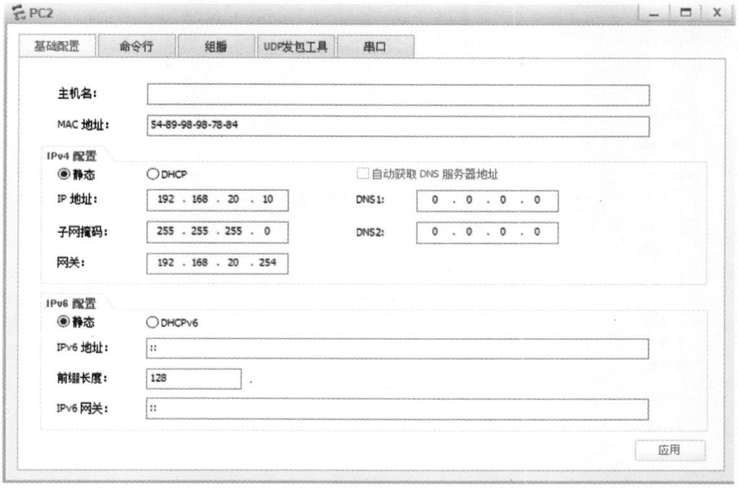

(b) PC2配置

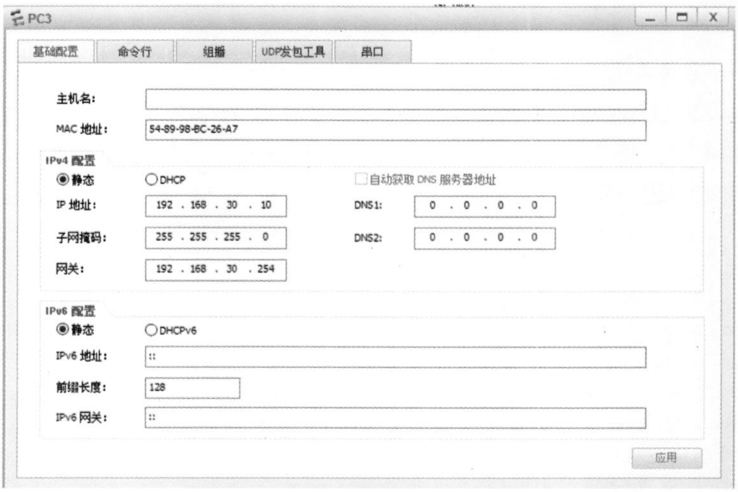

(c) PC3配置

图2-18-23

(d) PC4配置

(e) PC5配置

图2-18-23　PC配置

步骤3　vrrp 基本配置

在 sw1 和 sw2 上配置 vrrp 网关冗余协议。为了防止 vrrp 协议报文（心跳报文）所经过的链路不通，特配置一条心跳线，用于传递 vrrp 协议报文。但是配置心跳线后，网络中就形成环路，需要生成树进行解决。

1）基本配置

sw1:

```
[sw1]interface Vlanif 10
[sw1-Vlanif10]ip address 192.168.10.1 24
[sw1-Vlanif10]vrrp vrid 10 virtual-ip 192.168.10.254
[sw1-Vlanif10]vrrp vrid 10 priority 120
[sw1-Vlanif10]quit
[sw1]interface Vlanif 20
[sw1-Vlanif20]ip address 192.168.20.1 24
[sw1-Vlanif20]vrrp vrid 20 virtual-ip 192.168.20.254
```

```
[sw1-Vlanif20]vrrp vrid 20 priority 120
[sw1-Vlanif20]quit
[sw1]interface Vlanif 30
[sw1-Vlanif30]ip address 192.168.30.1 24
[sw1-Vlanif30]vrrp vrid 30 virtual-ip 192.168.30.254
[sw1-Vlanif30]quit
[sw1]interface Vlanif 40
[sw1-Vlanif40]ip address 192.168.40.1 24
[sw1-Vlanif40]vrrp vrid 40 virtual-ip 192.168.40.254
[sw1-Vlanif40]quit
[sw1]interface Vlanif 120
[sw1-Vlanif120]ip address 10.1.20.1
[sw1-Vlanif120]vrrp vrid 120 virtual-ip 10.1.20.254
[sw1-Vlanif40]quit
[sw1]quit
<sw1>save
```

sw2:
```
[sw2]interface Vlanif 10
[sw2-Vlanif10]192.168.10.2 24
[sw2-Vlanif10]vrrp vrid 10 virtual-ip 192.168.10.254
[sw2-Vlanif10]quit
[sw2]interface Vlanif 20
[sw2-Vlanif20]ip address 192.168.20.2 24
[sw2-Vlanif20]vrrp vrid 20 virtual-ip 192.168.20.254
[sw2-Vlanif20]quit
[sw2]interface Vlanif 30
[sw2-Vlanif30]ip address 192.168.30.2 24
[sw2-Vlanif30]vrrp vrid 30 virtual-ip 192.168.30.254
[sw2-Vlanif30]vrrp vrid 30 priority 120
[sw2-Vlanif30]quit
[sw2]interface Vlanif 40
[sw2-Vlanif40]ip address 192.168.40.2 24
[sw2-Vlanif40]vrrp vrid 40 virtual-ip 192.168.40.254
[sw2-Vlanif40]vrrp vrid 40 priority 105
[sw2-Vlanif40]quit
#配置 vrrp 网关地址与物理地址相同，sw2 就是 vlan 120 的 Master
[sw2]interface Vlanif 120
[sw2-Vlanif120]ip address 10.1.20.254 24
[sw2-Vlanif120]vrrp vrid 120 virtual-ip 10.1.20.254
[sw2-Vlanif120]vrrp vrid 120 priority 120
[sw2-Vlanif120]quit
[sw2]quit
<sw2>save
```

2）查看 vrrp 配置

sw1:
```
[sw1]display vrrp brief
VRID   State     Interface          Type      Virtual IP
--------------------------------------------------------
10     Master    Vlanif10           Normal    192.168.10.254
20     Master    Vlanif20           Normal    192.168.20.254
```

```
 30       Backup        Vlanif30                  Normal    192.168.30.254
 40       Backup        Vlanif40                  Normal    192.168.40.254
 120      Backup        Vlanif120                 Normal    10.1.20.254
--------------------------------------------------------------------------
Total:5     Master:2      Backup:3      Non-active:0
```

sw2:

```
[sw2]display vrrp brief
VRID   State         Interface                 Type      Virtual IP
--------------------------------------------------------------------------
 10       Backup        Vlanif10                  Normal    192.168.10.254
 20       Backup        Vlanif20                  Normal    192.168.20.254
 30       Master        Vlanif30                  Normal    192.168.30.254
 40       Master        Vlanif40                  Normal    192.168.40.254
 120      Master        Vlanif120                 Normal    10.1.20.254
--------------------------------------------------------------------------
Total:5     Master:3      Backup:2      Non-active:0
```

3）配置边缘端口

sw3:

```
[sw3]interface Ethernet 0/0/4
[sw3-Ethernet0/0/4]stp edged-port enable
[sw3-Ethernet0/0/4]quit
[sw3]interface Ethernet 0/0/5
[sw3-Ethernet0/0/5]stp edged-port enable
[sw3-Ethernet0/0/5]quit
[sw3]quit
<sw3>save
```

sw4:

```
[sw4]interface Ethernet 0/0/4
[sw4-Ethernet0/0/4]stp edged-port enable
[sw4-Ethernet0/0/4]quit
[sw4]interface Ethernet 0/0/5
[sw4-Ethernet0/0/5]stp edged-port enable
[sw4-Ethernet0/0/5]quit
[sw4]quit
<sw4>save
```

步骤4 Mstp 配置

sw1:

```
# 进入 MSTP 配置域视图
[sw1]stp region-configuration
# 配置域名为 fsm
[sw1-mst-region]region-name fsm
# 创建实例 1，并将 vlan 10 20 映射到实例 1 中
[sw1-mst-region]instance 1 vlan 10 20
# 创建实例 2，并将 vlan 30 40 120 映射到实例 2 中
[sw1-mst-region]instance 2 vlan 30 40 120
# 激活 MSTP 域
[sw1-mst-region]active region-configuration
Info: This operation may take a few seconds. Please wait for a moment...done.
[sw1-mst-region]quit
# 将交换机 sw1 为实例 1 的根桥
[sw1]stp instance 1 root primary
```

```
# 将交换机 sw1 为实例 2 的备份根桥
[sw1]stp instance 2 root secondary
[sw1]quit
<sw1>save
[sw2]stp region-configuration
[sw2-mst-region]region-name fsm
[sw2-mst-region]instance 1 vlan 10 20
[sw2-mst-region]instance 2 vlan 30 40 120
[sw2-mst-region]active region-configuration
Info: This operation may take a few seconds. Please wait for a moment...done.
[sw2-mst-region]quit
# 将交换机 sw2 为实例 2 的根桥
[sw2]stp instance 1 root secondary
# 将交换机 sw2 为实例 1 的备份根桥
[sw2]stp instance 2 root primary
[sw2]quit
<sw2>save
[sw3]stp region-configuration
[sw3-mst-region]region-name fsm
[sw3-mst-region]instance 1 vlan 10 20
[sw3-mst-region]instance 2 vlan 30 40 120
[sw3-mst-region]active region-configuration
Info: This operation may take a few seconds. Please wait for a moment...done.
[sw3-mst-region]quit
[sw3]quit
<sw3>save
[sw4]stp region-configuration
[sw4-mst-region]region-name fsm
[sw4-mst-region]instance 1 vlan 10 20
[sw4-mst-region]instance 2 vlan 30 40 120
[sw4-mst-region]active region-configuration
Info: This operation may take a few seconds. Please wait for a moment...done.
[sw4-mst-region]quit
[sw4]quit
<sw4>save
```

知识链接

sw1: 批量配置的办法

[sw1]port-group group-member GigabitEthernet 0/0/1 to GigabitEthernet 0/0/4

[sw1-GigabitEthernet0/0/1]port link-type trunk

[sw1-GigabitEthernet0/0/2]port link-type trunk

[sw1-GigabitEthernet0/0/3]port link-type trunk

[sw1-GigabitEthernet0/0/4]port link-type trunk

```
[sw1-port-group]port trunk allow-pass vlan 10 20 30 40 120
[sw1-GigabitEthernet0/0/1]port trunk allow-pass vlan 10 20 30 40 120
[sw1-GigabitEthernet0/0/2]port trunk allow-pass vlan 10 20 30 40 120
[sw1-GigabitEthernet0/0/3]port trunk allow-pass vlan 10 20 30 40 120
[sw1-GigabitEthernet0/0/4]port trunk allow-pass vlan 10 20 30 40 120
```

sw5:
```
[sw5]stp region-configuration
[sw5-mst-region]region-name fsm
[sw5-mst-region]instance 1 vlan 10 20
[sw5-mst-region]instance 2 vlan 30 40 120
[sw5-mst-region]active region-configuration
Info: This operation may take a few seconds. Please wait for a moment...done.
[sw5-mst-region]quit
[sw5]quit
<sw5>save
```

查看 Mstp 配置

sw1:
```
[sw1]display stp region-configuration
Oper configuration
 Format selector      :0
 Region name          :fsm
 Revision level       :0
 Instance    VLANs Mapped
 0          1 to 9, 11 to 19, 21 to 29, 31 to 39, 41 to 119, 121 to 4094
 1          10, 20, 30
 2          40, 120
```

依次查看其他交换机。

查询 Mstp 端口阻塞

sw1:
```
[sw1]display stp brief
 MSTID   Port                       Role   STP State     Protection
   0     GigabitEthernet0/0/1       ROOT   FORWARDING    NONE
   0     GigabitEthernet0/0/2       DESI   FORWARDING    NONE
   0     GigabitEthernet0/0/3       DESI   FORWARDING    NONE
   0     GigabitEthernet0/0/4       DESI   FORWARDING    NONE
   1     GigabitEthernet0/0/1       DESI   FORWARDING    NONE
   1     GigabitEthernet0/0/2       DESI   FORWARDING    NONE
   1     GigabitEthernet0/0/3       DESI   FORWARDING    NONE
   1     GigabitEthernet0/0/4       DESI   FORWARDING    NONE
   2     GigabitEthernet0/0/1       ROOT   FORWARDING    NONE
   2     GigabitEthernet0/0/2       DESI   FORWARDING    NONE
   2     GigabitEthernet0/0/3       DESI   FORWARDING    NONE
   2     GigabitEthernet0/0/4       DESI   FORWARDING    NONE
```

sw2:
```
[sw2]display stp brief
```

```
 MSTID  Port                      Role  STP State   Protection
    0   GigabitEthernet0/0/1      DESI  FORWARDING  NONE
    0   GigabitEthernet0/0/2      DESI  FORWARDING  NONE
    0   GigabitEthernet0/0/3      DESI  FORWARDING  NONE
    0   GigabitEthernet0/0/4      DESI  FORWARDING  NONE
    1   GigabitEthernet0/0/1      ROOT  FORWARDING  NONE
    1   GigabitEthernet0/0/2      DESI  FORWARDING  NONE
    1   GigabitEthernet0/0/3      DESI  FORWARDING  NONE
    1   GigabitEthernet0/0/4      DESI  FORWARDING  NONE
    2   GigabitEthernet0/0/1      DESI  FORWARDING  NONE
    2   GigabitEthernet0/0/2      DESI  FORWARDING  NONE
    2   GigabitEthernet0/0/3      DESI  FORWARDING  NONE
    2   GigabitEthernet0/0/4      DESI  FORWARDING  NONE
```

sw3:

```
[sw3]display stp brief
 MSTID  Port              Role  STP State   Protection
    0   Ethernet0/0/2     ALTE  DISCARDING  NONE
    0   Ethernet0/0/3     ROOT  FORWARDING  NONE
    0   Ethernet0/0/4     DESI  FORWARDING  NONE
    0   Ethernet0/0/5     DESI  FORWARDING  NONE
    1   Ethernet0/0/2     ROOT  FORWARDING  NONE
    1   Ethernet0/0/3     ALTE  DISCARDING  NONE
    1   Ethernet0/0/4     DESI  FORWARDING  NONE
    1   Ethernet0/0/5     DESI  FORWARDING  NONE
    2   Ethernet0/0/2     ALTE  DISCARDING  NONE
    2   Ethernet0/0/3     ROOT  FORWARDING  NONE
```

sw4:

```
[sw4]display stp brief
 MSTID  Port              Role  STP State   Protection
    0   Ethernet0/0/2     ROOT  FORWARDING  NONE
    0   Ethernet0/0/3     ALTE  DISCARDING  NONE
    0   Ethernet0/0/4     DESI  FORWARDING  NONE
    0   Ethernet0/0/5     DESI  FORWARDING  NONE
    1   Ethernet0/0/2     ALTE  DISCARDING  NONE
    1   Ethernet0/0/3     ROOT  FORWARDING  NONE
    1   Ethernet0/0/4     DESI  FORWARDING  NONE
    2   Ethernet0/0/2     ROOT  FORWARDING  NONE
    2   Ethernet0/0/3     ALTE  DISCARDING  NONE
    2   Ethernet0/0/5     DESI  FORWARDING  NONE
```

sw5:

```
[sw5]display stp brief
 MSTID  Port              Role  STP State   Protection
    0   Ethernet0/0/1     ROOT  FORWARDING  NONE
    0   Ethernet0/0/2     ALTE  DISCARDING  NONE
    0   Ethernet0/0/3     DESI  FORWARDING  NONE
    1   Ethernet0/0/1     ALTE  DISCARDING  NONE
    1   Ethernet0/0/2     ROOT  FORWARDING  NONE
    2   Ethernet0/0/1     ROOT  FORWARDING  NONE
    2   Ethernet0/0/2     ALTE  DISCARDING  NONE
    2   Ethernet0/0/3     DESI  FORWARDING  NONE
```

用红色圆圈代表阻塞实例1，用绿色圆圈代表阻塞实例2，见图2-18-24。

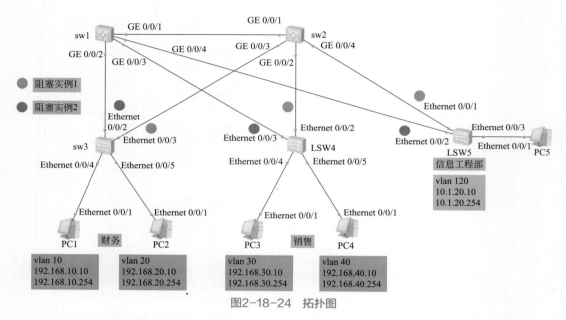

图2-18-24 拓扑图

查看网络情况：PC1(IP:192.168.10.10)-ping-PC2(IP:192.168.20.10)- 通

```
PC>ping 192.168.20.10
Ping 192.168.20.10: 32 data bytes, Press Ctrl_C to break
From 192.168.20.10: bytes=32 seq=1 ttl=127 time=78 ms
……省略……
--- 192.168.20.10 ping statistics ---
  5 packet(s) transmitted
  5 packet(s) received
  0.00% packet loss
  round-trip min/avg/max = 78/78/78 ms
```

PC1(IP:192.168.10.10)-ping-PC3(IP:192.168.30.10)- 通

```
PC>ping 192.168.30.10
Ping 192.168.30.10: 32 data bytes, Press Ctrl_C to break
From 192.168.30.10: bytes=32 seq=1 ttl=127 time=109 ms
……省略……
--- 192.168.30.10 ping statistics ---
  5 packet(s) transmitted
  5 packet(s) received
  0.00% packet loss
  round-trip min/avg/max = 93/103/110 ms
```

PC1(IP:192.168.10.10)-ping-PC4(IP:192.168.40.10)- 通

```
PC>ping 192.168.40.10
Ping 192.168.40.10: 32 data bytes, Press Ctrl_C to break
……省略……
--- 192.168.40.10 ping statistics ---
  5 packet(s) transmitted
  5 packet(s) received
  0.00% packet loss
  round-trip min/avg/max = 94/100/109 ms
```

PC1(IP:192.168.10.10)-ping-PC5(IP:10.1.20.10)- 通

```
PC>ping 10.1.20.10
Ping 10.1.20.10: 32 data bytes, Press Ctrl_C to break
From 10.1.20.10: bytes=32 seq=1 ttl=127 time=109 ms
……省略……
--- 10.1.20.10 ping statistics ---
  5 packet(s) transmitted
  5 packet(s) received
  0.00% packet loss
  round-trip min/avg/max = 94/103/125 ms
```

项目 19　使用 ftp /tftp 升级系统

ftp 协议是使用非常广泛的文件传输协议，控制连接使用 TCP 端口号 21，数据连接使用 TCP 端口号 20。tftp 简单文本传输协议，使用 UDP 端口号 69，只能提供文件下载（get）与上传 (put)。

项目简介

　　本项目在 eNSP 模拟器中创建路由器，将其配置为 ftp 服务器，主要实现功能就是给路由器系统打补丁。同时详细介绍了通过 3Daemon 软件配置 ftp/tftp 两种方式，在交换机中将配置文件上传（put）与下载（get）详细操作步骤。

一、路由器配置为 ftp 服务器

使用华为模拟器为路由器系统打补丁。路由器配置为 ftp 服务器，利用环回接口与模拟器主机通信。

1. 拓扑图

如图 2-19-1 所示。

图2-19-1　拓扑图

Cloud1（用于模拟器与主机之间通信）基本配置见图 2-19-2。电脑端环回接口地址：172.168.16.1/24，路由器地址：172.168.16.2/24。

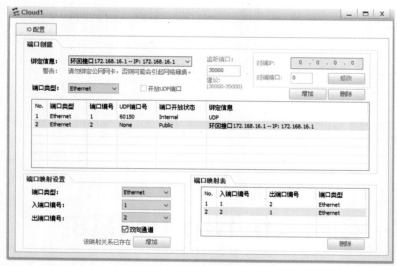

图2-19-2 Cloud1基本配置

2. 项目配置步骤

步骤1 AR1 路由器基本配置

```
<AR1>sys
Enter system view, return user view with Ctrl+Z.
# 路由器接口 IP 地址配置
[AR1]interface GigabitEthernet 0/0/0
[AR1-GigabitEthernet0/0/0]ip address 172.168.16.2 24
```

步骤2 AR1 路由器 ftp 服务配置

```
# 系统模式下 ftp 使能
[AR1]ftp server enable
[AR1]aaa
# 设置 ftp 用户登录用户名：fsmftp ；密码：a123456
[AR1-aaa]local-user fsmftp password cipher a123456
Info: Add a new user.
# 开启 ftp 服务
[AR1-aaa]local-user fsmftp service-type ftp
# 配置用户权限级别
[AR1-aaa]local-user fsmftp privilege level 15
# 配置访问目录范围
[AR1-aaa]local-user fsmftp ftp-directory flash:/
[AR1-aaa]quit
# 开启同时远程访问数量
[AR1]user-interface vty 0 4
# 配置使用用户名以及密码登录
[AR1-ui-vty0-4]authentication-mode aaa
[AR1-ui-vty0-4]quit
[AR1]quit
<AR1>save
```

步骤3 电脑端测试 ftp 功能

在键盘上操作 Win+R 输入 cmd，见图 2-19-3。

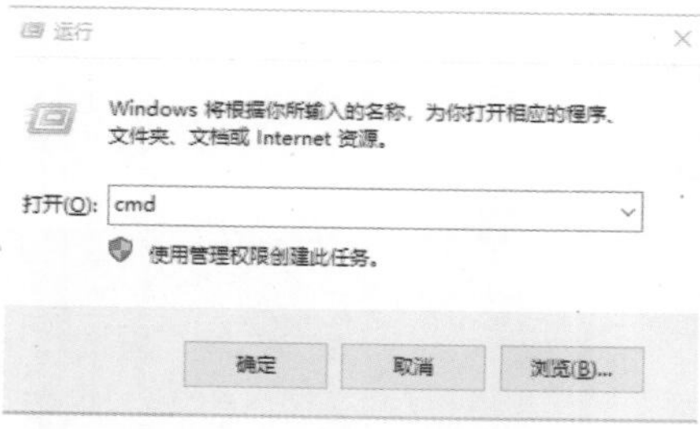

图2-19-3　Win+R 输入 cmd

然后输入：ftp 172.168.16.2，按照提示输入用户名以及密码。见图 2-19-4。

用户名：fsmftp，密码：a123456，登录后输入 dir，查询交换机文件。

从路由器中将 vrpcfg.zip 文件传到电脑（使用 get 命令），见图 2-19-5，文件保存路径在 C:\Users\fsm 文件中。

在物理主机查看刚从路由器下载的文件，见图 2-19-6。

```
C:\Users\fsm>ftp 172.168.16.2
连接到 172.168.16.2。
220 FTP service ready.
530 Please login with USER and PASS.
用户(172.168.16.2:(none)): fsmftp
331 Password required for fsmftp.
密码:
230 User logged in.
ftp> dir
200 Port command okay.
150 Opening ASCII mode data connection for *.
drwxrwxrwx   1 noone    nogroup          0 Aug 09 05:27 dhcp
-rwxrwxrwx   1 noone    nogroup     121802 May 26  2014 portalpage.zip
-rwxrwxrwx   1 noone    nogroup       2263 Aug 09 05:27 statemach.efs
-rwxrwxrwx   1 noone    nogroup     828482 May 26  2014 sslvpn.zip
-rwxrwxrwx   1 noone    nogroup        249 Aug 09 05:29 private-data.txt
drwxrwxrwx   1 noone    nogroup          0 Aug 11 03:48 .
-rwxrwxrwx   1 noone    nogroup        648 Aug 11 03:48 vrpcfg.zip
226 Transfer complete.
ftp: 收到 470 字节，用时 0.03秒 16.21千字节/秒。
ftp>
```

图2-19-4　登录ftp

```
ftp: 收到 470 字节，用时 0.04秒 10.68千字节/秒。
ftp> get vrpcfg.zip
200 Port command okay.
150 Opening ASCII mode data connection for vrpcfg.zip.
226 Transfer complete.
ftp: 收到 648 字节，用时 0.01秒 129.60千字节/秒。
ftp>
```

图2-19-5　将vrpcfg.zip 文件传到电脑

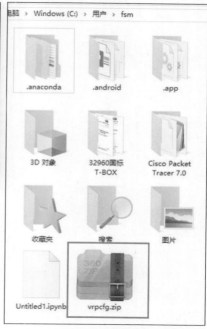

图2-19-6　物理主机查看从路由器下载的文件

步骤 4　在路由器中查询文件

如图 2-19-7 所示。

```
<AR1>dir
Directory of flash:/

  Idx  Attr     Size(Byte)  Date        Time(LMT)  FileName
    0  drw-              -  Aug 09 2021 05:27:55   dhcp
    1  -rw-        121,802  May 26 2014 09:20:58   portalpage.zip
    2  -rw-          2,263  Aug 09 2021 05:27:50   statemach.efs
    3  -rw-        828,482  May 26 2014 09:20:58   sslvpn.zip
    4  -rw-            249  Aug 09 2021 05:29:26   private-data.txt
    5  -rw-            648  Aug 11 2021 03:48:23   vrpcfg.zip

1,090,732 KB total (784,456 KB free)
```

图2-19-7　路由器中文件

将刚才传到电脑中的文件 vrpcfg.zip 重命名为 vrpcfg02.zip，见图 2-19-8。

通过 put 命令，将 vrpcfg02.zip 文件上传到路由器。见图 2-19-9。

在路由器 flash 中查询文件，见图 2-19-10，方框内已经显示上传的文件。

图 2-19-8　vrpcfg.zip 重命名为 vrpcfg02.zip

图 2-19-9　put vrpcfg02.zip 到路由器

图 2-19-10　查询文件

将 vrpcfg02.zip 删除（使用命令 delete），见图 2-19-11。

步骤 5　给路由器打补丁，当然也可以升级系统，此处以打补丁为例。先将补丁传到电脑用户目录下，见图 2-19-12。

将补丁文件下载到路由器 AR。见图 2-19-13。

通过 dir 命令查询下载到路由器的补丁文件，见图 2-19-14。

查看启动配置文件，见图 2-19-15。

升级补丁文件，输入"startup patch ar2220-v200r009sph025.pat"，见图 2-19-16。

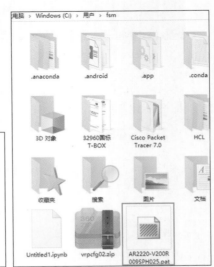

图 2-19-11　删除文件

图 2-19-12　路由器打补丁

```
ftp> put AR2220-V200R009SPH025.pat
200 Port command okay.
150 Opening ASCII mode data connection for AR2220-V200R009SPH025.pat.
226 Transfer complete.
ftp: 发送 7406208 字节，用时 59.06秒 125.40千字节/秒
ftp>
```

图2-19-13　补丁文件下载到路由器AR

```
<AR1>dir
Directory of flash:/

  Idx  Attr    Size(Byte)   Date         Time(LMT)   FileName
    0  drw-            -    Aug 09 2021  05:27:55    dhcp
    1  -rw-    7,406,208    Aug 11 2021  14:47:21    ar2220-v200r009sph025.pat
    2  -rw-      121,802    May 26 2014  09:20:58    portalpage.zip
    3  -rw-        2,263    Aug 09 2021  05:27:50    statemach.efs
    4  -rw-      828,482    May 26 2014  09:20:58    sslvpn.zip
    5  -rw-          249    Aug 09 2021  05:29:26    private-data.txt
    6  -rw-          648    Aug 11 2021  03:48:23    vrpcfg.zip

1,090,732 KB total (777,212 KB free)
<AR1>
```

图2-19-14　查询文件

```
<AR1>display startup
MainBoard:
  Startup system software:                     null
  Next startup system software:                null
  Backup system software for next startup:     null
  Startup saved-configuration file:            flash:/vrpcfg.zip
  Next startup saved-configuration file:       flash:/vrpcfg.zip
  Startup license file:                        null
  Next startup license file:                   null
  Startup patch package:                       null
  Next startup patch package:                  null
  Startup voice-files:                         null
  Next startup voice-files:                    null
<AR1>
```

图2-19-15　查看启动配置文件

```
<AR1>startup patch ar2220-v200r009sph025.pat
This operation will take several minutes, please wait......
Info: Succeeded in setting the file for booting system
<AR1>
```

图2-19-16　升级补丁文件

```
<AR1>startup patch ar2220-v200r009sph025.pat
This operation will take several minutes, please wait........
Info: Succeeded in setting the file for booting system
<AR1>
```

查看路由器升级补丁，输入"display startup"，见图 2-19-17。

```
<AR1>display startup
MainBoard:
  Startup system software:                     null
  Next startup system software:                null
  Backup system software for next startup:     null
  Startup saved-configuration file:            flash:/vrpcfg.zip
  Next startup saved-configuration file:       flash:/vrpcfg.zip
  Startup license file:                        null
  Next startup license file:                   null
  Startup patch package:                       null
  Next startup patch package:                  flash:/ar2220-v200r009sph025.pat
  Startup voice-files:                         null
  Next startup voice-files:                    null
<AR1>
```

图2-19-17　查看升级补丁

补丁文件生效，需要重启，重启后查看配置文件，见图 2-19-18。

```
<AR1>display startup
MainBoard:
 Startup system software:                      null
 Next startup system software:                 null
 Backup system software for next startup:      null
 Startup saved-configuration file:             flash:/vrpcfg.zip
 Next startup saved-configuration file:        flash:/vrpcfg.zip
 Startup license file:                         null
 Next startup license file:                    null
 Startup patch package:                        flash:/ar2220-v200r009sph025.pat
 Next startup patch package:                   flash:/ar2220-v200r009sph025.pat
 Startup voice-files:                          null
 Next startup voice-files:                     null
<AR1>
```

图 2-19-18　补丁文件生效，需要重启

二、在线使用交换机升级系统（使用 ftp 方法）

1. 在线使用交换机型号

型号为 S5130S-28P-EI，这台交换机已在正常使用，这里进行在线系统升级。

2. 升级步骤

步骤 1　查询交换机剩余空间，见图 2-19-19。

在向交换机传送升级文件之前，一定要查询剩余空间，看能否满足升级文件的空间大小需求。通过命令：dir flash:/ 来进行查询。

```
<DYJT-BQ-Access_test.4.36>dir flash:/
Directory of flash:
   0 -rw-      220684 Jan 01 2013 00:00:00   defaultfile.zip
   1 drw-           - Jan 01 2013 00:00:12   diagfile
   2 -rw-         735 Jul 24 2021 11:24:03   hostkey
   3 -rw-         837 Aug 09 2021 09:18:43   ifindex.dat
   4 -rw-           - Jan 01 2013 00:00:13   license
   5 drw-           - Jan 02 2013 00:00:11   logfile
   6 drw-           - Jan 01 2013 00:00:53   pki
   7 -rw-     6146048 Jan 01 2013 00:00:00   s5130s_ei-cmw710-boot-r6318p01.bin
   8 -rw-    49709056 Jan 01 2013 00:00:00   s5130s_ei-cmw710-system-r6318p01.bin
   9 drw-           - Jan 01 2013 00:00:12   seclog
  10 -rw-         591 Jul 24 2021 11:24:03   serverkey
  11 -rw-        3989 Aug 09 2021 09:18:45   startup.cfg
  12 -rw-      102771 Aug 09 2021 09:18:45   startup.mdb
  13 drw-           - Jan 01 2013 00:00:58   versionInfo

251904 KB total (191836 KB free)
```

图 2-19-19　交换机剩余空间

步骤 2　采用 3Daemon 软件，设置为 ftp 服务器。

1）选择设置 ftp 服务器（IP：192.168.9.121），见图 2-19-20。

2）选择上传 / 下载目录

在电脑桌面新建文件夹，命名为 3CDaemon，在 3CDaemon 软件中将其设置为上传 / 下载目录，见图 2-19-21。

3）设置用户名以及密码，见图 2-19-22。

图2-19-20 设置FTP服务器

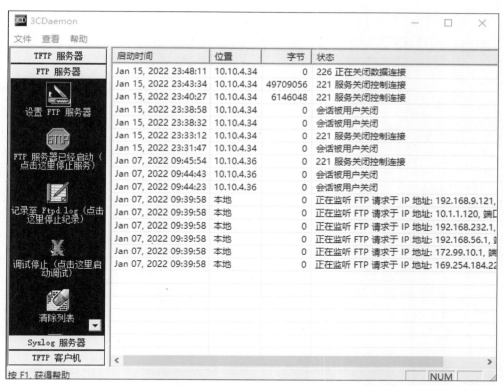

图2-19-21 上传/下载目录

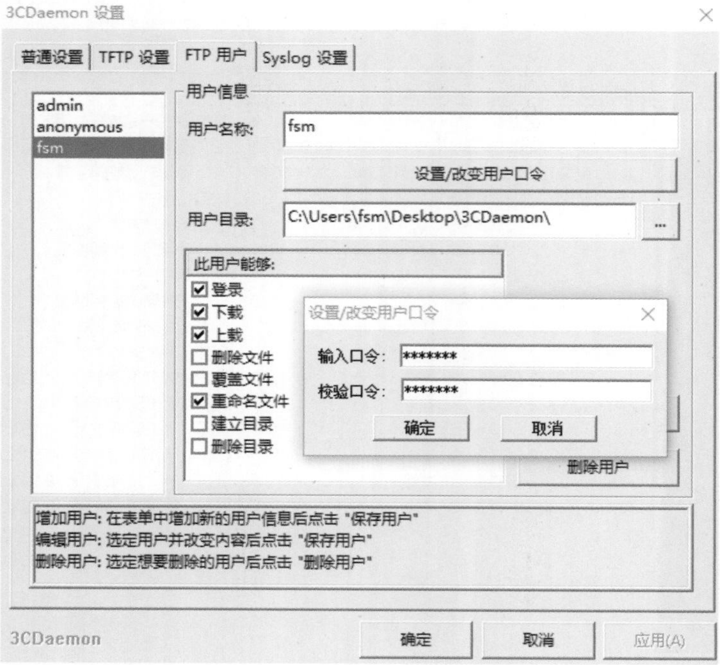

图2-19-22 设置用户名以及密码

步骤3 交换机升级系统。

将升级文件 S5130S_EI-CMW710-R6320.ipe 放至桌面 3CDaemon 文件中。

在交换机上访问 ftp 服务器，并将文件下载到交换机（使用 get 命令），见图 2-19-23。在传输文件中，可以实时观察文件传输状态，见图 2-19-24。

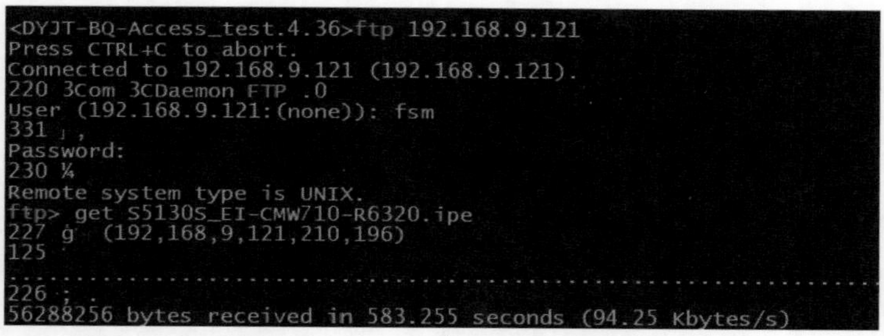

图2-19-23 将文件下载到交换机

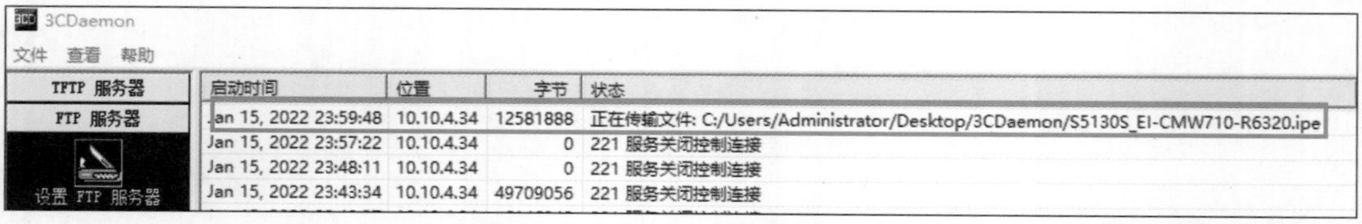

图2-19-24 观察文件传输状态

在交换机命令行输入"dir"，查看是否下载成功，见图 2-19-25。

在交换机命令行输入"boot-loader file flash:/S5130S_EI-CMW710-R6320.ipe slot 1 main"进行系统升级，见图 2-19-26。

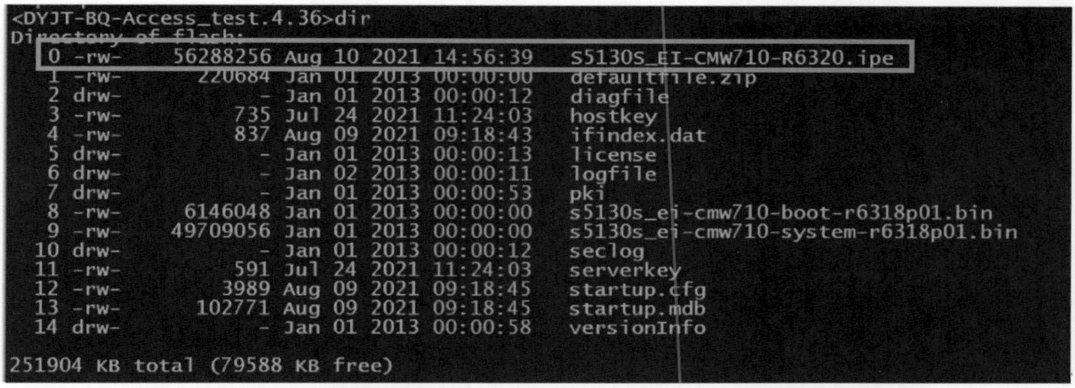

```
<DYJT-BQ-Access_test.4.36>dir
Directory of flash:
   0 -rw-     56288256 Aug 10 2021 14:56:39   S5130S_EI-CMW710-R6320.ipe
   1 -rw-       220684 Jan 01 2013 00:00:00   defaultfile.zip
   2 drw-            - Jan 01 2013 00:00:12   diagfile
   3 -rw-          735 Jul 24 2021 11:24:03   hostkey
   4 -rw-          837 Aug 09 2021 09:18:43   ifindex.dat
   5 drw-            - Jan 01 2013 00:00:13   license
   6 drw-            - Jan 02 2013 00:00:11   logfile
   7 drw-            - Jan 01 2013 00:00:53   pki
   8 -rw-      6146048 Jan 01 2013 00:00:00   s5130s_ei-cmw710-boot-r6318p01.bin
   9 -rw-     49709056 Jan 01 2013 00:00:00   s5130s_ei-cmw710-system-r6318p01.bin
  10 drw-            - Jan 01 2013 00:00:12   seclog
  11 -rw-          591 Jul 24 2021 11:24:03   serverkey
  12 -rw-         3989 Aug 09 2021 09:18:45   startup.cfg
  13 -rw-       102771 Aug 09 2021 09:18:45   startup.mdb
  14 drw-            - Jan 01 2013 00:00:58   versionInfo

251904 KB total (79588 KB free)
```

图2-19-25　输入“dir”，查看是否下载成功

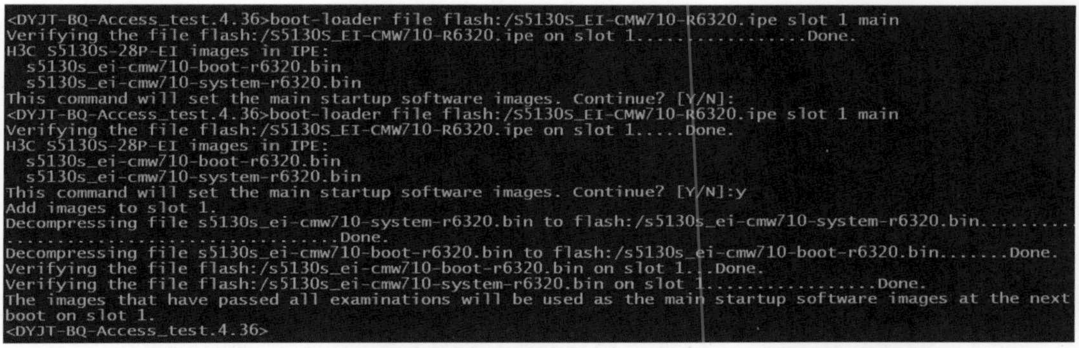

```
<DYJT-BQ-Access_test.4.36>boot-loader file flash:/S5130S_EI-CMW710-R6320.ipe slot 1 main
Verifying the file flash:/S5130S_EI-CMW710-R6320.ipe on slot 1.................Done.
H3C S5130S-28P-EI images in IPE:
  s5130s_ei-cmw710-boot-r6320.bin
  s5130s_ei-cmw710-system-r6320.bin
This command will set the main startup software images. Continue? [Y/N]:
<DYJT-BQ-Access_test.4.36>boot-loader file flash:/S5130S_EI-CMW710-R6320.ipe slot 1 main
Verifying the file flash:/S5130S_EI-CMW710-R6320.ipe on slot 1.....Done.
H3C S5130S-28P-EI images in IPE:
  s5130s_ei-cmw710-boot-r6320.bin
  s5130s_ei-cmw710-system-r6320.bin
This command will set the main startup software images. Continue? [Y/N]:y
Add images to slot 1.
Decompressing file s5130s_ei-cmw710-system-r6320.bin to flash:/s5130s_ei-cmw710-system-r6320.bin.........
................................Done.
Decompressing file s5130s_ei-cmw710-boot-r6320.bin to flash:/s5130s_ei-cmw710-boot-r6320.bin.......Done.
Verifying the file flash:/s5130s_ei-cmw710-boot-r6320.bin on slot 1...Done.
Verifying the file flash:/s5130s_ei-cmw710-system-r6320.bin on slot 1................Done.
The images that have passed all examinations will be used as the main startup software images at the next
boot on slot 1.
<DYJT-BQ-Access_test.4.36>
```

图2-19-26　升级系统

重启交换机完毕，输入“dis boot-loader”检查版本是否升级正常。图 2-19-27 所示系统已经升级到 6320 版本。

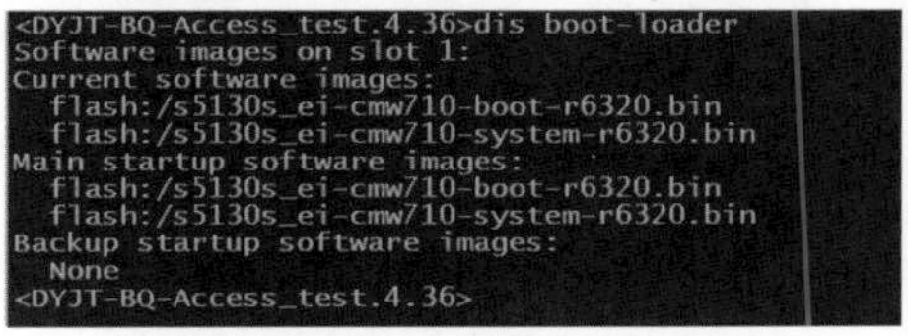

```
<DYJT-BQ-Access_test.4.36>dis boot-loader
Software images on slot 1:
Current software images:
  flash:/s5130s_ei-cmw710-boot-r6320.bin
  flash:/s5130s_ei-cmw710-system-r6320.bin
Main startup software images:
  flash:/s5130s_ei-cmw710-boot-r6320.bin
  flash:/s5130s_ei-cmw710-system-r6320.bin
Backup startup software images:
  None
<DYJT-BQ-Access_test.4.36>
```

图2-19-27　查看升级系统

三、华三真机在线更新指定配置文件（tftp 文件上传与下载）

项目目的：对于重要的交换机配置文件，建议按照周期保存到服务器，当要换新交换机或者恢复到指定时间点时，就非常便捷。

步骤 1　通过 ssh 远程访问一台在线交换机，将 startup.cfg 配置文件复制一份并命名 startup1.cfg，查看交换机所有文件，见图 2-19-28。

步骤 2　查看目前默认启动配置文件，输入“display startup”，默认为 startup.cfg，见图 2-19-29。

步骤 3　指定下次启动配置文件为 startup1.cfg，输入“startup saved-configuration startup1.cfg”，见图 2-19-30。

```
<DYJT-BQ-Access_BGJF.4.34>copy startup.cfg flash:/startup1.cfg
Copy flash:/startup.cfg to flash:/startup1.cfg? [Y/N]:y
Copying file flash:/startup.cfg to flash:/startup1.cfg... Done.
<DYJT-BQ-Access_BGJF.4.34>dir
Directory of flash:
   0 -rw-      220684 Jan 01 2013 00:00:00   defaultfile.zip
   1 drw-           - Jan 01 2013 00:00:12   diagfile
   2 -rw-         735 Jan 01 2013 00:09:47   hostkey
   3 -rw-         863 Jul 14 2021 07:18:24   ifindex.dat
   4 drw-           - Jan 01 2013 00:00:13   license
   5 -rw-           - Jan 02 2013 00:00:11   logfile
   6 drw-           - Jan 01 2013 00:00:47   pki
   7 -rw-     6146048 Jan 01 2013 00:00:00   s5130s_ei-cmw710-boot-r6318p01.bin
   8 -rw-    49709056 Jan 01 2013 00:00:00   s5130s_ei-cmw710-system-r6318p01.bin
   9 drw-           - Jan 01 2013 00:00:12   seclog
  10 -rw-         591 Jan 01 2013 00:09:47   serverkey
  11 -rw-        4265 Jul 14 2021 07:18:26   startup.cfg
  12 -rw-      103772 Jul 14 2021 07:18:27   startup.mdb
  13 -rw-        4265 Dec 15 2021 01:27:31   startup1.cfg
  14 drw-           - Jan 01 2013 00:00:52   versionInfo
  15 -rw-        4265 Dec 15 2021 01:26:13
251904 KB total (193832 KB free)
```

图2-19-28　copy配置文件，并查看

```
<DYJT-BQ-Access_BGJF.4.34>display startup
MainBoard:
 Current startup saved-configuration file: flash:/startup.cfg(*)
 Next main startup saved-configuration file: flash:/startup.cfg
 Next backup startup saved-configuration file: NULL
```

图2-19-29　查看启动文件

```
<DYJT-BQ-Access_BGJF.4.34>startup saved-configuration startup1.cfg
Please wait...... Done.
```

图2-19-30　指定启动配置文件

步骤4　保存之后再重启，见图2-19-31。

```
<DYJT-BQ-Access_BGJF.4.34>sa f
Validating file. Please wait...
Saved the current configuration to mainboard device successfully.
```

(a) 保存文件

```
<DYJT-BQ-Access_BGJF.4.34>reboot
Start to check configuration with next startup configuration file, please wait.........DONE!
This command will reboot the device. Continue? [Y/N]:y
Now rebooting, please wait......
```

(b) 重启系统

图2-19-31　保存文件并重启系统

步骤5　查看是否切换到新的配置文件，见图2-19-32。

```
<DYJT-BQ-Access_BGJF.4.34>display startup
MainBoard:
 Current startup saved-configuration file: flash:/startup1.cfg(*)
 Next main startup saved-configuration file: flash:/startup1.cfg
 Next backup startup saved-configuration file: NULL
```

图2-19-32　查看是否切换到新配置文件

步骤6　配置tftp服务器。

1）建立tftp服务器，见图2-19-33。

2）将startup1.cfg配置文件通过"put"上传至tftp 192.168.9.121服务器，方法见图2-19-34。

3）查看是否上传成功（进入192.168.9.121服务器），见图2-19-35。

4）将tftp-目录中的"starup1.cfg"复制，重命名为"starup5.cfg"，见图2-19-36。在交换机中输入"tftp 192.168.9.121 get startup5.cfg"，下载文件"startup5.cfg"到交换机，见图2-19-37，并输入"dir"查看。

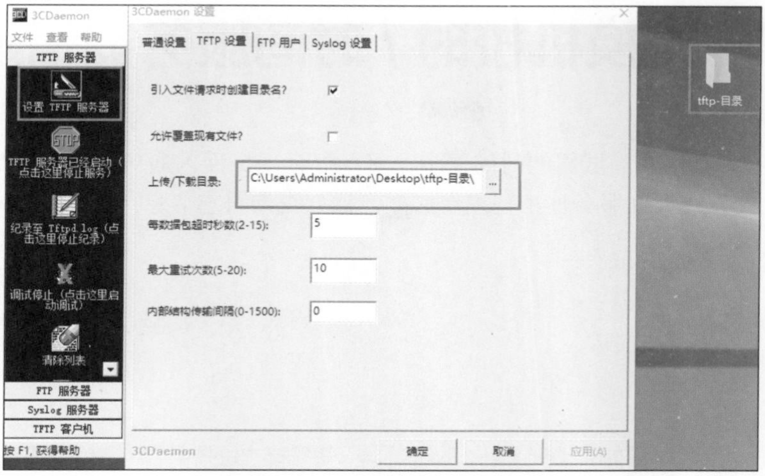

图2-19-33 设置tftp服务器

图2-19-34 "put"上传至tftp 192.168.9.121服务器

图2-19-35 上传成功

图2-19-36 复制"starup1.cfg"并重命名为"starup5.cfg"

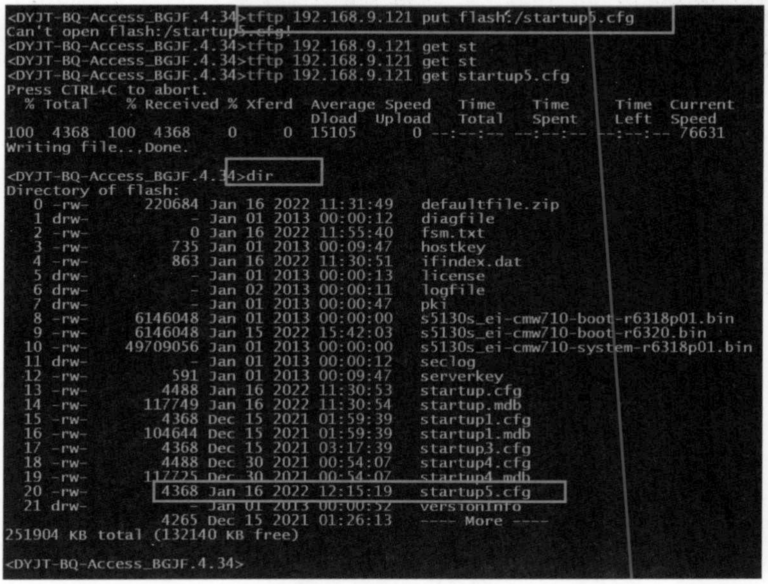

图2-19-37 下载文件"startup5.cfg"到交换机

项目 20　交换机校时/时间服务器搭建

ntp 时间服务器用于局域网的时间同步，可以保证局域网内服务器、交换机等设备时间保持一致性。企业中时间服务器搭建很有必要，比如查询日志信息、确保时间准确、摄像头录像时间准确等。

本项目重点介绍如何在 Linux 系统环境下搭建时间服务器，并讲解交换机以及 PC 如何校时，最后还介绍了摄像头校时方法。

一、Linux 环境下搭建时间服务器

（1）查询 Linux 操作系统 IP 地址，如图 2-20-1 所示。

操作命令：[root@localhost ~]# ip a

查询结果 IP：192.168.99.124

图2-20-1　查询IP地址

（2）确保服务器要能访问 Internet 外网，见图 2-20-2。

联系网络管理员开通服务器外网并进行测试，因为时间服务器需要通过外网与校时网站通信。开通外网后，通过 ping 百度网站，检查是否正常。

测试：[root@localhost ~]# ping www.baidu.com，查看结果网络正常。

图2-20-2　测试服务器能否访问外网

（3）查询 linux 操作系统是否安装 ntp 模块，见图 2-20-3。

操作命令：[root@localhost /]# rpm -qa |grep ntp

查询结果：没有安装

```
[root@localhost /]# rpm -qa |grep ntp
[root@localhost /]#
```

图2-20-3 查询linux操作系统是否安装ntp模块

(4) 安装 ntp 服务。

操作命令：[root@localhost /]# yum install ntp ntpdate -y

```
[root@localhost /]# yum install ntp ntpdate  -y
Loaded plugins: fastestmirror
Determining fastest mirrors
 * base: mirrors.aliyun.com
 * extras: mirrors.bfsu.edu.cn
 * updates: mirrors.bfsu.edu.cn
base
| 3.6 kB   00:00:00
http://mirrors.bfsu.edu.cn/centos/7.9.2009/extras/x86_64/repodata/repomd.xml: [Errno 12] Timeout on
http://mirrors.bfsu.edu.cn/centos/7.9.2009/extras/x86_64/repodata/repomd.xml: (28, 'Operation too slow. Less than 1000 bytes/sec transferred the last 30 seconds')
Trying other mirror.
Extras
| 2.9 kB   00:00:00
Updates
| 2.9 kB   00:00:00
(1/4): base/7/x86_64/group_gz
| 153 kB   00:00:00
(2/4): extras/7/x86_64/primary_db
| 243 kB   00:00:00
(3/4): updates/7/x86_64/primary_db
|  12 MB   00:00:00
……省略……
ntp.x86_64 0:4.2.6p5-29.el7.centos.2
ntpdate.x86_64 0:4.2.6p5-29.el7.centos.2
Dependency Installed:
  autogen-libopts.x86_64 0:5.18-5.el7
Complete!
```

(5) 再次查看安装 ntp 服务，见图 2-20-4。

操作命令：[root@localhost /]# rpm -qa |grep ntp

查询结果：服务已经安装到位

```
[root@localhost /]# rpm -qa |grep ntp
ntpdate-4.2.6p5-29.el7.centos.2.x86_64
ntp-4.2.6p5-29.el7.centos.2.x86_64
```

图2-20-4 查看安装ntp服务

(6) 在 etc 目录下查看配置文件，见图 2-20-5。

操作命令：[root@localhost etc]# ll |grep ntp

查询结果：ntp.conf 配置文件

```
[root@localhost etc]# ll |grep ntp
drwxr-xr-x.  3 root root       52 Nov  1 12:26 ntp
-rw-r--r--.  1 root root     2000 Nov 28  2019 ntp.conf
```

图2-20-5 查看配置文件

（7）进入 ntp.conf 文件，见图 2-20-6。

`[root@localhost ~]# vi /etc/ntp.conf`

[root@localhost ~]# vi /etc/ntp.conf

查看，内部已经内置四条时间服务器地址。

```
[root@localhost ~]# vi /etc/ntp.conf
# For more information about this file, see the man pages
# ntp.conf(5), ntp_acc(5), ntp_auth(5), ntp_clock(5), ntp_misc(5), ntp_mon(5).
driftfile /var/lib/ntp/drift

# Permit time synchronization with our time source, but do not
# permit the source to query or modify the service on this system.
restrict default nomodify notrap nopeer noquery

# Permit all access over the loopback interface.  This could
# be tightened as well, but to do so would effect some of
# the administrative functions.
restrict 127.0.0.1
restrict ::1

# Hosts on local network are less restricted.
#restrict 192.168.1.0 mask 255.255.255.0 nomodify notrap

# Use public servers from the pool.ntp.org project.
# Please consider joining the pool (http://www.pool.ntp.org/join.html).
server 0.centos.pool.ntp.org iburst
server 1.centos.pool.ntp.org iburst
server 2.centos.pool.ntp.org iburst
server 3.centos.pool.ntp.org iburst
```

图2-20-6　进入ntp.conf文件

在配置文件最后添加以下两条信息，目的是让 ntp Server 与自身同步，当内置的 Server 时间服务器都不可用时候，将使用 Local 时间作为 ntp 服务提供给客户端。

```
server 127.127.1.0 #local clock
fudge  127.127.1.0 stratum 10
```

（8）启动 ntpd 服务以及设置开机启动。

启动 ntpd：`systemctl start ntpd.service`
开机启动：`systemctl enable ntpd.service`

（9）通过命令"watch ntpq -p"查询 ntp Server 服务器与自身同步情况。见图 2-20-7。

```
[root@localhost ~]# watch ntpq -p
Every 2.0s: ntpq -p

     remote           refid      st t when poll reach   delay   offset  jitter
==============================================================================
+a.chl.la        131.188.3.222    2 u  227  256  377  175.465  -14.969  12.698
+ntp6.flashdance 192.36.143.130   2 u  221  256  367  174.100  -18.716  16.686
-makaki.miuku.ne 210.23.25.77     2 u  510  256  376  215.982  -59.824  57.703
*time.neu.edu.cn .PTP.            1 u  313  256  376   79.824   -6.442  12.655
```

图2-20-7　查询ntp Server 服务器与自身同步情况

◇ remote：远程 ntp 服务器的 IP 地址或域名，"+"表示优先，"*"表示次优先。
◇ refid：远程 ntp 服务器的上层服务器的 IP 地址或域名。
◇ st：远程 ntp 服务器所在的层数。
◇ t：本地 ntp 服务器与远程 ntp 服务器的通信方式，u：单播。
◇ when：上一次校正时间与现在时间的差值。
◇ poll：下次更新在多少秒后。

✧ reach：是一种衡量前 8 次查询是否成功的位掩码值，377 表示都成功，0 表示不成功。
✧ delay：网络延时。
✧ offset：时间补偿。
✧ jitter：查询偏差的分布值。

注意 当时间服务器重新启动后，在该时间内客户端输入命令同步时间，将会提示“no server suitable for synchronization found”，稍后即可正常。

关于防火墙（firewalld）的命令，如表 2-20-1。

表 2-20-1　防火墙（firewalld）的命令

systemctl status firewalld	查询防火墙状态
systemctl enable firewalld	开机启用
systemctl disable firewalld	永久关闭
systemctl stop firewalld	关闭
systemctl start firewalld	开启防火墙

校时服务器需要关闭防火墙，或者开启防火墙，放通 udp 123 端口，否则终端无法校时。通过命令 systemctl status firewalld，查询防火墙状态，图 2-20-8 所示防火墙处于运行状态（running）。

```
[root@localhost ~]# systemctl status firewalld
● firewalld.service - firewalld - dynamic firewall daemon
   Loaded: loaded (/usr/lib/systemd/system/firewalld.service; enabled; vendor preset: enabled)
   Active: active (running) since Mon 2021-11-01 20:58:27 CST; 4s ago
     Docs: man:firewalld(1)
 Main PID: 6950 (firewalld)
   CGroup: /system.slice/firewalld.service
           └─ 6950 /usr/bin/python -Es /usr/sbin/firewalld --nofork --nopid

Nov 01 20:58:27 localhost.localdomain systemd[1]: Starting firewalld - dynamic firewall daemon...
Nov 01 20:58:27 localhost.localdomain systemd[1]: Started firewalld - dynamic firewall daemon.
```

图2-20-8　防火墙处于运行状态

systemctl stop firewalld 关闭防火墙，查询防火墙状态，已经关闭状态（dead）。见图 2-20-9。

```
[root@localhost ~]# systemctl stop firewalld
[root@localhost ~]# systemctl status firewalld
● firewalld.service - firewalld - dynamic firewall daemon
   Loaded: loaded (/usr/lib/systemd/system/firewalld.service; disabled; vendor preset: enabled)
   Active: inactive (dead)
     Docs: man:firewalld(1)

Nov 01 21:35:06 localhost.localdomain systemd[1]: Stopping firewalld - dynamic firewall daemon...
Nov 01 21:35:07 localhost.localdomain systemd[1]: Stopped firewalld - dynamic firewall daemon.
Nov 01 21:35:07 localhost.localdomain systemd[1]: Starting firewalld - dynamic firewall daemon...
Nov 01 21:35:07 localhost.localdomain systemd[1]: Started firewalld - dynamic firewall daemon.
Nov 01 21:35:10 localhost.localdomain systemd[1]: Stopping firewalld - dynamic firewall daemon...
Nov 01 21:35:11 localhost.localdomain systemd[1]: Stopped firewalld - dynamic firewall daemon.
Nov 01 21:35:11 localhost.localdomain systemd[1]: Starting firewalld - dynamic firewall daemon...
Nov 01 21:35:11 localhost.localdomain systemd[1]: Started firewalld - dynamic firewall daemon.
Nov 01 21:36:26 localhost.localdomain systemd[1]: Stopping firewalld - dynamic firewall daemon...
Nov 01 21:36:27 localhost.localdomain systemd[1]: Stopped firewalld - dynamic firewall daemon.
```

图2-20-9　查询防火墙状态为已经关闭状态

二、终端校时

1. 交换机校时

```
# 设置时区
[DYJT-BQ-Access_BG1F.4.38]clock timezone beijing add 8
[DYJT-BQ-Access_BG1F.4.38]dis clock
17:07:31.271 beijing Mon 11/01/2021
Time Zone : beijing add 08:00:00
[DYJT-BQ-Access_BG1F.4.38]clock protocol none
[DYJT-BQ-Access_BG1F.4.38]quit
# 将交换机时间修改为 07:50:52 01/21/2021
<DYJT-BQ-Access_BG1F.4.38>clock datetime 07:50:52 01/21/2021
# 查询时间
<DYJT-BQ-Access_BG1F.4.38>display clock
07:50:59.228 beijing Thu 01/21/2021
Time Zone : beijing add 08:00:00
<DYJT-BQ-Access_BG1F.4.38>sys
System View: return to User View with Ctrl+Z.
# 交换机开启 ntp 服务
[DYJT-BQ-Access_BG1F.4.38] ntp-service enable
# 配置校时服务器地址
[DYJT-BQ-Access_BG1F.4.38] ntp-service unicast-server 192.168.99.124
# 配置校时服务为 ntp
[DYJT-BQ-Access_BG1F.4.38]clock protocol ntp
# 查询交换机配置关于 ntp 命令
[DYJT-BQ-Access_BG1F.4.38]dis cur | include ntp
clock protocol ntp
ntp-service enable
ntp-service unicast-server 192.168.99.124
# 查询时间，已经校时正确
<DYJT-BQ-Access_BG1F.4.38>dis clock
17:37:45.703 beijing Mon 11/01/2021
Time Zone : beijing add 08:00:00
```

2. Windows 系统校时配置（Windows 10 为例）

步骤1 打开控制面板，见图 2-20-10。

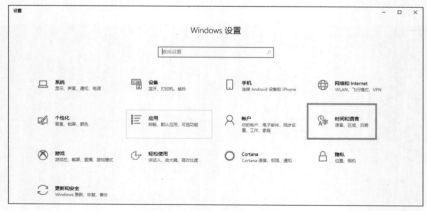

图2-20-10 打开控制面板

步骤2 选择时间和语言，见图2-20-11。

自动设置时间关闭 - 修改时间（错误时间），更改时间为 21:29。

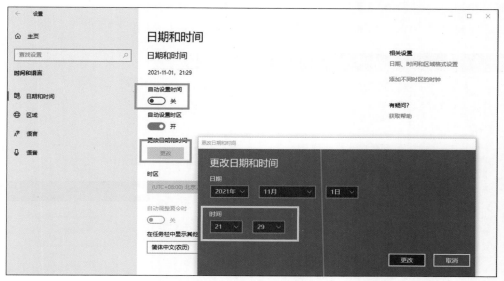

图2-20-11 修改时间

步骤3 选择不同时区的时钟，服务器里填写 192.168.99.124，立即更新，显示时钟在 2021-11-01 21:53，与 192.168.99.124 同步成功。见图 2-20-12。

图2-20-12 填写服务器地址，同步时间

3. linux 操作系统校时

格式：ntpdate -u 192.168.99.124

-u，指定使用无特权的端口发送数据包。

```
[root@localhost ~]# ntpdate -u 192.168.99.124
 1 Nov 22:01:21 ntpdate[337251]: step time server 192.168.99.124 offset 234.699259 sec
```

4. 摄像头校时

企业中摄像头的数量成百上千，为确保摄像头时间准确性，一般在摄像头中配置时间服务器进行校对。图 2-20-13 为一款摄像头 ntp 校时配置界面（配置 - 系统配置 - 服务器地址 - 校时时间间隔），点击"测试"，弹出"测试成功"，保存即可。见图 2-20-14。

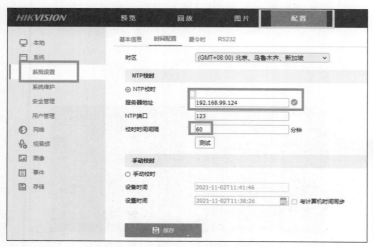

图2-20-13　ntp校时配置界面

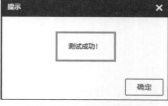

图2-20-14　测试成功

项目 21　实战 VPN 两种配置（L2TP/IPsec VPN）办法

通过 L2TP VPN 隧道协议实现互联网访问公司网络要求，L2TP VPN 主要实现出差用户通过互联网 PC 访问公司内部数据，而 IPsec VPN 通过建立双向安全联盟机制，能形成安全互通的数据加密隧道，实现处于不同区域的两个内网之间通过互联网数据互通。本项目通过 L2TP/IPsec VPN 两种方式进行配置。

一、H3C GR2200 路由器 L2TP VPN 配置方法

1. H3C GR2200 路由器配置

步骤 1　选择 VPN-L2TP VPN，见图 2-21-1。

步骤 2　选择 L2TP 服务端，见图 2-21-2。

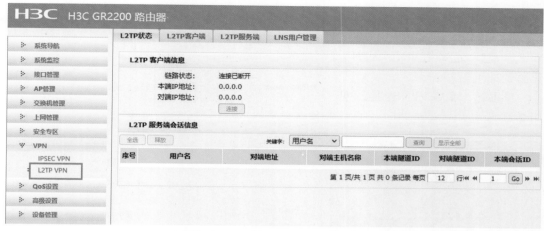

图2-21-1 选择VPN-L2TP VPN

勾选启用LNS。路由器上开启L2TP时，配置L2TP服务器地址池，为拨号进来的远端PC分配IP地址。点击"应用"。

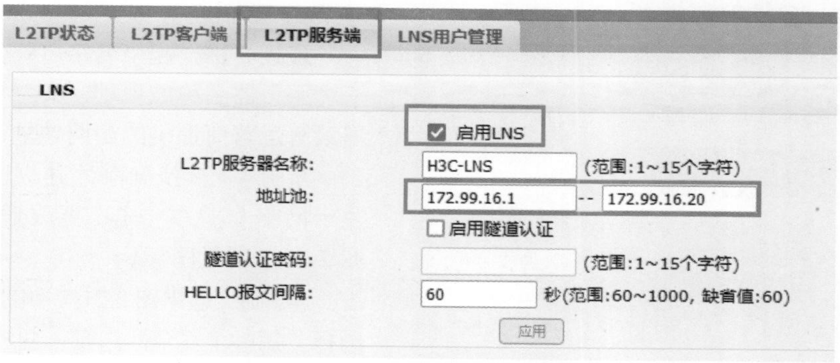

图2-21-2 选择L2TP服务端

步骤3 配置管理用户见图2-21-3。
也就是拨号进来需要验证用户名以及密码。

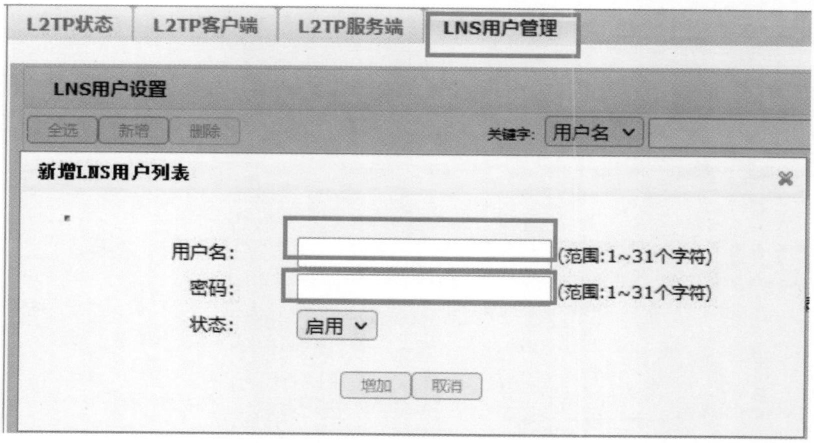

图2-21-3 配置管理用户

2. 电脑端设置步骤

以 Windows 10 操作系统为例。

步骤1　通知栏点击"VPN"。见图 2-21-4。

步骤2　添加 VPN 连接，见图 2-21-5。

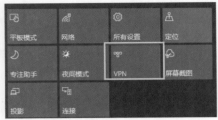

图2-21-4　点击"VPN"

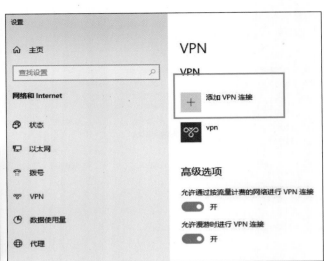

图2-21-5　添加VPN连接

步骤3　填写基本信息，见图 2-21-6。

VPN 提供商：选择 Windows（内置）

连接名称：起个名字

服务器名称或地址：60.**.**.**（这里填写运营商提供的公网地址）。

步骤4　查找到刚才建立的 VPN，路径：网络与共享中心 - 更改适配器设置 - 图 2-21-7 所示图标。

步骤5　选中图 2-21-7 中的图标，右键选择"属性 - 安全"选项，见图 2-21-8。

VPN 类型选择"使用 IPsec 的第 2 层隧道协议（L2TP/IPsec）"，允许使用协议下打上对勾，其他默认。

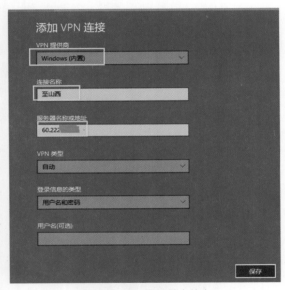

图2-21-6　填写基本信息

图2-21-7　查找到刚才建立的VPN

3. 连接上网

步骤1 选择 VPN "至山西"，填写用户名以及密码，见图 2-21-9。

步骤2 查看状态 - 显示已连接，见图 2-21-10。

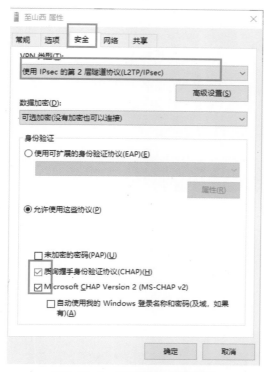

图2-21-8 选择属性-安全选项

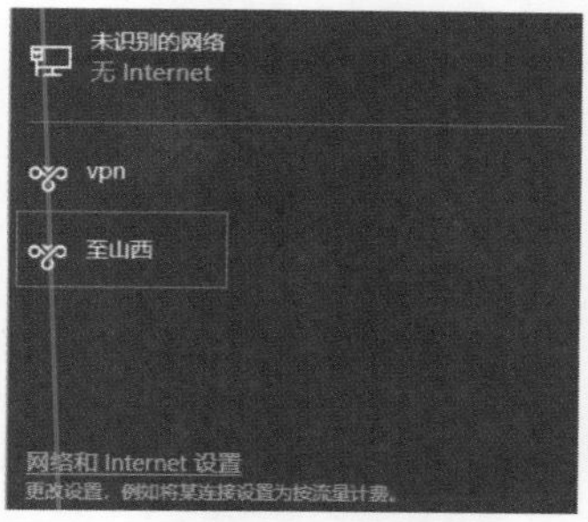

图2-21-9 选择VPN "至山西"

图2-21-10 显示连接状态

步骤3 查看网卡信息，IP：172.99.16.1，见图 2-21-11。

测试网络已通（IP：192.168.1.127 为公司服务器地址），见图 2-21-12。

图2-21-11 网卡信息

图2-21-12 网络测试

二、IPsec VPN 配置

下面介绍如何在 H3C GR2200 路由器与天清汉马 VPN 两个设备之间建立 IPsec VPN。

1. H3C GR2200 路由器方面配置

步骤1 创建虚接口，见图 2-21-13。

路径：IPsec VPN- 虚接口 - 新增 -WAN1。

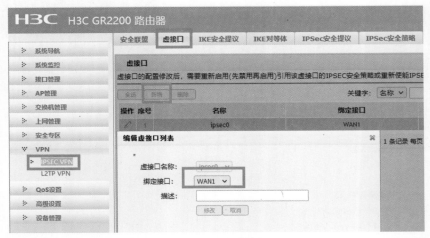

图2-21-13 创建虚接口

步骤2 IKE 安全提议设置，见图 2-21-14。

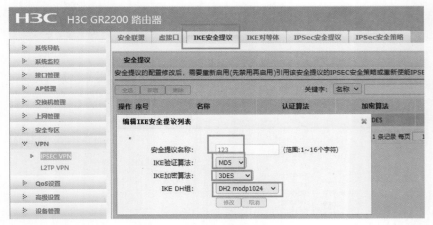

图2-21-14 KE安全提议设置

配置信息见表 2-21-1。

表 2-21-1 配置信息

安全提议命名	IKE 验证算法	IKE 加密算法	IKE DH 组
123	MD5	3DES	modp1024

步骤3 IKE 对等体配置见图 2-21-15。

对端地址：填写对方 VPN 设备公网地址，预共享秘钥（PSK）：两个设备一致。

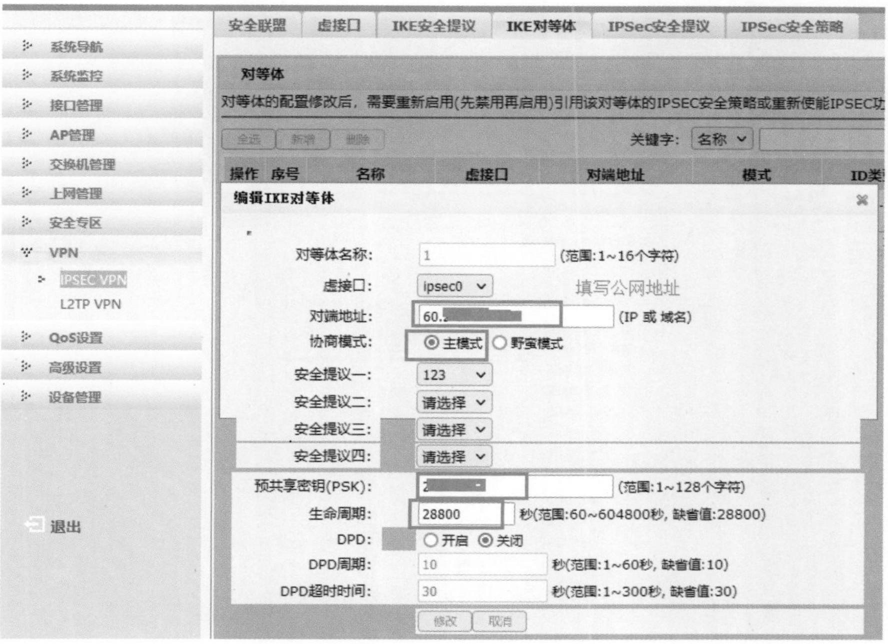

图2-21-15　IKE对等体配置

步骤4　IPsec 安全提议，见图 2-21-16。

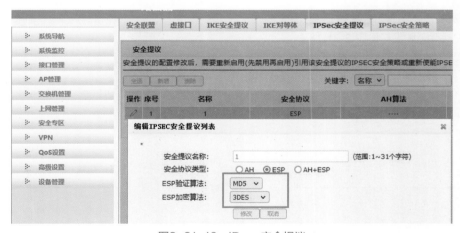

图2-21-16　IPsec安全提议

IPsec 安全提议配置见表 2-21-2。

表 2-21-2　IPsec 安全提议配置

安全协议类型	ESP 验证算法	ESP 加密算法
ESP	MD5	3DES

步骤5　IPsec 安全策略，见图 2-21-17。

启用 IPsec 功能选项打上对勾，本地子网 IP/ 掩码（192.168.1.0/255.255.255.0），对端子网 IP/ 掩码（10.1.0.0/255.255.255.0）。

255

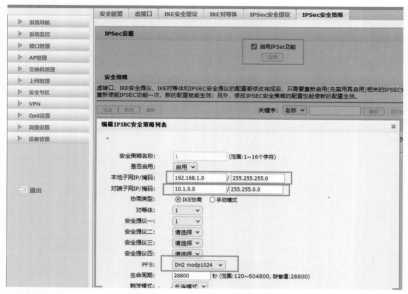

图2-21-17　IPsec安全策略

2. 天清汉马 VPN 安全网关系统 V3.0（SAG-6000-2000）

步骤1　新建 VPN 规则，填写本端与对端保护网络地址。见图 2-21-18。

步骤2　IKE 配置，见图 2-21-19。

填写对端地址，对方设备运营商公网地址。

步骤3　网关隧道配置，见图 2-21-20。

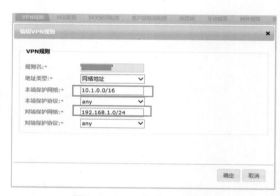

图2-21-18　新建VPN规则

图2-21-19　IKE配置

图2-21-20　网关隧道配置

3. 查看状态

步骤1 在天清汉马 VPN 安全网关系统 V3.0（SAG-6000-2000）设备上查看，点击"隧道监控"-"网关隧道"，状态已经显示"已经建立"见图 2-21-21。

（a）点击"隧道监控" 　　　　　　　　（b）显示已经建立

图2-21-21　查看状态

步骤2 在 H3C GR2200 路由器查看，安全联盟中可以看到数据双向数据流，状态正常，见图 2-21-22。

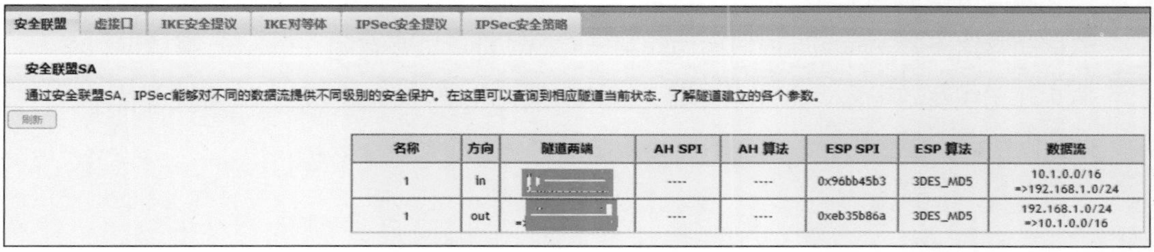

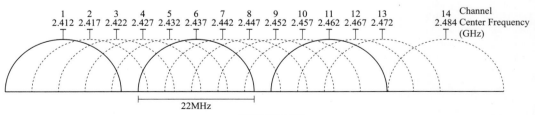

图2-21-22　互通数据流量

项目 22　构建无线局域网

随着智能互联终端的普及，企业网络中有越来越多的数据需要经过无线网络进行传输，因此在企业网络中，进行无线网络设计必不可少。

无线控制器（Wireless Access Point Controller）是一种网络设备，简称 AC，用来集中化控制无线 AP，是无线网络的核心。无线射频频率范围是 2.4GHz 频段（2.4～2.4835GHz）和 5GHz 频段（频率范围是 5.150～5.350GHz 和 5.725～5.850GHz）。

各个国家 2.4G 开放的信道不一样，我国开放 1～13 信道，每个信号占用 22MHz 带宽，在这 13 个信道中找出 3 个独立信道，见图中 1、6、11 三个独立的信道。见图 2-22-1。

802.11b频段带宽示意图

图2-22-1　1、6、11三个独立的信道

在无线 AP 网络部署中，采用蜂窝式布局，为了避免相邻 AP 同频干扰，蜂窝式布局相邻的 AP 使用不交叠的独立信道（1、6、11 三个独立的信道），能有效地解决同频干扰。见图 2-22-2。

AC 在网络部署中有两种方式，一种是串接到网络链路中，还有一种就是旁挂式，旁挂式部署居多。

图2-22-2 解决同频干扰

 项目简介

项目重点讲解无线 AP 与 AC 通信，以及无线 AP 在 AC 中认证方式，同时介绍移动终端通过无线 wifi 连接局域网采用密码认证，以及在同一个 ssid 覆盖下，终端实现无线漫游配置。企业中部署无线网一般把生产与办公分开，本项目重点介绍办公网络可以上因特网，而生产网络无法上因特网的设置方法。

一、拓扑图

如图 2-22-3 所示。

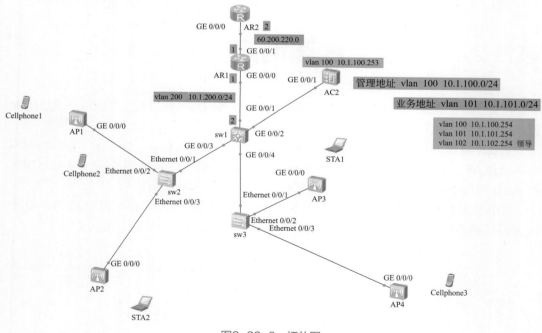

图2-22-3 拓扑图

基本要求

① AP 管理 vlan 100；网段：10.1.100.0/24。

② Sw1-AR1 网段：10.1.200.0/24；AR1-AR2 网段：60.200.220.0/24。
③ 生产业务终端 vlan101 网段：10.1.101.0/24，生产网 AP1- AP2 负责。
领导业务终端 vlan 102 网段：10.1.102.0/24，领导办公使用 AP3- AP4 负责。
④ AC 采用旁挂式本地转发。
⑤ AR2 设置环回接口（IP：1.1.1.1/32），模拟外网地址。
⑥ AC 管理 AP1-AP2，使用 MAC 地址认证，AC 管理 AP3-AP4 使用 sn 地址认证。
⑦ 所有终端使用密码认证，密码：a123456，ssid: fsm。

二、基本配置

1. 交换机 sw1 配置

1）基本配置（vlan、网关、DHCP）

```
<Huawei>sys
<Huawei>system-view
Enter system view, return user view with Ctrl+Z.
# 关闭信息
[Huawei]un in en
Info: Information center is disabled.
[Huawei]sysname sw1
# 创建 vlan
[sw1]vlan batch 100 101
Info: This operation may take a few seconds. Please wait for a moment...done.
# 创建 vlan 100 网关
[sw1]interface Vlanif 100
[sw1-Vlanif100]ip address 10.1.100.254 24
# 标识 vlan 100 的用途，TO-AP-GL（vlan 100 用于管理 AP）
[sw1-Vlanif100]description TO-AP-GL
[sw1-Vlanif100]quit
# 创建 vlan 101 网关
[sw1]interface Vlanif 101
[sw1-Vlanif101]ip address 10.1.101.254 24
# 标识 vlan 101 的用途，生产业务
[sw1-Vlanif101]description TO-AP-shengchanyewu
[sw1-Vlanif101]quit
# 全局开启 dhcp 服务
[sw1]dhcp enable
Info: The operation may take a few seconds. Please wait for a moment.done.
# 地址池命令为 AP-GL
[sw1]ip pool AP-GL
# 配置网关
[sw1-ip-pool-ap-gl]gateway-list 10.1.100.254
# 配置地址范围 10.1.100.0/24
[sw1-ip-pool-ap-gl]network 10.1.100.0 mask 24
# 全局地址池给 DHCP 客户端使用
[sw1-Vlanif100]dhcp select global
# 配置生产业务使用 IP 地址池
[sw1]ip pool AP-shengchanyewu
Info:It's successful to create an IP address pool.
```

配置网关
[sw1-ip-pool-ap-shengchanyewu]gateway-list 10.1.101.254
配置地址范围 10.1.101.0/24
[sw1-ip-pool-ap-shengchanyewu]network 10.1.101.0 mask 24
配置 DNS1
[sw1-ip-pool-ap-shengchanyewu]dns-list 114.114.114.114
配置 DNS2
[sw1-ip-pool-ap-shengchanyewu]dns-list 8.8.8.8
[sw1-ip-pool-ap-shengchanyewu]quit
[sw1]interface Vlanif 101
全局的地址池给 DHCP 客户端使用
[sw1-Vlanif101]dhcp select global
[sw1-Vlanif101]quit

2）sw1 接口配置

[sw1]interface GigabitEthernet 0/0/2
[sw1-GigabitEthernet0/0/2]port link-type trunk
[sw1-GigabitEthernet0/0/2]port trunk allow-pass vlan 100 101
查询配置
[sw1-GigabitEthernet0/0/2]dis this
#
interface GigabitEthernet0/0/2
port link-type trunk
port trunk allow-pass vlan 100 to 101
#
return
[sw1-GigabitEthernet0/0/2]quit
[sw1]int GigabitEthernet 0/0/3
[sw1-GigabitEthernet0/0/3]port link-type trunk
[sw1-GigabitEthernet0/0/3] port trunk allow-pass vlan 100 to 101
查询配置
[sw1-GigabitEthernet0/0/3]dis this
#
interface GigabitEthernet0/0/3
port link-type trunk
port trunk allow-pass vlan 100 to 101
#
return
[sw1-GigabitEthernet0/0/3]quit

2. 交换机 sw2 配置

sw2 交换机接口配置如下。

[sw2-Ethernet0/0/1] port link-type trunk
[sw2-Ethernet0/0/1]port trunk allow-pass vlan 100 to 101
查询配置
[sw2-Ethernet0/0/1]dis this
#
interface Ethernet0/0/1
port link-type trunk
port trunk allow-pass vlan 100 to 101
#
return

```
[sw2-Ethernet0/0/1]quit
[sw2]interface Ethernet 0/0/2
[sw2-Ethernet0/0/2] port link-type trunk
#默认pvid是vlan1，修改pvid vlan 100。
[sw2-Ethernet0/0/2] port trunk pvid vlan 100
[sw2-Ethernet0/0/2] port trunk allow-pass vlan 100 to 101
[sw2-Ethernet0/0/2]quit
[sw2]interface  Ethernet 0/0/3
[sw2-Ethernet0/0/3] port link-type trunk
[sw2-Ethernet0/0/3] port trunk pvid vlan 100
[sw2-Ethernet0/0/3] port trunk allow-pass vlan 100 to 101
[sw2-Ethernet0/0/3]quit
[sw2]
```

3. 无线 AC 控制器配置

1）基本配置

```
[AC]vlan batch 100 101
Info: This operation may take a few seconds. Please wait for a moment...done.
[AC]interface  GigabitEthernet 0/0/1
[AC-GigabitEthernet0/0/1]port link-type trunk
[AC-GigabitEthernet0/0/1]port trunk allow-pass vlan 100 101
[AC-GigabitEthernet0/0/1]quit
[AC]interface Vlanif 100
```

2）设备配置 IP 管理地址

```
[AC-Vlanif100]ip address 10.1.100.253 24
[AC-Vlanif100]quit
[AC]interface  Vlanif 101
[AC-Vlanif101]ip address 10.1.101.253 24
[AC-Vlanif101]quit
```

测试 AC-ping-sw1（IP: 10.1.100.254）- 通

```
[AC]ping 10.1.100.254
    PING 10.1.100.254: 56  data bytes, press CTRL_C to break
    Reply from 10.1.100.254: bytes=56 Sequence=1 ttl=255 time=30 ms
    Reply from 10.1.100.254: bytes=56 Sequence=2 ttl=255 time=30 ms
……省略……
--- 10.1.100.254 ping statistics ---
  5 packet(s) transmitted
  5 packet(s) received
  0.00% packet loss
  round-trip min/avg/max = 10/22/30 ms
[AC]quit
```

3）AP 连接 AC 认证方式（采用 MAC 认证）

```
[AC]wlan
# 配置模式选择
[AC-wlan-view]ap auth-mode ?
  mac-auth   MAC authenticated mode, default authenticated mode
  no-auth    No authenticated mode
  sn-auth    SN authenticated mode
# 选择  mac-auth 认证
[AC-wlan-view]ap auth-mode mac-auth
```

```
[AC-wlan-view]quit
```

4）配置 AC 源地址

配置 AC 源地址用于 AC 与 AP 之间建立隧道通信。

```
[AC]capwap source ip-address 10.1.100.253
```

5）配置 sw1 交换机 DHCP 排除地址

```
# 排除 AC 源地址（保留两个 AC 地址，一个备用，本例使用 IP 地址 10.1.100.253,10.1.100.252 备用）
[sw1-ip-pool-ap-gl]excluded-ip-address 10.1.100.252 10.1.100.253
# 排除业务地址
[sw1-ip-pool-ap-shengchanyewu]excluded-ip-address 10.1.101.252 10.1.101.253
```

6）打开 AP1 与 AP2，查看能否获取 IP 地址

```
#AP 在命令行下是无法修改 AP 名字的
# 查询 AP1 获取的 IP 地址 - 已经获取 10.1.100.251
[Huawei]display ip interface brief
……省略
Interface                    IP Address/Mask        Physical    Protocol
NULL0                        unassigned             up          up(s)
Vlanif1                      10.1.100.251/24        up          up
测试 AP1(IP:10.1.100.251)-ping-AC(IP:10.1.100.253)-通
[Huawei]ping 10.1.100.253
  PING 10.1.100.253: 56   data bytes, press CTRL_C to break
    Reply from 10.1.100.253: bytes=56 Sequence=1 ttl=255 time=180 ms
    Reply from 10.1.100.253: bytes=56 Sequence=2 ttl=255 time=70 ms
……省略……
--- 10.1.100.253 ping statistics ---
  5 packet(s) transmitted
  5 packet(s) received
  0.00% packet loss
  round-trip min/avg/max = 40/82/180 ms
# 查询 AP2 获取的 IP 地址 -10.1.100.250
[Huawei]display ip interface  brief
……省略……
Interface                    IP Address/Mask        Physical    Protocol
NULL0                        unassigned             up          up(s)
Vlanif1                      10.1.100.250/24        up          up
测试 AP2(IP:10.1.100.250)-ping-AC(IP:10.1.100.253)-通
[Huawei]ping 10.1.100.253
  PING 10.1.100.253: 56   data bytes, press CTRL_C to break
    Reply from 10.1.100.253: bytes=56 Sequence=1 ttl=255 time=210 ms
……省略……
--- 10.1.100.253 ping statistics ---
  5 packet(s) transmitted
  5 packet(s) received
  0.00% packet loss
  round-trip min/avg/max = 30/70/210 ms
```

7）AP1 与 AP2-MAC 认证

两个 AP 设备的 MAC 以及 SN 截图见图 2-22-4 和图 2-22-5。

图2-22-4　AP1-MAC地址以及SN

图2-22-5　AP2-MAC地址以及SN

```
[AC]wlan
[AC-wlan-view]ap auth-mode mac-auth
#AC 上绑定 AP1 设备的 MAC 地址
[AC-wlan-view]ap-id 1 ap-mac 00E0-FCC2-43D0
# 修改 AP1 的名字为 AP-shengchan-1
[AC-wlan-ap-1]ap-name AP-shengchan-1
[AC-wlan-ap-1]quit
```

```
#AC 上绑定 AP2 设备的 MAC 地址
[AC-wlan-view]ap-id 2 ap-mac 00E0-FC2E-2080
# 修改 AP2 的名字为 AP-shengchan-2
[AC-wlan-ap-2]ap-name AP-shenchan-2
[AC-wlan-ap-2]quit
[AC-wlan-view]quit
[AC]quit
<AC>save
```

AC 设备上查询 AP1 与 AP2 上线情况（idle 显示说明还没有上线），见图 2-22-6。

```
[AC]display ap all
Info: This operation may take a few seconds. Please wait for a moment.done.
Total AP information:
idle : idle             [2]
------------------------------------------------------------------------
---------
ID   MAC            Name           Group    IP  Type        State STA Uptime
------------------------------------------------------------------------
1    00e0-fcc2-43d0 AP-shengchan-1 default   -   -           idle  0   -
2    00e0-fc2e-2080 AP-shenchan-2  default   -   -           idle  0   -
---------
Total: 2
[AC]
```

图2-22-6　在AC设备上查询AP1与AP2上线情况（状态idle）

大约 2min 后，AP 连接状态（idle）变为 nor（正常），显示获取地址，同时 AP 设备 IP 地址已经同步。见图 2-22-7。

```
[AC]display ap all
Info: This operation may take a few seconds. Please wait for a moment.done.
Total AP information:
nor  : normal           [2]
------------------------------------------------------------------------
------------------
ID   MAC            Name           Group    IP            Type      State ST
A Uptime
------------------------------------------------------------------------
1    00e0-fcc2-43d0 AP-shengchan-1 default  10.1.100.251  AP3030DN   nor  0
  28S
2    00e0-fc2e-2080 AP-shenchan-2  default  10.1.100.250  AP3030DN   nor  0
  28S
------------------------------------------------------------------------
------------------
Total: 2
```

图2-22-7　在AC设备上查询AP1与AP2上线情况（状态nor）

再次查询 AP1 与 AP2，发现名字已经修改，分别为 <AP-shengchan-1>、<AP-shenchan-2>。
查询 capwap 情况：
capwap 是一个通用的隧道协议，完成 AP 发现 AC 等基本协议功能。

```
[AC]display capwap configuration
----------------------------------------------------------------
 Source interface                          : -
 Source ip-address                         : 10.1.100.253
 Echo interval(seconds)                    : 25
 Echo times                                : 6
 Control priority(server to client)        : 7
 Control priority(client to server)        : 7
```

```
Control-link DTLS encrypt                          : disable
DTLS PSK value                                     : ******
PSK mandatroy match switch                         : disable
Control-link inter-controller DTLS encrypt         : disable
Inter-controller DTLS PSK value                    : ******
IPv6 status                                        : disable
Message-integrity PSK value                        : ******
Message-integrity check switch                     : enable
```

8）建立AP组

\# 新建组名为shengchan-group（生产组）

[AC-wlan-view]ap-group name shengchan-group
Info: This operation may take a few seconds. Please wait for a moment.done.

\# 查询新建AP组中域控制模板全部信息

[AC-wlan-ap-group-shengchan-group]dis regulatory-domain-profile all

```
Profile name    Reference
default         2
```

Total: 1

\# 查询新建AP组域管理模板default信息（本例采用默认default）

[AC-wlan-ap-group-shengchan-group]display regulatory-domain-profile name default

```
Profile name              : default
Country code              : CN    // 中国
2.4G dca channel-set      : 1,6,11
5G dca bandwidth          : 20mhz
5G dca channel-set        : 149,153,157,161,165
Wideband switch           : disable
```

\# 新建组调用默认域

[AC-wlan-ap-group-shengchan-group]regulatory-domain-profile default
Warning: Modifying the country code will clear channel, power and antenna gain configurations of the radio and reset the AP. Continue?[Y/N]:y
[AC-wlan-ap-group-shengchan-group]

\# 进入AP-shengchan-1

[AC-wlan-view]ap-name AP-shengchan-1

\#AP-shengchan-1 绑定到 shengchan-group 组

[AC-wlan-ap-1]ap-group shengchan-group
Warning: This operation may cause AP reset. If the country code changes, it will
    clear channel, power and antenna gain configurations of the radio, Whether to c
    ontinue? [Y/N]:y
Info: This operation may take a few seconds. Please wait for a moment..done.
[AC-wlan-ap-1]

查询AP1已经在组里，见图2-22-8。

```
[AC-wlan-view]dis ap all
Info: This operation may take a few seconds. Please wait for a moment.done.
Total AP information:
fault: fault            [1]
nor  : normal           [1]
--------------------------------------------------------------------------------
ID   MAC              Name            Group          IP            Type
State STA Uptime
--------------------------------------------------------------------------------
1    00e0-fcc2-43d0  AP-shengchan-1   shengchan-group -             AP3030DN
fault 0
2    00e0-fc2e-2080  AP-shenchan-2    default        10.1.100.250  AP3030DN
nor   0   18H:6M:40S
--------------------------------------------------------------------------------
Total: 2
```

图2-22-8　查询AP1已经在组里

[AC-wlan-view]ap-name AP-shenchan-2
[AC-wlan-ap-2]ap-group shengchan-group
Warning: This operation may cause AP reset. If the country code changes, it will clear channel, power and antenna gain configurations of the radio, Whether to continue? [Y/N]:
[AC-wlan-ap-2]

查询两个 AP 已经都加到 shengchan-group，见图 2-22-9。

```
[AC-wlan-view]display ap all
Info: This operation may take a few seconds. Please wait for a moment.done.
Total AP information:
nor  : normal           [2]
--------------------------------------------------------------------------------
ID   MAC              Name            Group          IP            Type
State STA Uptime
--------------------------------------------------------------------------------
1    00e0-fcc2-43d0  AP-shengchan-1  shengchan-group 10.1.100.251  AP3030DN
nor   0   3M:19S
2    00e0-fc2e-2080  AP-shenchan-2   shengchan-group 10.1.100.250  AP3030DN
nor   0   33S
--------------------------------------------------------------------------------
Total: 2
```

图2-22-9　AP1与AP2全部加到shengchan-group

9）建立安全服务模板（终端电脑与 AP 连接的安全模板）
建议安全模板名称 fsm-sec-1
[AC-wlan-view]security-profile name fsm-sec-1
配置安全策略，并设置终端接入密码为 a1234567.
[AC-wlan-sec-prof-fsm-sec-1]security wpa-wpa2 psk pass-phrase a1234567 aes
配置 ssid 模板 fsm-2.4G
[AC-wlan-view]ssid-profile name fsm-2.4G
#ssid 模板中，设置 ssid 名称为 wifi-2.4G
[AC-wlan-ssid-prof-fsm-2.4G]ssid wifi-2.4G
Info: This operation may take a few seconds, please wait.done.
[AC-wlan-ssid-prof-fsm-2.4G]QUIT
#ssid 模板中，设置模版名称为 fsm-5G
[AC-wlan-view]ssid-profile name fsm-5G
#ssid 模板中，设置 ssid 名称为 wifi-5G
[AC-wlan-ssid-prof-fsm-5G]ssid wifi-5G

```
Info: This operation may take a few seconds, please wait.done.
[AC-wlan-ssid-prof-fsm-5G]quit
[AC-wlan-view]
```

10）创建 vap 模板（无线配置模板的称呼，里面包含了 SSID 等参数）

```
[AC-wlan-view]vap-profile name fsm-vap-1
# 配置业务数据为直接转发模式
[AC-wlan-vap-prof-fsm-vap-1]forward-mode direct-forward
# 引用安全模板 fsm-sec-1
[AC-wlan-vap-prof-fsm-vap-1]security-profile fsm-sec-1
Info: This operation may take a few seconds, please wait.done.
# vap1 引用对应 fsm-5G wifi 模板
[AC-wlan-vap-prof-fsm-vap-1]ssid-profile fsm-5G
Info: This operation may take a few seconds, please wait.done.
[AC-wlan-vap-prof-fsm-vap-1]quit
[AC-wlan-view]
# 配置 vap 1 业务 vlan
[AC-wlan-vap-prof-fsm-vap-1]service-vlan vlan-id 101
Info: This operation may take a few seconds, please wait.done.
# 查询 vap 1 所有信息
[AC-wlan-vap-prof-fsm-vap-1]dis this
#
  service-vlan vlan-id 101
  ssid-profile fsm-5G
  security-profile fsm-sec-1
#
return
[AC-wlan-vap-prof-fsm-vap-1]
# 创建 vap2 模板
[AC-wlan-view]vap-profile name fsm-vap-2
# 转发模式
[AC-wlan-vap-prof-fsm-vap-2]forward-mode ?
  direct-forward  Direct forward
  softgre         Softgre profile
  tunnel          Tunnel
[AC-wlan-vap-prof-fsm-vap-2]forward-mode direct-forward
[AC-wlan-vap-prof-fsm-vap-2]security-profile fsm-sec-1
Info: This operation may take a few seconds, please wait.done.
# vap2 引用对应 fsm-2.4G wifi 模板
[AC-wlan-vap-prof-fsm-vap-2]ssid-profile fsm-2.4G
Info: This operation may take a few seconds, please wait.done.
[AC-wlan-vap-prof-fsm-vap-2]service-vlan vlan-id 101
Info: This operation may take a few seconds, please wait.done.
[AC-wlan-vap-prof-fsm-vap-2]dis this
#
  service-vlan vlan-id 101
  ssid-profile fsm-2.4G
  security-profile fsm-sec-1
#
return
[AC-wlan-vap-prof-fsm-vap-2]
```

11）配置无线 AP1 与 AP2

配置 AP1
进入 AP1
[AC-wlan-view]ap-name AP-shengchan-1
配置 AP1，引用 vap-1 模板，射频 1 对应的 5G 射频信号
[AC-wlan-ap-1]vap-profile fsm-vap-1 wlan 1 radio 1
配置 AP1，引用 vap-2，射频 0 对应的 2.4G 射频信号
[AC-wlan-ap-1]vap-profile fsm-vap-2 wlan 1 radio 0
查询无线 AP1 的配置情况
[AC-wlan-ap-1]dis this
#
  ap-name AP-shengchan-1
  ap-group shengchan-group
  radio 0
    vap-profile fsm-vap-2 wlan 1
  radio 1
    vap-profile fsm-vap-1 wlan 1
#
return
[AC-wlan-ap-1]
配置 AP2
[AC-wlan-view]ap-name AP-shenchan-2
[AC-wlan-ap-2]vap-profile fsm-vap-1 wlan 1 radio 1
[AC-wlan-ap-2]vap-profile fsm-vap-2 wlan 1 radio 0
[AC-wlan-ap-2]

配置完毕（图 2-22-10），发现 AP1-AP2 已经开始发射无线射频信号。

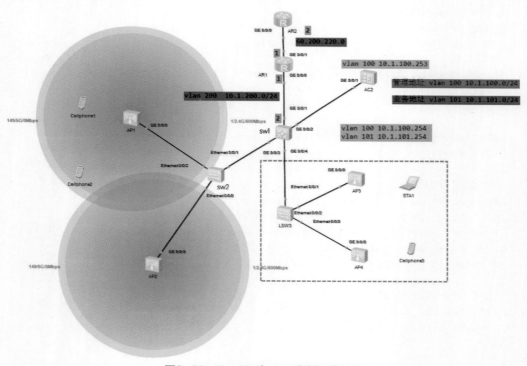

图2-22-10　AP1与AP2发射无线信号

配置 Cellphone1 与无线 AP 连接，不要忘了密码是 a1234567。AP1 连接 wifi-2.4G，见图 2-22-11 和图 2-22-12。

图 2-22-11　Cellphone1 接收到 wifi-2.4G　wifi-5G

图 2-22-12　AP1 连接 wifi-2.4G

测试 Cellphone1-ping-vlan 101 网关（10.1.101.254）- 通，见图 2-22-13。

269

```
Ping 10.1.101.254: 32 data bytes, Press Ctrl_C to break
From 10.1.101.254: bytes=32 seq=1 ttl=255 time=172 ms
From 10.1.101.254: bytes=32 seq=2 ttl=255 time=140 ms
From 10.1.101.254: bytes=32 seq=3 ttl=255 time=156 ms
From 10.1.101.254: bytes=32 seq=4 ttl=255 time=125 ms
```

图2-22-13　网络测试

AP1 连接 wifi-5G，见图 2-22-14。

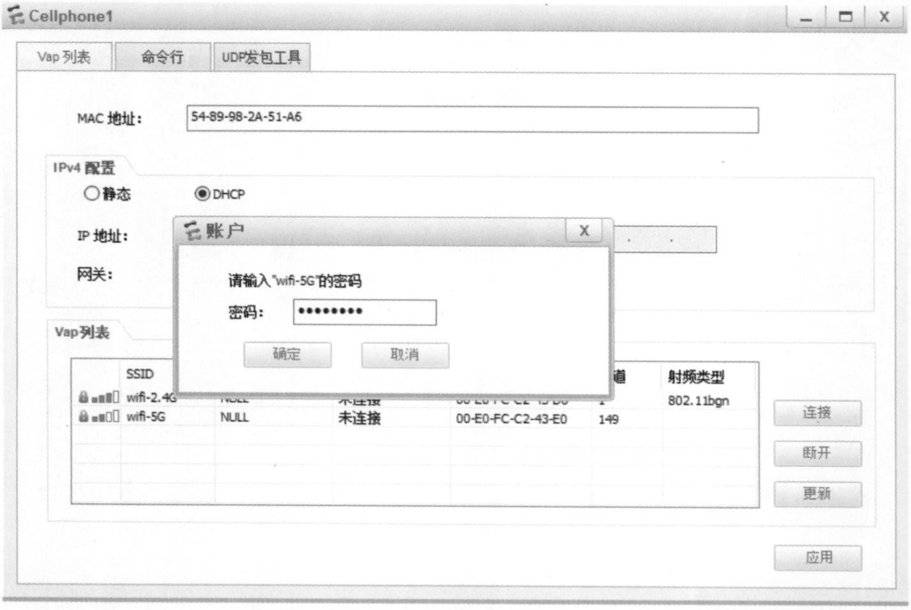

图2-22-14　AP1 连接wifi-5G

测试 Cellphone1-ping-vlan 101 网关（10.1.101.254）- 通，见图 2-22-15。

```
STA>ipconfig
Link local IPv6 address............: ::
IPv6 address.......................: :: / 128
IPv6 gateway.......................: ::
IPv4 address.......................: 10.1.101.250
Subnet mask........................: 255.255.255.0
Gateway............................: 10.1.101.254
Physical address...................: 54-89-98-2A-51-A6
DNS server.........................: 114.114.114.114
                                     8.8.8.8

STA>ping 10.1.10.1.254
host 10.1.10.1.254 unreachable

STA>ping 10.1.101.254
Ping 10.1.101.254: 32 data bytes, Press Ctrl_C to break
From 10.1.101.254: bytes=32 seq=1 ttl=255 time=141 ms
From 10.1.101.254: bytes=32 seq=2 ttl=255 time=140 ms
```

图2-22-15　网络测试

Cellphone2 连接网络，见图 2-22-16。

测试 Cellphone2-ping-vlan 101 网关（10.1.101.254）- 通，见图 2-22-17。

测试：Cellphone1-ping- Cellphone2- 通，见图 2-22-18。

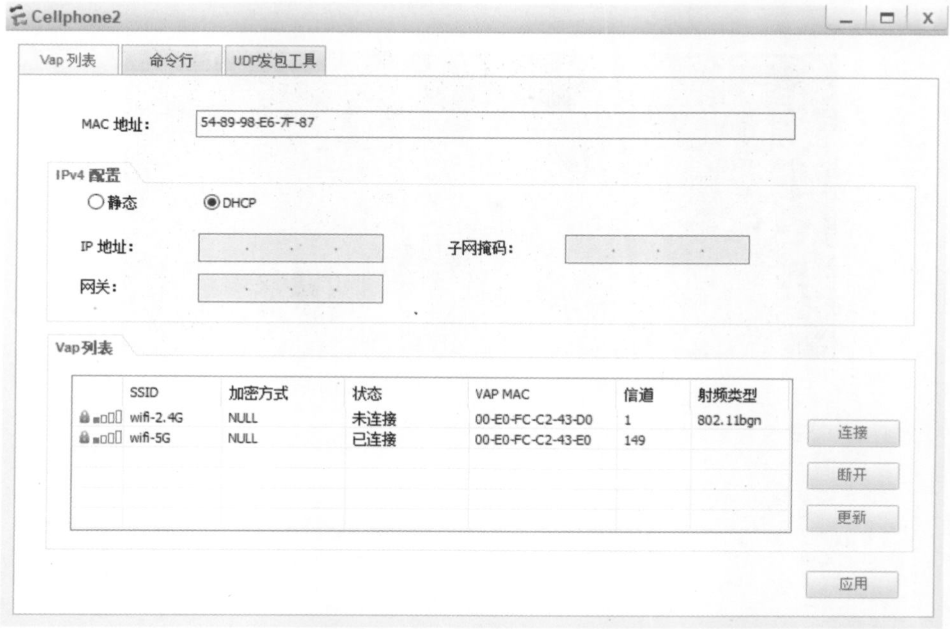

图2-22-16　Cellphone2 连接wifi-5G

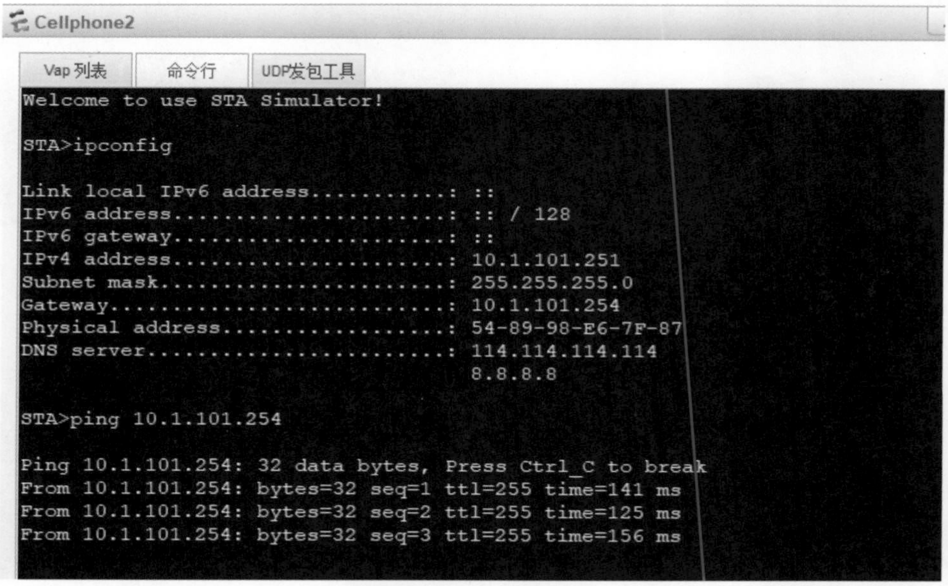

图2-22-17　网络测试

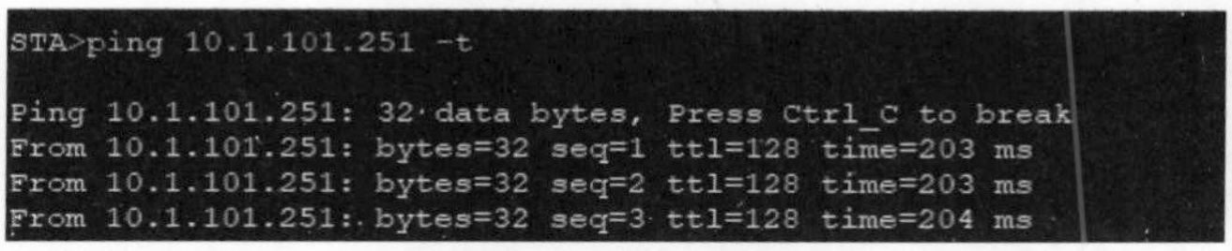

图2-22-18　网络测试

在 AP2 无线辐射范围内，增加工作站 STA2，并 ping-vlan 101 网关（10.1.101.254）测试 - 通，见图 2-22-19。

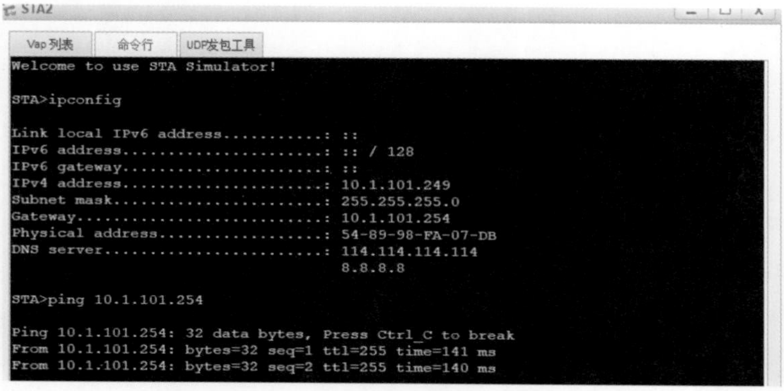

图2-22-19　自动获取IP地址并测试网络

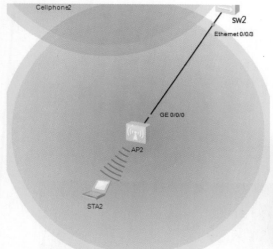

见图 2-22-20，增加的 STA2 工作站已经联通。

测试 STA2 从 AP2 迁移到 AP1，信号暂时中断，由于使用的是相同的 ssid 以及密码，达到无线漫游的目的。见图 2-22-21。

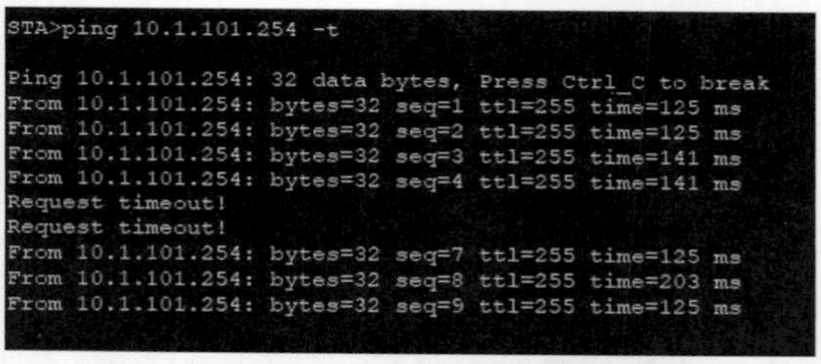

图2-22-20　STA2连接到AP2　　　　图2-22-21　STA2 无线漫游

三、管理人员的无线网络配置

配置情况与上述类似，注释不再体现。

1. 交换机 sw1 配置

增加 vlan 102 以及配置 vlan 102 网关，以及 port g 0/0/2，g 0/0/4 增加 vlan。

```
[sw1]vlan 102
[sw1-vlan102]quit
[sw1]interface Vlanif 102
[sw1-Vlanif102]ip address 10.1.102.254 24
[sw1-Vlanif102]quit
[sw1]interface GigabitEthernet 0/0/2
[sw1-GigabitEthernet0/0/2]port link-type trunk
[sw1-GigabitEthernet0/0/2]port trunk allow-pass vlan
[sw1-GigabitEthernet0/0/2]port trunk allow-pass vlan 102
[sw1-GigabitEthernet0/0/2]quit
[sw1]interface  GigabitEthernet 0/0/4
[sw1-GigabitEthernet0/0/4]port link-type trunk
[sw1-GigabitEthernet0/0/4]port trunk allow-pass vlan 100 101 102
```

sw1-配置 dhcp
[sw1]ip pool AP-lingdaoyewu
Info:It's successful to create an IP address pool.
[sw1-ip-pool-ap-lingdaoyewu]gateway-list 10.1.102.254
[sw1-ip-pool-ap-lingdaoyewu]network 10.1.102.0 mask 24
[sw1-ip-pool-ap-lingdaoyewu]dns-list 114.114.114.114
[sw1-ip-pool-ap-lingdaoyewu]dns-list 8.8.8.8
[sw1-ip-pool-ap-lingdaoyewu]excluded-ip-address 10.1.102.252 10.1.102.253
[sw1-ip-pool-ap-lingdaoyewu]quit
[sw1]interface Vlanif 102
[sw1-Vlanif102]dhcp select global
[sw1-Vlanif102]quit
[sw1]quit
<sw1>sa

2. AC 增加 vlan 102，并且配置 IP 地址

<AC>sys
Enter system view, return user view with Ctrl+Z.
[AC]vlan 102
Info: This operation may take a few seconds. Please wait for a moment...done.
[AC-vlan102]quit
[AC]interface GigabitEthernet 0/0/1
[AC-GigabitEthernet0/0/1]port link-type trunk
[AC-GigabitEthernet0/0/1]port trunk allow-pass vlan 102
[AC-GigabitEthernet0/0/1]quit
[AC]interface Vlanif 102
[AC-Vlanif102]ip address 10.1.102.253 24
[AC-Vlanif102]quit
[AC]quit
<AC>sa

3. sw3 配置

<Huawei>system-view
Enter system view, return user view with Ctrl+Z.
[Huawei]sys
[Huawei]sysname sw3
[sw3]vlan batch 100 102
Info: This operation may take a few seconds. Please wait for a moment...done.
[sw3]interface Ethernet 0/0/1
#
[sw3-Ethernet0/0/1]port link-type trunk
[sw3-Ethernet0/0/1]port trunk allow-pass vlan 100 102
[sw3-Ethernet0/0/1]quit
#
[sw3]interface Ethernet 0/0/2
[sw3-Ethernet0/0/2]port link-type trunk
[sw3-Ethernet0/0/2]port trunk allow-pass vlan 100 102
[sw3-Ethernet0/0/2]quit
#
[sw3]interface Ethernet 0/0/3
[sw3-Ethernet0/0/3]port link-type trunk
[sw3-Ethernet0/0/3]port trunk allow-pass vlan 100 102
[sw3-Ethernet0/0/3]

```
[sw3-Ethernet0/0/3]dis this
#
interface Ethernet0/0/3
port link-type trunk
port trunk allow-pass vlan 100 102
#
return
[sw3-Ethernet0/0/3]
#
```

修改 pvid

```
[sw3]interface Ethernet 0/0/2
[sw3-Ethernet0/0/2]port trunk pvid vlan 100
[sw3-Ethernet0/0/2]quit
[sw3]interface  Ethernet 0/0/3
[sw3-Ethernet0/0/3]port trunk pvid vlan 100
[sw3-Ethernet0/0/3]dis this
#
interface Ethernet0/0/3
port link-type trunk
port trunk pvid vlan 100
port trunk allow-pass vlan 100 102
#
return
[sw3-Ethernet0/0/3]
```

4. AP3 与 AP4 获取地址查询

AP3

```
<Huawei>dis ip interface brief
*down: administratively down
……省略……
Interface                        IP Address/Mask      Physical    Protocol
NULL0                            unassigned           up          up(s)
Vlanif1                          10.1.100.249/24      up          up
```

AP4

Ap4 获取地址查询

```
<Huawei>dis ip interface  brief
……省略……
Interface                        IP Address/Mask      Physical    Protocol
NULL0                            unassigned           up          up(s)
Vlanif1                          10.1.100.248/24      up          up
<Huawei>
```

测试 Ap3 与 AC 通信 - 通

```
<Huawei>ping 10.1.100.253
  PING 10.1.100.253: 56  data bytes, press CTRL_C to break
    Reply from 10.1.100.253: bytes=56 Sequence=1 ttl=255 time=200 ms
    Reply from 10.1.100.253: bytes=56 Sequence=2 ttl=255 time=50 ms
……省略……
--- 10.1.100.253 ping statistics ---
  5 packet(s) transmitted
  5 packet(s) received
  0.00% packet loss
  round-trip min/avg/max = 30/78/200 ms
```

```
AP4 与 AC 通信 - 通
<Huawei>ping   10.1.100.253
  PING 10.1.100.253: 56   data bytes, press CTRL_C to break
    Reply from 10.1.100.253: bytes=56 Sequence=1 ttl=255 time=150 ms
…………省略……
--- 10.1.100.253 ping statistics ---
  5 packet(s) transmitted
  5 packet(s) received
  0.00% packet loss
  round-trip min/avg/max = 50/80/150 ms
```

继续 AP 认证，图 2-22-22 是 AP3 和 AP4 的配置。

(a) AP3-MAC以及SN地址

(b) AP4-MAC以及SN地址

图2-22-22　AP3和AP4的配置

AP3 sn 认证配置
```
[AC-wlan-view]ap auth-mode sn-auth
[AC-wlan-view]ap-id 5 ap-sn 2102354483105B01BB75
[AC-wlan-ap-5]ap-name AP-lingdao-5
[AC-wlan-ap-5]quit
```

AP3 名字已经修改，见图 2-22-23。

```
<AP-lingdao-5>
```

图2-22-23　AP3 名字已经修改

AP4 sn 认证
```
[AC-wlan-view]ap auth-mode sn-auth
[AC-wlan-view]ap-id 3 ap-sn 21023544831049141C67
[AC-wlan-ap-3]ap-name AP-lingdao-1
[AC-wlan-ap-3]quit
```

AP4 名字已经修改，见图 2-22-24。

```
<AP-lingdao-1>
```

图2-22-24　AP4 名字已经修改

AC 查看上线信息：蓝色框标注的是原来的 AP1 和 AP2 的上线信息，AP3 与 AP4 用红色框标注。见图 2-22-25。

```
[AC]display ap all
Info: This operation may take a few seconds. Please wait for a moment.done.
Total AP information:
nor  : normal              [4]
--------------------------------------------------------------------------------
-----------------------------------
ID   MAC            Name           Group           IP            Type
State STA Uptime
-----------------------------------
1    00e0-fcc2-43d0 AP-shengchan-1 shengchan-group 10.1.100.251 AP3030DN
nor  2    5H:54M:36S
2    00e0-fc2e-2080 AP-shenchan-2  shengchan-group 10.1.100.250 AP3030DN
nor  1    5H:51M:50S
3    00e0-fc10-0890 AP-lingdao-1   default         10.1.100.248 AP3030DN
nor  0    1H:45M:20S
5    00e0-fc46-3c00 AP-lingdao-5   default         10.1.100.249 AP3030DN
nor  0    9M:24S
-----------------------------------
Total: 4
[AC]
```

图2-22-25　AP上线情况

AC-ping- AP3 - 通、AC-ping- AP4 - 通
```
[AC]ping 10.1.100.249
  PING 10.1.100.249: 56  data bytes, press CTRL_C to break
    Reply from 10.1.100.249: bytes=56 Sequence=1 ttl=255 time=70 ms
………省略……
--- 10.1.100.249 ping statistics ---
  5 packet(s) transmitted
  5 packet(s) received
  0.00% packet loss
```

```
    round-trip min/avg/max = 30/52/70 ms
[AC]ping 10.1.100.248
  PING 10.1.100.248: 56  data bytes, press CTRL_C to break
    Reply from 10.1.100.248: bytes=56 Sequence=1 ttl=255 time=50 ms
…………省略……
--- 10.1.100.248 ping statistics ---
  5 packet(s) transmitted
  5 packet(s) received
  0.00% packet loss
  round-trip min/avg/max = 40/54/70 ms
[AC]
```

查看 capwap，见图 2-22-26。

```
[AC]display capwap configuration
--------------------------------------------------------------
 Source interface                              : -
 Source ip-address                             : 10.1.100.253
 Echo interval(seconds)                        : 25
 Echo times                                    : 6
 Control priority(server to client)            : 7
 Control priority(client to server)            : 7
 Control-link DTLS encrypt                     : disable
 DTLS PSK value                                : ******
 PSK mandatroy match switch                    : disable
 Control-link inter-controller DTLS encrypt    : disable
 Inter-controller DTLS PSK value               : ******
 IPv6 status                                   : disable
 Message-integrity PSK value                   : ******
 Message-integrity check switch                : enable
--------------------------------------------------------------
[AC]
```

图 2-22-26　查看 capwap

5. 建立 AP 组

[AC]wlan
新建组名字为 lingdao-group
[AC-wlan-view]ap-group name lingdao-group
Info: This operation may take a few seconds. Please wait for a moment.done.
调用默认域 default
[AC-wlan-ap-group-lingdao-group]regulatory-domain-profile default
Warning: Modifying the country code will clear channel, power and antenna gain configurations of the radio and reset the AP. Continue?[Y/N]:y
[AC-wlan-ap-group-lingdao-group]

6. 将 AP3 与 AP4 加到领导组

[AC-wlan-view]ap-name AP-lingdao-5
[AC-wlan-ap-5]ap-group lingdao-group
Warning: This operation may cause AP reset. If the country code changes, it will clear channel, power and antenna gain configurations of the radio, Whether to continue? [Y/N]:y
Info: This operation may take a few seconds. Please wait for a moment..done.
[AC-wlan-ap-5]quit
[AC-wlan-view]ap-name AP-lingdao-1
[AC-wlan-ap-3]ap-group lingdao-group

```
Warning: This operation may cause AP reset. If the country code changes, it
will clear channel, power and antenna gain configurations of the radio, Whether
to continue? [Y/N]:y
    Info: This operation may take a few seconds. Please wait for a moment..
done.
    [AC-wlan-ap-3]quit
```

查询 AP 是否加入。如图 2-22-27 中所示，红色方框内显示已经加入 lingdao-group。

```
[AC]display ap all
Info: This operation may take a few seconds. Please wait for a moment.done.
Total AP information:
nor  : normal           [4]
---------------------------------------------------------------------------
---------------------------
ID   MAC            Name           Group           IP           Type
State STA Uptime
---------------------------------------------------------------------------
---------------------------
1    00e0-fcc2-43d0 AP-shengchan-1 shengchan-group 10.1.100.251 AP3030DN
nor  2   6H:7M:59S
2    00e0-fc2e-2080 AP-shenchan-2  shengchan-group 10.1.100.250 AP3030DN
nor  1   6H:5M:13S
3    00e0-fc10-0890 AP-lingdao-1   lingdao-group   10.1.100.248 AP3030DN
nor  0   12S
5    00e0-fc46-3c00 AP-lingdao-5   lingdao-group   10.1.100.249 AP3030DN
nor  0   57S
---------------------------------------------------------------------------
---------------------------
Total: 4
```

图2-22-27　AP3、AP4加入lingdao-group

7. 新建安全模板

```
# 新建安全模板 fsm-sec-2
[AC-wlan-view]security-profile name fsm-sec-2
# 采用 wpa2 加密方式，密码为 a123456
[AC-wlan-sec-prof-fsm-sec-2]security wpa2 psk pass-phrase a1234567 aes
```

8. 配置 ssid 模板

直接调用以上配置 fsm-2.4G 与 fsm-5G 的两个模板。

9. 创建 vap 模板

```
# 创建 fsm-vap-3
[AC-wlan-view]vap-profile name fsm-vap-3
[AC-wlan-vap-prof-fsm-vap-3]forward-mode direct-forward
[AC-wlan-vap-prof-fsm-vap-3]security-profile fsm-sec-2
Info: This operation may take a few seconds, please wait.done.
[AC-wlan-vap-prof-fsm-vap-3]ssid-profile fsm-5G
Info: This operation may take a few seconds, please wait.done.
# 配置领导终端使用的 vlan 102
[AC-wlan-vap-prof-fsm-vap-3]service-vlan vlan-id 102
Info: This operation may take a few seconds, please wait.done.
[AC-wlan-vap-prof-fsm-vap-3]
# 查询 fsm-vap-3 配置情况
[AC-wlan-vap-prof-fsm-vap-3]dis this
```

```
 #
   service-vlan vlan-id 102
   ssid-profile fsm-5G
   security-profile fsm-sec-2
 #
 return
[AC-wlan-vap-prof-fsm-vap-3]
```
创建 fsm-vap-4
```
[AC-wlan-view]vap-profile name fsm-vap-4
[AC-wlan-vap-prof-fsm-vap-4]security-profile fsm-sec-2
Info: This operation may take a few seconds, please wait.done.
[AC-wlan-vap-prof-fsm-vap-4]ssid-profile fsm-2.4G
Info: This operation may take a few seconds, please wait.done.
[AC-wlan-vap-prof-fsm-vap-4]service-vlan vlan-id 102
Info: This operation may take a few seconds, please wait.done.
```
查询 fsm-vap-4 配置信息
```
[AC-wlan-vap-prof-fsm-vap-4]dis this
 #
   service-vlan vlan-id 102
   ssid-profile fsm-2.4G
   security-profile fsm-sec-2
 #
 return
[AC-wlan-vap-prof-fsm-vap-4]
```
配置 AP3 参数
```
[AC-wlan-view]ap-name AP-lingdao-5
```
配置 AP3，引用 vap-3 射频 1 对应的 5G 射频信号
```
[AC-wlan-ap-5]vap-profile fsm-vap-3 wlan 1 radio 1
```
配置 AP3，引用 vap-4，射频 0 对应的 2.4G 射频信号
```
[AC-wlan-ap-5]vap-profile fsm-vap-4 wlan 1 radio 0
```
查询 AP3 的配置信息
```
[AC-wlan-ap-5]dis this
 #
   ap-name AP-lingdao-5
   ap-group lingdao-group
   radio 0
    vap-profile fsm-vap-4 wlan 1
   radio 1
    vap-profile fsm-vap-3 wlan 1
 #
 return
[AC-wlan-ap-5]
```
配置 AP4 参数
```
[AC-wlan-view]ap-name AP-lingdao-1
[AC-wlan-ap-3]vap-profile fsm-vap-3 wlan 1 radio 1
[AC-wlan-ap-3]vap-profile fsm-vap-4 wlan 1 radio 0
[AC-wlan-ap-3]quit
[AC-wlan-view]
```

AP1-AP4 已经发射信号拓扑图见图 2-22-28。

将 STA1 终端连接 AP3，连接 wifi-5G 信号，见图 2-22-29。

测试 STA1 与 vlan 102（IP:10.1.102.254）- 通，见图 2-22-30。

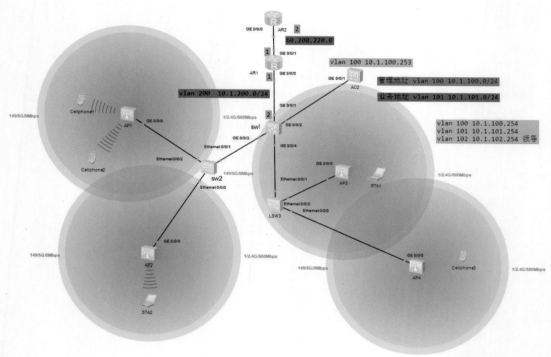

图2-22-28 拓扑图

图2-22-29 STA1终端连接AP3 – wifi-5G

将 Cellphone3 终端连接 AP4 设置，连接 wifi-5G 信号，见图 2-22-31。
测试 Cellphone3 与 vlan 102 (IP:10.1.102.254)- 通，见图 2-22-32。
STA1（IP:10.1.102.250）与 Cellphone3（IP:10.1.101.251）- 通，见图 2-22-33。

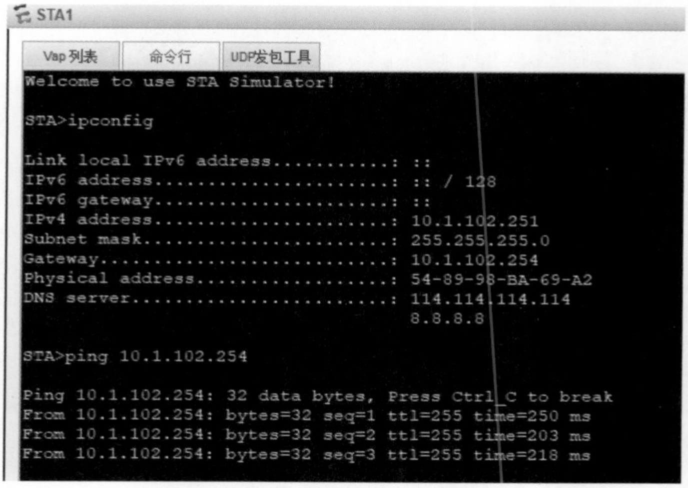

图2-22-30 测试网络

图2-22-31 Cellphone3 终端连接AP4 -wifi-5G

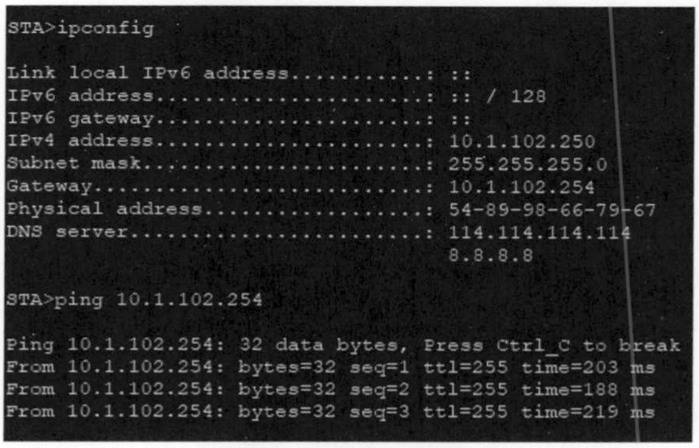

图2-22-32 网络测试

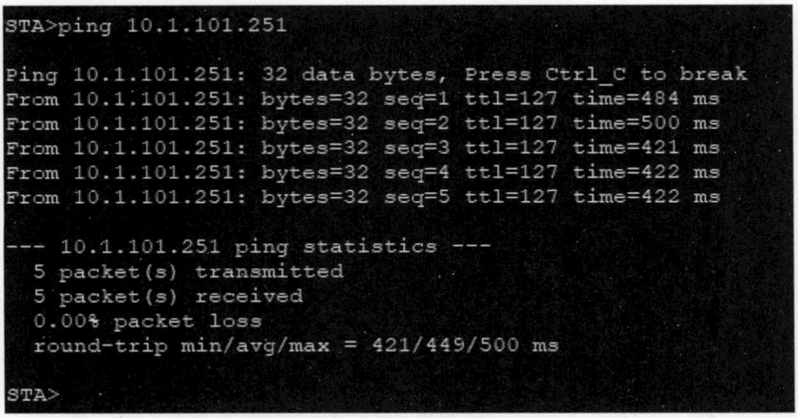

```
STA>ping 10.1.101.251

Ping 10.1.101.251: 32 data bytes, Press Ctrl_C to break
From 10.1.101.251: bytes=32 seq=1 ttl=127 time=484 ms
From 10.1.101.251: bytes=32 seq=2 ttl=127 time=500 ms
From 10.1.101.251: bytes=32 seq=3 ttl=127 time=421 ms
From 10.1.101.251: bytes=32 seq=4 ttl=127 time=422 ms
From 10.1.101.251: bytes=32 seq=5 ttl=127 time=422 ms

--- 10.1.101.251 ping statistics ---
  5 packet(s) transmitted
  5 packet(s) received
  0.00% packet loss
  round-trip min/avg/max = 421/449/500 ms

STA>
```

图2-22-33　网络测试

四、实现管理者的网段能访问互联网

要求：IP：1.1.1.1 模拟互联网地址，管理者的 10.1.102.0 可以上互联网，而 10.1.101.0/24 生产网段不能上互联网。

基本配置如下。

sw1

```
<sw1>system-view
Enter system view, return user view with Ctrl+Z.
# 创建vlan200
[sw1]vlan 200
[sw1-vlan200]quit
[sw1]interface Vlanif 200
[sw1-Vlanif200]ip address 10.1.200.2 24
[sw1]interface GigabitEthernet 0/0/1
[sw1-GigabitEthernet0/0/1]port link-type access
[sw1-GigabitEthernet0/0/1]port default vlan 200
[sw1-GigabitEthernet0/0/1]quit
```

AR1

```
[AR1]interface GigabitEthernet 0/0/0
[AR1-GigabitEthernet0/0/0]ip address 10.1.200.1 24
[AR1]interface  GigabitEthernet 0/0/1
[AR1-GigabitEthernet0/0/1]ip address 60.200.220.1 24
```

AR2

```
<Huawei>system-view
Enter system view, return user view with Ctrl+Z.
[Huawei]sysname AR2
[AR2]interface GigabitEthernet 0/0/0
[AR2-GigabitEthernet0/0/0]ip address 60.200.220.2 24
[AR2-GigabitEthernet0/0/0]quit
# 关闭信息中心
[AR2]un in en
Info: Information center is disabled.
[AR2]interface LoopBack 0
```

```
# 配置环回地址，模拟外网百度地址
[AR2-LoopBack0]ip address 1.1.1.1 32
[AR2-LoopBack0]quit
```
测试AR2-ping-1.1.1.1-通
```
[AR2]ping 1.1.1.1
  PING 1.1.1.1: 56  data bytes, press CTRL_C to break
    Reply from 1.1.1.1: bytes=56 Sequence=1 ttl=255 time=30 ms
……省略……
--- 1.1.1.1 ping statistics ---
  5 packet(s) transmitted
  5 packet(s) received
  0.00% packet loss
  round-trip min/avg/max = 1/6/30 ms
```

sw1 交换机
```
# 配置默认路由
[sw1]ip route-static 0.0.0.0 0 10.1.200.1
```

AR1 路由器配置
```
# 配置 NAT 地址转换
[AR1]acl 2000
[AR1-acl-basic-2000]rule 5 permit source 10.1.102.0 0.0.0.255
[AR1-acl-basic-2000]quit
[AR1]interface  GigabitEthernet 0/0/1
[AR1-GigabitEthernet0/0/1]nat outbound 2000
[AR1-GigabitEthernet0/0/1]quit
# 配置静态路由
[AR1]ip route-static 0.0.0.0 0 60.200.220.2
# 配置明细路由
[AR1]ip route-static 10.1.102.0 24 10.1.200.2
```

查询 AR1 路由表，见图 2-22-34。

```
[AR1]display ip routing-table
Route Flags: R - relay, D - download to fib
------------------------------------------------------------------------------
Routing Tables: Public
          Destinations : 12      Routes : 12

Destination/Mask      Proto   Pre  Cost      Flags  NextHop         Interface

        0.0.0.0/0    Static  60   0            RD   60.200.220.2    GigabitEthernet
0/0/1
     10.1.102.0/24   Static  60   0            RD   10.1.200.2      GigabitEthernet
0/0/0
     10.1.200.0/24   Direct  0    0            D    10.1.200.1      GigabitEthernet
0/0/0
     10.1.200.1/32   Direct  0    0            D    127.0.0.1       GigabitEthernet
0/0/0
     10.1.200.255/32 Direct  0    0            D    127.0.0.1       GigabitEthernet
0/0/0
     60.200.220.0/24 Direct  0    0            D    60.200.220.1    GigabitEthernet
0/0/1
     60.200.220.1/32 Direct  0    0            D    127.0.0.1       GigabitEthernet
0/0/1
     60.200.220.255/32 Direct 0   0            D    127.0.0.1       GigabitEthernet
0/0/1
      127.0.0.0/8    Direct  0    0            D    127.0.0.1       InLoopBack0
      127.0.0.1/32   Direct  0    0            D    127.0.0.1       InLoopBack0
127.255.255.255/32   Direct  0    0            D    127.0.0.1       InLoopBack0
255.255.255.255/32   Direct  0    0            D    127.0.0.1       InLoopBack0
```

图2-22-34　AR1路由表

测试 Cellphone3 与 AR1（IP：10.1.200.1）- 通，测试 Cellphone3 与 AR2（IP：60.200.220.2）- 通，测试 Cellphone3 与 IP：1.1.1.1- 通，完成领导无线网络（10.1.102.0/24），可以上外网，见图 2-22-35。

而 cellphone2 无法 ping 通 1.1.1.1（生产使用的终端无法连接外网），如图 2-22-36 所示。AR1 路由器 NAT 会话信息，见图 2-22-37。

图2-22-35　网络测试

图2-22-36　网络测试

图2-22-37　AR1路由器NAT会话信息

项目23　交换机配置文件自动备份

交换机配置文件备份的重要性不言而喻，大中型企业中交换机达到数百台，当交换机故障后，如果有原来的配置文件，对于处理故障能达到事半功倍的效果。尤其是汇聚层以及核心层交换机，配置文件比较多，不断探索交换机备份方式以及无人值守自动化，很有必要。

交换机备份配置，首先要确保交换机时间准确，那就需要配置 ntp 自动校时服务器，然后交换机配置定时采集，将配置文件复制并重新命名为新配置文件，发送到指定 tftp 服务器,同时需要将交换机上备份的新配置文件删除,否则随着时间的延长,交换机储存空间就满了,无法运行。同时在 Windows server 服务器上配置"任务计划程序"调动批处理文件,按天进行配置文件备份。

主要通过建立 tftp 服务器,将交换机配置文件上传至 tftp 服务器,在 Windows server 服务器上配置计划,执行脚本文件,最终实现交换机配置文件能够按天保存到指定文件夹中。

一、配置交换机自动校时功能

基本命令:
```
clock timezone beijing add 08:00:00   // 设置时区
clock protocol ntp    // 设置 ntp 校时
ntp-service enable    // 启动 ntp 校时
ntp-service unicast-server 192.168.99.124 // 配置校时服务器
```
配置实例:
```
// 设置时区
[Access]clock timezone beijng add 8
// 查询交换机配置关于 ntp 的命令,已经配置完毕
[Access]display current-configuration | include ntp
clock protocol ntp
ntp-service enable
ntp-service unicast-server 192.168.99.124
```

二、配置交换机自动备份

1. 设置 tftp 服务器

采用 3CDaemon 软件,实现 tftp 功能。设置步骤如下。

步骤1 点击"设置 tftp"服务器,弹出对话框,"普通设置"默认选择即可。见图 2-23-1。

步骤2 tftp 设置,"允许覆盖现有文件"打上对勾,配置交换机上传文件位置。见图 2-23-2。

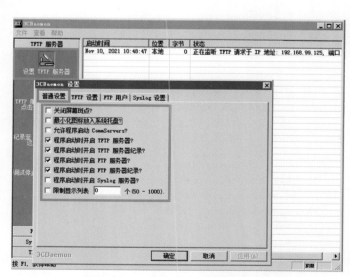

图2-23-1 点击"设置tftp"服务器

285

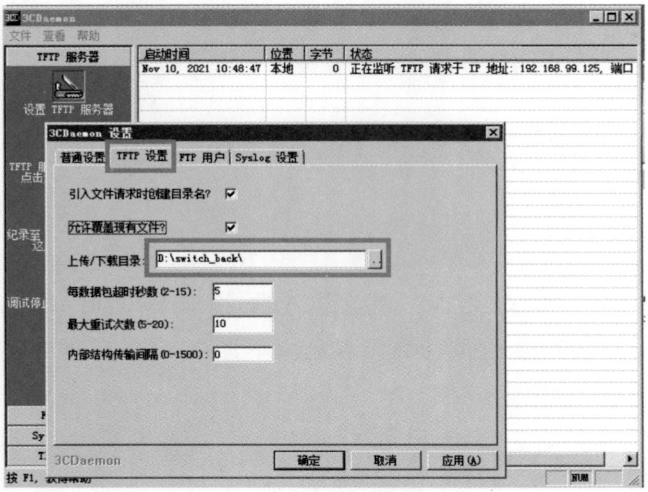

图2-23-2 "允许覆盖现有文件"打上对勾

2. 交换机自动上传备份文件命令步骤

步骤1 配置定时执行命令

```
scheduler schedule switch_back   // switch_back 为交换机备份工作名
user-role network-operator
user-role network-admin
job switch_back
time repeating at 00:00    // 执行时间备份配置时间
```

步骤2 执行动作

```
scheduler job switch_back   //执行工作内容
// copy flash 目录下 startup.cfg 文件,并重新命名(示例: 4_178.cfg)
command 1 copy flash:/startup.cfg flash:/4_178.cfg
// 询问时,自动选择"y"
command 2 y
// 将 4_178.cfg(示例)上传至 tftp 服务器 IP:192.168.99.125
command 3 tftp 192.168.99.125 put flash:/4_178.cfg
// 删除交换机 copy 的配置文件
command 4 delete flash:/4_178.cfg
// 询问时,自动选择"y"
command 5 y
```

步骤3 交换机配置实战

为了便于演示,时间定在21:30,复制步骤1与2的配置命令(不包含注释)在交换上执行。

```
[Access]scheduler schedule switch_back
[Access]user-role network-operator
[Access]user-role network-admin
[Access]job switch_back
[Access]time repeating at 21:50
[Access]
[Access]  scheduler job switch_back
[Access] command 1 copy flash:/startup.cfg flash:/4_178.cfg
```

```
[Access] command 2 y
[Access] command 3 tftp 192.168.99.125 put flash:/4_178.cfg
[Access] command 4 delete flash:/4_178.cfg
[Access] command 5 y
[Access]quit
[Access]
```

在server-192.168.99.125服务器上查看，交换机上传配置文件见图2-23-3。
文件夹：switch_back，文件名：4_178，时间：2021/11/10 21:50。

图2-23-3　查询tftp服务器上传的备份文件

通过以上操作，完全可以实现交换机配置文件实现备份，但是tftp启用了覆盖功能，只能得到前一天的配置文件。

三、按天备份配置

交换机配置文件实现按天备份到指定的文件夹，具体操作步骤如下。
步骤1　创建任务计划名称，见图2-23-4和图2-23-5。

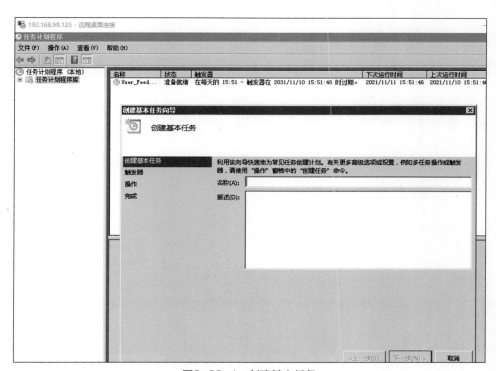

图2-23-4　创建基本任务

步骤2 触发器选择"每天"。见图 2-23-6。

步骤3 配置开始时间见图 2-23-7。

步骤4 操作-启动程序，选择执行脚本。见图 2-23-8 和图 2-23-9。

步骤5 双击计划名称 switch，见图 2-23-10。

常规选择"不管用户是否登录都要运行"，打上对勾。见图 2-23-11。

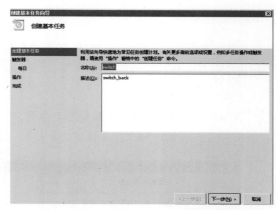

图2-23-5 创建几乎名称

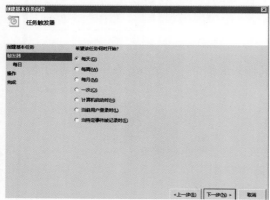

图2-23-6 触发器选择

图2-23-7 配置开始时间

图2-23-8 启动程序

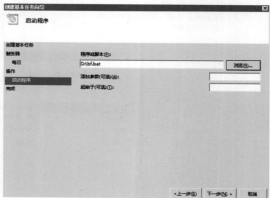

图2-23-9 执行脚本

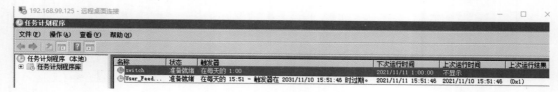

图2-23-10 双击计划名称switch

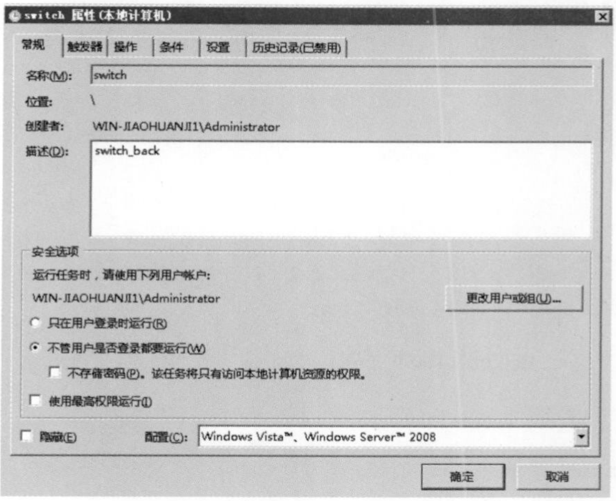

图2-23-11

附备份脚本：

```
@echo off
setlocal enabledelayedexpansion\
rem yyyymmdd
set yyyymmdd=%date:~0,10%
// 将 D 盘 switch_back 文件夹，按照日期备份到 D 盘 switch_all_back 文件夹中。
echo D|xcopy D:\switch_back D:\switch_all_back\%yyyymmdd%
```

Windows server 服务器中查看已经按天备份的交换机配置文件，见图2-23-12。

2021-10-30	2021-10-30 2:00	文件夹
2021-10-31	2021-10-31 2:00	文件夹
2021-11-01	2021-11-1 2:00	文件夹
2021-11-02	2021-11-2 2:00	文件夹
2021-11-03	2021-11-3 2:00	文件夹
2021-11-04	2021-11-4 2:00	文件夹
2021-11-05	2021-11-5 2:00	文件夹
2021-11-06	2021-11-6 2:00	文件夹
2021-11-07	2021-11-7 2:00	文件夹

图2-23-12　按天备份的交换机配置文件

项目 24　文件管理系统

目前网络设备中的操作系统大多数是 Linux 系统底层，掌握基本操作命令，便于文件操作。比如在交换机升级配置文件，需要从服务器上下载（get）新配置文件，当升级完毕，通过命令删除（delete）原来的配置文件。

1. 显示目录下的文件信息（命令：dir）

```
<Huawei>dir
Directory of flash:/
  IdxAttr      Size(Byte)  Date         Time(LMT)   FileName
    0  drw-            -  Sep 17 2021  08:57:12    dhcp
    1  -rw-      121,802  May 26 2014  09:20:58    portalpage.zip
    2  -rw-        2,263  Sep 17 2021  08:57:08    statemach.efs
    3  -rw-      828,482  May 26 2014  09:20:58    sslvpn.zip
1,090,732 KB total (784,464 KB free)
```

以上查询的信息，存在设备的 flash 中。

2. 新建文件目录（命令：mkdir）

```
<Huawei>mkdir fsm
Info: Create directory flash:/fsm......Done
<Huawei>dir
Directory of flash:/
  IdxAttr      Size(Byte)  Date         Time(LMT)   FileName
    0  drw-            -  Sep 17 2021  08:57:12    dhcp
    1  -rw-      121,802  May 26 2014  09:20:58    portalpage.zip
    2  -rw-        2,263  Sep 17 2021  08:57:08    statemach.efs
    3  -rw-      828,482  May 26 2014  09:20:58    sslvpn.zip
    4  drw-            -  Sep 17 2021  09:02:33    fsm
1,090,732 KB total (784,460 KB free)
```

以上字母简要信息，"d"代表目录，"r"代表允许可读权限，"w"代表允许可写权限。

```
# 进入新建的目录 fsm
<Huawei>cd fsm
# 查看当前目录
<Huawei>pwd
flash:/fsm
```

3. 删除目录（命令：rmdir）

```
<Huawei>rmdir fsm
Remove directory flash:/fsm? (y/n)[n]:y
%Removing directory flash:/fsm...Done!
# 进入 dhcp 目录
<Huawei>cd dhcp
<Huawei>dir
Directory of flash:/dhcp/
  IdxAttr      Size(Byte)  Date         Time(LMT)   FileName
    0  -rw-           98  Sep 17 2021  08:57:12    dhcp-duid.txt
1,090,732 KB total (784,464 KB free)
```

4. 查看文件命令（命令：more）

```
<Huawei>more dhcp-duid.txt
*Huawei DHCP DUID*
*time* 2021-09-17 16:57:12
*version* 1
#DUID_LL: 0003000100E0FCDA522D
*end*
```

5. 复制文件（命令：copy）

```
<Huawei>cd dhcp
<Huawei>pwd
flash:/dhcp
<Huawei>copy dhcp-duid.txt fsm.txt
Copy flash:/dhcp/dhcp-duid.txt to flash:/dhcp/fsm.txt? (y/n)[n]:y
100%    complete
Info: Copied file flash:/dhcp/dhcp-duid.txt to flash:/dhcp/fsm.txt...Done
<Huawei>pwd
flash:/dhcp
<Huawei>dir
Directory of flash:/dhcp/
  Idx   Attr     Size(Byte)    Date        Time(LMT)   FileName
   0    -rw-             98    Sep 18 2021 00:31:01    dhcp-duid.txt
   1    -rw-             98    Sep 18 2021 00:35:15    fsm.txt
1,090,732 KB total (784,452 KB free)
<Huawei>more fsm.txt
*Huawei DHCP DUID*
*time* 2021-09-18 08:31:01
*version* 1
#DUID_LL: 0003000100E0FC303BA5
*end*
```

以上是复制到同一目录中时，如何将 dhcp 目录中的 dhcp-duid.txt 文件复制到 fsm 目录中。将 dhcp 目录中的 dhcp-duid.txt 文件复制到 flash:/fsm/fsm.txt（目录是 fsm，并且文件名是 fsm.txt）。

```
<Huawei>copy dhcp-duid.txt flash:/fsm/fsm.txt
Copy flash:/dhcp/dhcp-duid.txt to flash:/fsm/fsm.txt? (y/n)[n]:y
100%    complete
Info: Copied file flash:/dhcp/dhcp-duid.txt to flash:/fsm/fsm.txt...Done
# 返回上级目录（命令：cd ..）
<Huawei>cd ..
# 查看目录
<Huawei>pwd
flash:
<Huawei>cd fsm
<Huawei>pwd
flash:/fsm
<Huawei>dir
Directory of flash:/fsm/
  Idx   Attr     Size(Byte)    Date        Time(LMT)   FileName
   0    -rw-             98    Sep 18 2021 00:57:30    fsm.txt
1,090,732 KB total (784,448 KB free)
```

6. 删除文件（命令：delete）

当删除目录 fsm 的时候，提示错误，是因为这个目录中还有文件，不是空目录。经过查看存在 fsm.txt 文件，需要将 fsm.txt 文件删除，方可删除 fsm 目录。

```
<Huawei>rmdir fsm
Remove directory flash:/fsm? (y/n)[n]:y
%Removing directory flash:/fsm...
Error: Directory is not empty
```

```
<Huawei>cd fsm
<Huawei>pwd
flash:/fsm
<Huawei>dir
Directory of flash:/fsm/
  Idx   Attr    Size(Byte)    Date        Time(LMT)    FileName
    0   -rw-            98    Sep 18 2021 00:57:30    fsm.txt
1,090,732 KB total (784,452 KB free)
```
删除 fsm.txt 文件
```
<Huawei>delete fsm.txt
Delete flash:/fsm/fsm.txt? (y/n)[n]:y
Info: Deleting file flash:/fsm/fsm.txt...succeed.
```
退回根目录（flash）
```
<Huawei>cd ..
<Huawei>pwd
flash:
```
删除目录
```
<Huawei>rmdir fsm
Remove directory flash:/fsm? (y/n)[n]:y
%Removing directory flash:/fsm...Done!
```
查看确认
```
<Huawei>dir
Directory of flash:/
  Idx   Attr    Size(Byte)    Date        Time(LMT)    FileName
    0   drw-             -    Sep 18 2021 00:35:15    dhcp
    1   -rw-       121,802    May 26 2014 09:20:58    portalpage.zip
    2   -rw-         2,263    Sep 18 2021 00:30:56    statemach.efs
    3   -rw-       828,482    May 26 2014 09:20:58    sslvpn.zip
1,090,732 KB total (784,460 KB free)
```

7. 文件移动（命令：move）

将 dhcp 目录中文件 fsm.txt 移到根目录。

查看 dhcp 目录中文件
```
<Huawei>dir
Directory of flash:/dhcp/
  Idx   Attr    Size(Byte)    Date        Time(LMT)    FileName
    0   -rw-            98    Sep 18 2021 00:31:01    dhcp-duid.txt
    1   -rw-            98    Sep 18 2021 00:35:15    fsm.txt
1,090,732 KB total (784,460 KB free)
<Huawei>pwd
flash:/dhcp
```
#dhcp 目录中文件 fsm.txt 移动到根目录
```
<Huawei>move fsm.txt flash:/
Move flash:/dhcp/fsm.txt to flash:/fsm.txt? (y/n)[n]:y
%Moved file flash:/dhcp/fsm.txt to flash:/fsm.txt.
```
返回到根目录，查看文件
```
<Huawei>cd ..
<Huawei>dir
Directory of flash:/
  Idx   Attr    Size(Byte)    Date        Time(LMT)    FileName
    0   drw-             -    Sep 18 2021 01:11:51    dhcp
```

```
    1    -rw-         121,802   May 26 2014 09:20:58   portalpage.zip
    2    -rw-              98   Sep 18 2021 00:35:15   fsm.txt
    3    -rw-           2,263   Sep 18 2021 00:30:56   statemach.efs
    4    -rw-         828,482   May 26 2014 09:20:58   sslvpn.zip
1,090,732 KB total (784,460 KB free)
```
查看 dhcp 目录中 fsm.txt 文件已经不在了。
```
<Huawei>dir
Directory of flash:/dhcp/
  Idx  Attr    Size(Byte)    Date       Time(LMT)   FileName
    0    -rw-              98   Sep 18 2021 00:31:01   dhcp-duid.txt
1,090,732 KB total (784,460 KB free)
Rename   重命名
Reset recycle-bin   fsm.txt   删除回收站中的 fsm 文件
```

8. 文件重新命名（命令：remane）

将根目录下 fsm1.zip 重新命名为 fsm2.zip。

查看文件
```
<Huawei>dir
Directory of flash:/
  Idx  Attr    Size(Byte)    Date       Time(LMT)   FileName
    0    drw-               -   Sep 18 2021 03:56:15   dhcp
    1    -rw-         121,802   May 26 2014 09:20:58   portalpage.zip
    2    -rw-           2,263   Sep 18 2021 03:50:16   statemach.efs
    3    -rw-         828,482   May 26 2014 09:20:58   sslvpn.zip
    4    -rw-             249   Sep 18 2021 03:51:04   private-data.txt
    5    drw-               -   Sep 18 2021 03:54:51   fsm
    6    -rw-             516   Sep 18 2021 03:54:06   fsm1.zip
    7    -rw-             520   Sep 18 2021 03:53:39   vrpcfg.zip
1,090,732 KB total (784,444 KB free)
```
重新命名
```
<Huawei>rename fsm1.zip  fsm2.zip
Rename flash:/fsm1.zip to flash:/fsm2.zip? (y/n)[n]:y
Info: Rename file flash:/fsm1.zip to flash:/fsm2.zip ......Done
```
再次查看文件，fsm1.zip 已经修改位 fsm2.zip.
```
<Huawei>dir
Directory of flash:/
  Idx  Attr    Size(Byte)    Date       Time(LMT)   FileName
    0    drw-               -   Sep 18 2021 03:56:15   dhcp
    1    -rw-         121,802   May 26 2014 09:20:58   portalpage.zip
    2    -rw-           2,263   Sep 18 2021 03:50:16   statemach.efs
    3    -rw-         828,482   May 26 2014 09:20:58   sslvpn.zip
    4    -rw-             249   Sep 18 2021 03:51:04   private-data.txt
    5    drw-               -   Sep 18 2021 03:54:51   fsm
    6    -rw-             516   Sep 18 2021 03:54:06   fsm2.zip
    7    -rw-             520   Sep 18 2021 03:53:39   vrpcfg.zip
1,090,732 KB total (784,444 KB free)
```

9. 如何彻底清除回收站的文件

查询文件，并且将文件 fsm2.zip 删除。

```
<sw1>dir
```

```
Directory of flash:/
  Idx  Attr   Size(Byte)   Date         Time(LMT)   FileName
   0   drw-            -   Sep 18 2021  03:56:15    dhcp
   1   -rw-      121,802   May 26 2014  09:20:58    portalpage.zip
   2   -rw-        2,263   Sep 18 2021  03:50:16    statemach.efs
   3   -rw-      828,482   May 26 2014  09:20:58    sslvpn.zip
   4   -rw-          249   Sep 18 2021  03:51:04    private-data.txt
   5   drw-            -   Sep 18 2021  03:54:51    fsm
   6   -rw-          516   Sep 18 2021  03:54:06    fsm2.zip
   7   -rw-          532   Sep 18 2021  05:33:33    vrpcfg.zip
1,090,732 KB total (784,444 KB free)
```

删除文件

```
<sw1>delete fsm2.zip
Delete flash:/fsm2.zip? (y/n)[n]:y
Info: Deleting file flash:/fsm2.zip...succeed.
```

再次查看文件,fsm2.zip 文件已经不存在。

```
<sw1>dir
Directory of flash:/
  Idx  Attr   Size(Byte)   Date         Time(LMT)   FileName
   0   drw-            -   Sep 18 2021  03:56:15    dhcp
   1   -rw-      121,802   May 26 2014  09:20:58    portalpage.zip
   2   -rw-        2,263   Sep 18 2021  03:50:16    statemach.efs
   3   -rw-      828,482   May 26 2014  09:20:58    sslvpn.zip
   4   -rw-          249   Sep 18 2021  03:51:04    private-data.txt
   5   drw-            -   Sep 18 2021  03:54:51    fsm
   6   -rw-          532   Sep 18 2021  05:33:33    vrpcfg.zip
1,090,732 KB total (784,440 KB free)
```

通过命令(dir /all),查询全部文件,包含回收站的文件,回收站文件在 [ ] 内。

```
<sw1>dir /all
Directory of flash:/
  Idx  Attr   Size(Byte)   Date         Time(LMT)   FileName
   0   drw-            -   Sep 18 2021  03:56:15    dhcp
   1   -rw-      121,802   May 26 2014  09:20:58    portalpage.zip
   2   -rw-        2,263   Sep 18 2021  03:50:16    statemach.efs
   3   -rw-      828,482   May 26 2014  09:20:58    sslvpn.zip
   4   -rw-          249   Sep 18 2021  03:51:04    private-data.txt
   5   drw-            -   Sep 18 2021  03:54:51    fsm
   6   -rw-          532   Sep 18 2021  05:33:33    vrpcfg.zip
   7   -rw-          516   Sep 18 2021  05:37:12    [fsm2.zip]
1,090,732 KB total (784,440 KB free)
```

通过命令 reset recycle-bin fsm2.zip 将 fsm2.zip 文件从回收站中删除。

```
<sw1>reset recycle-bin fsm2.zip
Squeeze flash:/fsm2.zip? (y/n)[n]:y
Clear file from flash will take a long time if needed...Done.
%Cleared file flash:/fsm2.zip.
```

再次查看,已经看不到 fsm2.zip 文件。

```
<sw1>dir /all
Directory of flash:/
  Idx  Attr   Size(Byte)   Date         Time(LMT)   FileName
```

```
0       drw-          -         Sep 18 2021 03:56:15    dhcp
1       -rw-    121,802         May 26 2014 09:20:58    portalpage.zip
2       -rw-      2,263         Sep 18 2021 03:50:16    statemach.efs
3       -rw-    828,482         May 26 2014 09:20:58    sslvpn.zip
4       -rw-        249         Sep 18 2021 03:51:04    private-data.txt
5       drw-          -         Sep 18 2021 03:54:51    fsm
6       -rw-        532         Sep 18 2021 05:33:33    vrpcfg.zip
1,090,732 KB total (784,444 KB free)
```

10. 查看没有保存配置文件的命令

```
# 首先修改交换机名字
[sw1]sysname fsmsw
# 查询没有保存的配置文件
[fsmsw]display current-configuration
[V200R003C00]
#
sysname fsmsw
#
snmp-agent local-engineid 800007DB03000000000000
snmp-agent
#
clock timezone China-Standard-Time minus 08:00:00
……省略……
# 查询保存的配置文件,重点查看保存的文件,交换机的名字还是原来的 sw1,因为上面进行交换机名字修改后,没有进行保存 save,文件还保存在内存中。
[fsmsw]dis saved-configuration
[V200R003C00]
#
sysname sw1
#
snmp-agent local-engineid 800007DB03000000000000
snmp-agent
……省略……
```

项目 25　中大型企业网络规划

　　某大型企业新厂区规划网络,接到网络项目需求后,前期必须到现场了解网络规模,包括生产以及办公终端数量,安防、门禁、考勤等安装位置以及数量。根据前期调研结果,初步在模拟器上绘制拓扑图,配置相关命令,进行网络通信测试,见图 2-25-1。

　　为了确保网络稳定与畅通,还需要进行设备以及链路冗余,结合前期调研,分为三大部分实施:①规划 3 个汇聚点(负责办公以及生产网络),汇聚层采用两台设备堆叠(IRF)配置,与核心之间采用 4 条链路聚合;②安防、考勤、门禁等设备(汇聚层)无冗余备份;③数据中心网络设备与链路采用冗余设计。见图 2-25-2。

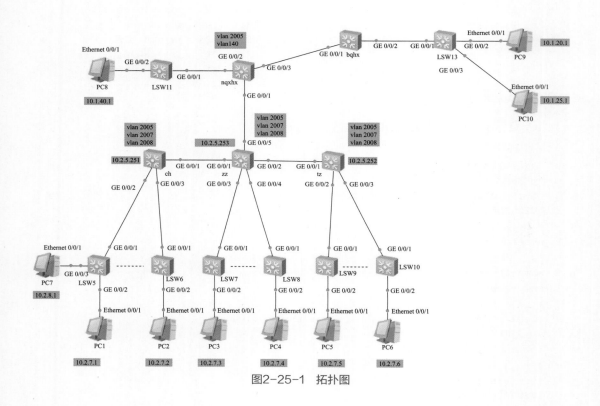

图2-25-1 拓扑图

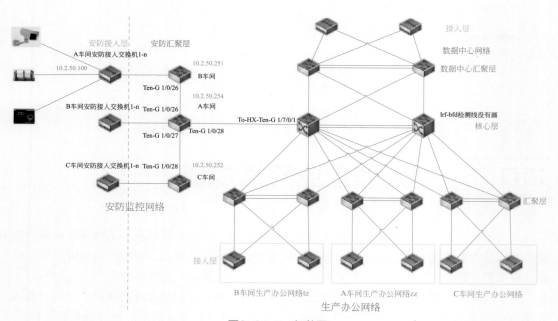

图2-25-2 拓扑图

一、网络设备选型

根据生产规模、办公点位，以及门禁、考勤等终端数量，确定交换机选型表 2-25-1。

表 2-25-1　设备选型

核心	H3C S10508X
汇聚（生产与办公）	H3C S5560X-30F-EI
汇聚（安防）	H3C S5560X-30F-EI
汇聚（数据中心）	H3C S6520X-54QC-EI
接入（生产）	H3C S5130S-52TP-EI
接入（安防）	H3C S5130S-28P-PWR-EI

二、堆叠与聚合接口规划

1. 核心交换机堆叠接口，如表 2-25-2 所示。

表 2-25-2　核心交换机堆叠接口

Master	Slave
Ten-GigabitEthernet1/5/0/36	Ten-GigabitEthernet2/5/0/36
Ten-GigabitEthernet1/7/0/36	Ten-GigabitEthernet2/7/0/36

核心交换机 IRF 堆叠 MAD-BFD 检测 port 接口，如表 2-25-3。

表 2-25-3　MAD-BFD 检查的接口

interface GigabitEthernet1/3/0/48	interface GigabitEthernet2/3/0/48	vlan 4094

2. A 车间汇聚交换机聚合与堆叠接口

A 车间汇聚交换机聚合接口与核心交换机相连的聚合接口，如表 2-25-4。

表 2-25-4　A 车间聚合接口与核心相连聚合接口

核心聚合组	A 车间聚合组
Ten-G 1/7/0/45	Ten-GigabitEthernet1/0/25
Ten-G 1/7/0/48	Ten-GigabitEthernet2/0/25
Ten-G 2/7/0/45	Ten-GigabitEthernet1/0/27
Ten-G 2/7/0/48	Ten-GigabitEthernet2/0/27

A 车间堆叠接口，如表 2-25-5。

表 2-25-5　A 车间堆叠接口

Master	Slave
Ten-GigabitEthernet1/0/26	Ten-GigabitEthernet2/0/26
Ten-GigabitEthernet1/0/28	Ten-GigabitEthernet2/0/28

3. B 车间汇聚交换机聚合与堆叠接口

B 车间汇聚交换机聚合接口与核心交换机相连的聚合接口如表 2-25-6 所示。

表 2-25-6　B 车间汇聚交换机聚合接口与核心交换机相连的聚合接口

核心聚合组	B 车间
Ten-G 1/7/0/44	Ten-GigabitEthernet1/0/25
Ten-G 1/7/0/47	Ten-GigabitEthernet2/0/25
Ten-G 2/7/0/44	Ten-GigabitEthernet1/0/27
Ten-G 2/7/0/47	Ten-GigabitEthernet2/0/27

B 车间堆叠接口如表 2-25-7 所示。

表 2-25-7　B 车间堆叠接口

Master	Slave
Ten-GigabitEthernet1/0/26	Ten-GigabitEthernet2/0/26
Ten-GigabitEthernet1/0/28	Ten-GigabitEthernet2/0/28

4. C 车间汇聚交换机聚合与堆叠接口

C 车间汇聚交换机聚合接口与核心交换机相连的聚合接口如表 2-25-8 所示。

表 2-25-8　C 车间汇聚交换机聚合接口与核心交换机相连的聚合接口

核心聚合组	C 车间
Ten-G 1/7/0/43	Ten-GigabitEthernet1/0/25
Ten-G 1/7/0/46	Ten-GigabitEthernet2/0/25
Ten-G 2/7/0/43	Ten-GigabitEthernet1/0/27
Ten-G 2/7/0/46	Ten-GigabitEthernet2/0/27

C 车间堆叠接口如表 2-25-9 所示。

表 2-25-9　C 车间堆叠接口

Master	Slave
Ten-GigabitEthernet1/0/26	Ten-GigabitEthernet2/0/26
Ten-GigabitEthernet1/0/28	Ten-GigabitEthernet2/0/28

5. 数据中心汇聚交换机聚合与堆叠接口。

数据中心交换机聚合接口与核心交换机相连的聚合接口如表 2-25-10 所示。

表 2-25-10　数据中心汇聚交换机聚合与堆叠接口

核心聚合组	server
Ten-G 1/7/0/41	Ten-GigabitEthernet1/0/46
Ten-G 1/7/0/42	Ten-GigabitEthernet2/0/45
Ten-G 2/7/0/41	Ten-GigabitEthernet2/0/46
Ten-G 2/7/0/42	Ten-GigabitEthernet1/0/45

数据中心汇聚交换机堆叠接口如表 2-25-11 所示。

表 2-25-11　数据中心汇聚交换机堆叠接口

Master	Slave
Ten-GigabitEthernet1/0/47	Ten-GigabitEthernet2/0/47
Ten-GigabitEthernet1/0/48	Ten-GigabitEthernet2/0/48

6. 安防汇聚交换机与核心交换机连接接口

安防汇聚交换机与核心交换机连接接口如表 2-25-12 所示。

表 2-25-12　安防汇聚交换机与核心交换机连接接口

核心	安防
Ten-G 1/7/0/1	Ten-GigabitEthernet1/0/28

网络配置方面，各个汇聚与核心交换机之间采用 ospf 动态协议，办公与生产使用同一汇聚划分不同 vlan 来区分使用网络环境。接入交换机管理地址网关配置在汇聚层，业务层网络的网关配置在汇聚层。

三、网段划分

1. 核心交换机与汇聚交换机（采用 ospf 动态路由协议）互联地址

如表 2-25-13 所示。

表 2-25-13　互联地址

核心	汇聚
10.2.201.1	A 车间汇聚 10.2.201.2
10.2.202.1	B 车间汇聚 10.2.202.2
10.2.203.1	C 车间汇聚 10.2.203.2
10.2.204.1	数据中心汇聚 10.2.204.2
10.2.205.1	监控汇聚 10.2.205.2

2. 安防、门禁、考勤业务网段、vlan 规划

如表 2-25-14 所示。

表 2-25-14　安防、门禁、考勤业务网段、vlan 规划

业务划分	vlan	网段
交换机管理	2050	10.2.50.0/24
考勤机	2060	10.2.60.0/24
监控	2070	10.2.70.0/24
门禁	2080	10.2.80.0/24
监控	2090	10.2.90.0/24

3. A 车间网段规划

如表 2-25-15 所示。

表 2-25-15　A 车间网段规划

业务划分	vlan	网段
交换机管理	2001	10.2.1.0/24
办公	2011	10.2.11.0/24
生产	2015	10.2.15.0/24

4. B 车间网段规划

如表 2-25-16 所示。

表 2-25-16　B 车间网段规划

业务划分	vlan	网段
交换机管理	2002	10.2.2.0/24
办公	2021	10.2.21.0/24
生产	2025	10.2.25.0/24

5. C 车间网段规划

如表 2-25-17 所示。

表 2-25-17　C 车间网段规划

业务划分	vlan	网段
交换机管理	2003	10.2.3.0/24
办公	2031	10.2.31.0/24
生产	2035	10.2.35.0/24

6. 数据中心网段规划

如表 2-25-18 所示。

表 2-25-18　数据中心网段规划

业务划分	vlan	网段
交换机管理	2050	10.2.50.0/24
应用服务器	2051	10.2.51.0/24
数据服务器	2052	10.2.52.0/24
应用服务器	2053	10.2.53.0/24

四、核心交换机配置

1. 核心交换机

见图 2-25-3，型号：H3C S10508X。

2. 主要配置

```
<HX>dis cur
#
version 7.1.070, Release 7585P05
#
mdc Admin id 1    // 默认配置。
#
sysname HX    // 命名
#
forward-path-detection enable    // 用来开启转发通道自动检测功能。默认开启。
#
 irf domain 10    // 配置堆叠域 ID，在堆叠线检测时使用。
irf mac-address persistent always    //用来配置 irf 的桥 MAC 地址永久保留不改变
 irf auto-update enable    // 开启 irf 系统启动文件的自动加载功能，默认配置。
 irf auto-merge enable    // 开启 irf 合并自动重启功能，缺省配置
 undo irf link-delay    // 配置 irf 链路 down 延迟上报时间为 0
 irf member 1 priority 10    // 配置A设备 irf 优先级是 10，member 编号默认 1。
 irf member 2 priority 1    // B设备 irf 默认优先级是 1，member 编号 2。
 irf member 1 description   master    //标识 member 1 是主设备
 irf member 2 description   Slave    //标识 member 2 是从设备
 irf mode normal    //堆叠模式为普通
#
ospf 2 router-id 10.2.202.1 //配置 router-id
 default-route-advertise //将缺省路由通告到普通 ospf 区域（默认路由下发）
 import-route static    // 引入静态路由
 bandwidth-reference 10000
 area 0.0.0.0    //进入区域 0
  description B workshop    //标识车间
  network 10.2.202.0 0.0.0.3    // 宣告网段（可使用 IP 地址 10.2.202.1-10.2.202.2）
 area 0.0.0.1    //进入区域 1
  description A workshop    //标识车间
  network 10.2.201.0 0.0.0.3
 area 0.0.0.3    //进入区域 3
  description C workshop    //标识车间
  network 10.2.203.0 0.0.0.3
 area 0.0.0.4    //进入区域 4
  description SERVER
  network 10.2.204.0 0.0.0.3
 area 0.0.0.5    //进入区域 5
  description JK
  network 10.2.205.0 0.0.0.3
#
ip unreachables enable    // 用来开启设备的 ICMP 目的不可达报文的发送功能
ip ttl-expires enable    // 用来开启设备的 ICMP 超时报文的发送功能
#
dhcp enable    //dhcp 使能
#
lldp global enable    // 开启交换机邻居关系
#
system-working-mode standard    //系统工作模式
password-recovery enable    //配置密码恢复功能使能
```

图2-25-3 核心交换机

```
#
vlan 1
#
vlan 4094
#
irf-port 1/2      //将堆叠接口加入逻辑堆叠组
port group interface Ten-GigabitEthernet1/5/0/36 mode extended
port group interface Ten-GigabitEthernet1/7/0/36 mode extended
#
irf-port 2/1      //将堆叠接口加入逻辑堆叠组
port group interface Ten-GigabitEthernet2/5/0/36 mode extended
port group interface Ten-GigabitEthernet2/7/0/36 mode extended
#
interface Route-Aggregation11     //聚合组11配置（A车间）
description A workshop   //标识A车间
ip address 10.2.201.1 255.255.255.252 // 配置IP地址
#
interface Route-Aggregation12    //聚合组12配置（B车间）
description B workshop //标识B车间
ip address 10.2.202.1 255.255.255.252 // 配置IP地址
#
interface Route-Aggregation13    //聚合组13配置
description C workshop //标识C车间
ip address 10.2.203.1 255.255.255.252 // 配置IP地址
#
interface Route-Aggregation14    //聚合组14配置
description TO_SERVER        //标识数据中心
ip address 10.2.204.1 255.255.255.252 //配置IP地址
#
interface Vlan-interface4094      //以下几条是irf 堆叠线mad bfd检测配置
mad bfd enable
mad ip address 172.16.31.1 255.255.255.252 member 1
mad ip address 172.16.31.2 255.255.255.252 member 2
#
interface GigabitEthernet1/5/0/1    // 配置与另一个园区互联地址（对端IP：1.1.1.1）
port link-mode route
ip address 1.1.1.2 255.255.255.252
#
……省略……
#
interface GigabitEthernet1/3/0/47
port link-mode bridge
#
interface GigabitEthernet1/3/0/48   // irf 堆叠线BFD检测线port
port link-mode bridge
port access vlan 4094
undo stp enable    //该接口关闭stp。(BDF检测与stp不能同时存在)
#
……省略……
#
interface GigabitEthernet1/5/0/48
```

```
port link-mode bridge
#
interface GigabitEthernet2/3/0/1
port link-mode bridge
#
……省略……
#
interface GigabitEthernet2/3/0/47
port link-mode bridge
#
interface GigabitEthernet2/3/0/48   // irf 堆叠线 BFD 检测线 port
port link-mode bridge
port access vlan 4094
undo stp enable    // 关闭该接口 stp 功能
#
interface GigabitEthernet2/5/0/1
port link-mode bridge
#
……省略……
#
interface Ten-GigabitEthernet1/7/0/1    // 至安防车间 A 汇聚交换机
port link-mode route
description anfang A
ip address 10.2.205.1 255.255.255.252
#
interface Ten-GigabitEthernet1/7/0/41   // 至数据中心汇聚
port link-mode route
port link-aggregation group 14
#
interface Ten-GigabitEthernet1/7/0/42   // 至数据中心汇聚
port link-mode route
port link-aggregation group 14
#
interface Ten-GigabitEthernet1/7/0/43   // 至 C 车间汇聚
port link-mode route
port link-aggregation group 13
#
interface Ten-GigabitEthernet1/7/0/44   // 至 B 车间汇聚
port link-mode route
port link-aggregation group 12
#
interface Ten-GigabitEthernet1/7/0/45   // 至 A 车间汇聚
port link-mode route
port link-aggregation group 11
#
interface Ten-GigabitEthernet1/7/0/46   // 至 C 车间汇聚
port link-mode route
port link-aggregation group 13
#
interface Ten-GigabitEthernet1/7/0/47  // 至 B 车间汇聚
```

```
 port link-mode route
 port link-aggregation group 12
#
 interface Ten-GigabitEthernet1/7/0/48   //至A车间汇聚
 port link-mode route
 port link-aggregation group 11
#
 interface Ten-GigabitEthernet2/7/0/39   //暂时不用
 port link-mode route
 port link-aggregation group 16
#
 interface Ten-GigabitEthernet2/7/0/41   //至数据中心汇聚
 port link-mode route
 port link-aggregation group 14
#
 interface Ten-GigabitEthernet2/7/0/42   //至数据中心汇聚
 port link-mode route
 port link-aggregation group 14
#
 interface Ten-GigabitEthernet2/7/0/43   //至C车间汇聚
 port link-mode route
 port link-aggregation group 13
#
 interface Ten-GigabitEthernet2/7/0/44   //至B车间汇聚
 port link-mode route
 port link-aggregation group 12
#
 interface Ten-GigabitEthernet2/7/0/45   //至A车间汇聚
 port link-mode route
 port link-aggregation group 11
#
 interface Ten-GigabitEthernet2/7/0/46   //至C车间汇聚
 port link-mode route
 port link-aggregation group 13
#
 interface Ten-GigabitEthernet2/7/0/47   //至B车间汇聚
 port link-mode route
 port link-aggregation group 12
#
 interface Ten-GigabitEthernet2/7/0/48   //至A车间汇聚
 port link-mode route
 port link-aggregation group 11
#
 interface Ten-GigabitEthernet1/5/0/29
 port link-mode bridge
#
……省略……
#
 interface Ten-GigabitEthernet1/7/0/38
 port link-mode bridge
#
```

```
interface Ten-GigabitEthernet2/5/0/29
 port link-mode bridge
#
……省略……
#
 scheduler logfile size 16      // 设置schedule日志文件的大小
#
 line vty 0 6
  authentication-mode scheme
  user-role network-admin
  user-role network-operator
  protocol inbound ssh
  idle-timeout 0 0
#
 line vty 7 63
  authentication-mode scheme
  user-role network-operator
#
 ip route-static 0.0.0.0 0 1.1.1.1      // 配置默认静态路由（企业总出口设置在另一个园区）
#
 info-center source default console level error
#
 ssh server enable    // 远程ssh访问使能
#
 ntp-service enable    // ntp 校时使能
 ntp-service refclock-master 1    // 配置ntp主时钟
#
 domain system
#
 domain default enable system
#
 role name level-0
  description Predefined level-0 role
#
……省略……
 role name level-14
  description Predefined level-14 role
#
 user-group system
#
 local-user admin class manage      // 用户名admin
  password hash
$h$6$N3xjPQURZGbSPMLl$WPtYMb8aTBY1Coskd3juFBHY3p20cWnqCTKc03yZkMbsCxyub-
MEQIJ8785TakOlBctkIPPdXT0usOwQEdDopkg==
  service-type ssh terminal
  authorization-attribute user-role network-admin
  authorization-attribute user-role network-operator
#
 local-user fsm class manage    // 配置远程登录了用户名fsm
  password hash
$h$6$Wxa/X1kEULjjrA9c$TPmV/P80MeNOE3p6oNgkjGVT2lSJzKesp8YY0BtoatPdYUC6eP-
JkKDX7u1xoFxELy5iGsS5PwAnLeSgsMa6+jg==
```

```
service-type ssh terminal
authorization-attribute user-role network-admin
authorization-attribute user-role network-operator
#
Return
```

以上是主要配置。

五、A 车间汇聚交换机配置（主要配置）

1. 设备型号

型号：H3C S5560X-30F-EI，见图 2-25-4。

2. 主要配置

图2-25-4　汇聚交换机

```
<A- 201.2>dis cur
#
version 7.1.070, Release 1119P11
#
sysname A- 201.2
#
clock protocol none
#
irf mac-address persistent timer
irf auto-update enable
undo irf link-delay
irf member 1 priority 10
irf member 2 priority 1
#
ospf 2
 bandwidth-reference 10000
 area 0.0.0.1
  network 10.2.1.0 0.0.0.255    // A 车间交换机管理地址
  network 10.2.11.0 0.0.0.255   // A 车间办公 IP 地址段
  network 10.2.15.0 0.0.0.255   // A 车间生产 IP 地址段
  network 10.2.201.0 0.0.0.3    // 与核心互联地址段
#
lldp global enable    // 邻居关系使能
#
password-recovery enable    // 启用密码恢复功能
#
vlan 1
#
vlan 2001 // A 车间交换机管理地址 vlan
#
vlan 2011 // A 车间 办公网络 vlan
#
vlan 2015 // B 车间 生产网络 vlan
#
irf-port 1/2    // A 车间堆叠 port
 port group interface Ten-GigabitEthernet1/0/26
```

```
port group interface Ten-GigabitEthernet1/0/28
#
irf-port 2/1
port group interface Ten-GigabitEthernet2/0/26
port group interface Ten-GigabitEthernet2/0/28
#
stp instance 0 root primary
stp global enable
#
interface Bridge-Aggregation1
port link-type trunk
undo port trunk permit vlan 1
port trunk permit vlan 2001 2011 2015
#
interface Bridge-Aggregation2
port link-type trunk
undo port trunk permit vlan 1
port trunk permit vlan 2001 2011 2015
#
……省略……
#
interface Bridge-Aggregation23
port link-type trunk
undo port trunk permit vlan 1
port trunk permit vlan 2001 2011 2015
#
interface Route-Aggregation12   // 配置与核心互联IP地址
description HX
ip address 10.2.201.2 255.255.255.252
ospf bfd enable
bfd min-transmit-interval 200
bfd min-receive-interval 200
bfd detect-multiplier 3
#
interface NULL0
#
interface Vlan-interface2001  // 配置A车间交换机管理地址网关
ip address 10.2.1.254 255.255.255.0
#
interface Vlan-interface2011 // 配置A车间办公网关
description bangong
ip address 10.2.11.254 255.255.255.0
#
interface Vlan-interface2015  // 配置A车间生产网关
description shengchan
ip address 10.2.15.254 255.255.255.0
#
interface GigabitEthernet1/0/1
port link-mode bridge
port link-type trunk
undo port trunk permit vlan 1
```

```
    port trunk permit vlan 2001 2011 2015
    port link-aggregation group 1
#
……省略……
#
    interface GigabitEthernet2/0/1
    port link-mode bridge
    port link-type trunk
    undo port trunk permit vlan 1
    port trunk permit vlan 2001 2011 2015
    port link-aggregation group 1
#
……省略……
#
    interface GigabitEthernet2/0/23
    port link-mode bridge
    port link-type trunk
    undo port trunk permit vlan 1
    port trunk permit vlan 2001 2011 2015
    combo enable fiber
    port link-aggregation group 23
#
    line vty 0 63
    authentication-mode scheme
    user-role network-operator
#
    ssh server enable
#
    ntp-service enable              // ntp 校时使能
    ntp-service unicast-server 192.168.99.124  // 配置校时服务器
#
    user-group system
#
    local-user fsm class manage    // 配置 ssh 远程访问用户名
    password hash $h$6$n9lnJ78fJYSU4ZdH$vGqdQMJP1c0m4weXSzKe/aRMS4hkY4/wCDwNey-
hlorLtjikYeDQDGlVLOdkFtJgc62Zk5ZD0uHWkuBQt6eSIrg==
    service-type ssh terminal
    authorization-attribute user-role network-admin
    authorization-attribute user-role network-operator
#
Return
```

以上是 A 车间汇聚交换机配置。

B 与 C 车间汇聚交换机配置与 A 类似，不再赘述。

六、安防系统规划

该系统包括厂区监控、门禁、考勤系统网络。按照厂区位置分为三个汇聚：A 车间安防汇聚（兼安防核心）、B 车间安防汇聚、C 车间安防汇聚。A 车间安防汇聚与厂区总核心交换机连接。安防系统无设备与链路冗余设计。

1. 拓扑图

见图 2-25-5。

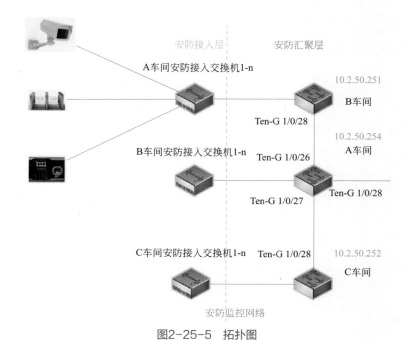

图2-25-5　拓扑图

2. vlan 以及网段规划

vlan 划分表格，如表 2-25-19。

表 2-25-19　vlan 划分表格

业务划分	vlan	网段
交换机管理	2050	10.2.50.0/24
考勤机	2060	10.2.60.0/24
监控	2070	10.2.70.0/24
门禁	2080	10.2.80.0/24
监控	2090	10.2.90.0/24

3. 交换机配置

1）汇聚交换机型号，见图 2-25-6。

型号：H3C S5560X-30F-EI

2）三台汇聚安防交换机配置分析

图2-25-6　安防汇聚交换机

```
<A-50.254>dis cur
#
version 7.1.070, Release 1119P11
#
sysname A-50.254        //命名
#
clock timezone beijing add 08:00:00 // 时区设置
```

```
clock protocol ntp   // 配置ntp校时
#
irf mac-address persistent timer//IRF桥MAC地址的保留时间6min
irf auto-update enable
undo irf link-delay
irf member 1 priority 1
```

> **知识点**
>
> irf mac-address persistent 命令用来指定 irf 桥 MAC 地址的保留时间。timer：指定 irf 桥 MAC 地址保留时间，如配置了 MAC 地址保留时间为 6min，当 Master 离开 irf 时，irf 桥 MAC 地址 6min 内不变化；如果 Master 设备在 6min 内重新又加入 irf，则 irf 桥 MAC 不会变化。如果 6min 后 Master 设备没有回到 irf，则会使用新选举的 Master 设备的桥 MAC 作为 irf 桥 MAC。

```
#
ospf 1
#
ospf 2
bandwidth-reference 10000  //配置带宽
area 0.0.0.0               // 在区域0宣告以下网段
 network 10.2.50.0 0.0.0.255
 network 10.2.60.0 0.0.0.255
 network 10.2.70.0 0.0.0.255
 network 10.2.80.0 0.0.0.255
 network 10.2.90.0 0.0.0.255
 network 10.2.205.0 0.0.0.3   //宣告与核心交换机互联地址网段
area 0.0.0.5
#
lldp global enable   // 邻居协议使能
#
loopback-detection global enable vlan 1 to 4094
#
password-recovery enable // 启用密码恢复功能
#
vlan 1
#
vlan 2050
#
vlan 2060
#
vlan 2070
#
vlan 2080
#
vlan 2090
#
stp instance 0 root primary
stp global enable
```

```
#
interface NULL0
#
interface Vlan-interface2050    // 配置交换机管理地址网关
description GL
ip address 10.2.50.254 255.255.255.0
#
interface Vlan-interface2060    // 配置考勤机管理地址网关
description KQ
ip address 10.2.60.254 255.255.255.0
#
interface Vlan-interface2070    // 配置监控网段网关
description JK
ip address 10.2.70.254 255.255.255.0
#
interface Vlan-interface2080    // 配置门禁网段网关
description MJ
ip address 10.2.80.254 255.255.255.0
#
interface Vlan-interface2090    // 配置监控网段网关
description JK
ip address 10.2.90.254 255.255.255.0
#
interface GigabitEthernet1/0/1       // 配置为 trunk，连接接入交换机
port link-mode bridge                // 接口模式为二层模式
port link-type trunk
port trunk permit vlan 1 2050 2060 2070 2080 2090
#
……省略……
#
interface GigabitEthernet1/0/18    // 该接口允许单个 vlan 通过
port link-mode bridge
port access vlan 2070
combo enable copper    // copper：表示该 Combo 接口的电口被激活，使用双绞线连接
#
interface GigabitEthernet1/0/19
port link-mode bridge
port access vlan 2070
combo enable copper
#
……省略……
#
interface M-GigabitEthernet0/0/0
#
interface Ten-GigabitEthernet1/0/26    // 与 B 车间安防汇聚交换机互联
port link-mode bridge
port link-type trunk
port trunk permit vlan 1 2050 2060 2070 2080 2090
#
interface Ten-GigabitEthernet1/0/27    // 与 C 车间安防汇聚交换机互联
port link-mode bridge
port link-type trunk
```

```
port trunk permit vlan 1 2050 2060 2070 2080 2090
#
interface Ten-GigabitEthernet1/0/28  // 接口配置三层模式，配置与核心交换机互联 IP 地址
port link-mode route
description to-HX
ip address 10.2.205.2 255.255.255.252    // 主核心 IP 配置 10.2.205.1
#
……省略……
#
ip route-static 0.0.0.0 0 10.2.50.254   // 配置默认路由
#
snmp-agent
snmp-agent local-engineid 800063A28000BED548B25C00000001
#
ssh server enable     //ssh 远程服务打开
#
ntp-service enable //ntp 使能服务
ntp-service unicast-server 192.168.99.124 // 配置校时服务器
#
……省略……
#
role name level-14
description Predefined level-14 role
#
user-group system
#
local-user fsm class manage   // 配置 ssh 远程访问用户名
password hash
$h$6$fGjKaVTWI4uxn4yd$ItgKFMiJtCq9O7IloUbkdhs5kYiqxKAkz-BRIK3rSE2kUaVOL+chwmufKMDBMsdN1ON9PlWhY2fPEmDKyHqQRug==
service-type ssh
authorization-attribute user-role network-admin
authorization-attribute user-role network-operator
#
return
```

B 与 C 车间安防汇聚交换机配置忽略。

3）poe 接入交换机配置

① 设备型号，见图 2-25-7。

型号：H3C S5130S-28P-PWR-EI

② 接入交换机主要配置（安防接入交换机举例）

```
[JK_50.100]dis cur
#
version 7.1.070, Release 6126P20
#
sysname JK_50.100
#
clock protocol none
#
irf mac-address persistent timer
irf auto-update enable
undo irf link-delay
```

图 2-25-7　设备型号

```
 irf member 1 priority 1
#
lldp global enable
#
password-recovery enable
#
vlan 1
#
vlan 2070
#
vlan 2050
#
stp global enable
#
interface NULL0
#
interface Vlan-interface2050    // 配置该交换机管理地址
 ip address 10.2.50.24 255.255.255.0
#
interface GigabitEthernet1/0/1
 port access vlan 2070
 poe enable
#
……省略……
#
interface GigabitEthernet1/0/25
#
interface GigabitEthernet1/0/26
#
interface GigabitEthernet1/0/27   // 与 B 安防汇聚交换机连接
 port link-type trunk
 port trunk permit 1 vlan 2050 2070
#
interface GigabitEthernet1/0/28
#
scheduler logfile size 16
#
……省略……
ip route-static 0.0.0.0 0 10.2.50.254    // 配置默认路由
#
ssh server enable
#
……省略……
#
local-user fsm class manage    // 配置 ssh 远程访问用户名
 password hash
$h$6$nCaIe/BmxTGz/4fW$YoS8xyZf0smO/xI2CBTu4Ft9ZXS+MRIgMkQ+gDNsQiPLKCsnqxl-JkKFt0sGRWS9J0kafc6sOcw7d8fVzUzPPmg==
 service-type ssh
 authorization-attribute user-role network-admin
 authorization-attribute user-role network-operator
#
Return
```

项目 26　防火墙基础配置

防火墙（Firewall）技术是通过有机结合各类用于安全管理与筛选的软件和硬件设备，帮助计算机网络在其内、外网之间构建一道相对隔绝的保护屏障，以保护用户资料与信息安全性的一种技术。

防火墙技术的基本功能是划分网络边界与加固内部网络安全，及时发现并处理计算机网络运行时可能存在的安全风险、数据传输等问题，其中处理措施包括隔离与保护，同时可对计算机网络安全当中的各项操作实施记录与检测，以确保计算机网络运行的安全性，保障用户资料与信息的完整性。

项目简介

本项目帮助读者了解防火墙基础知识，掌握防火墙远程配置办法。防火墙一般部署在企业边界，在安全策略中按需配置允许与拒绝的数据流量，比如在防火墙安全策略配置上一般都封禁"勒索病毒"涉及 135、137、138、139、445 高危端口。

一、拓扑图

实现功能如图 2-26-1 所示，电脑端配置环回接口，通过 CRT 软件远程进行配置防火墙，在防火墙上配置安全策略，实现 PC1 与 PC3 之间无法互访。

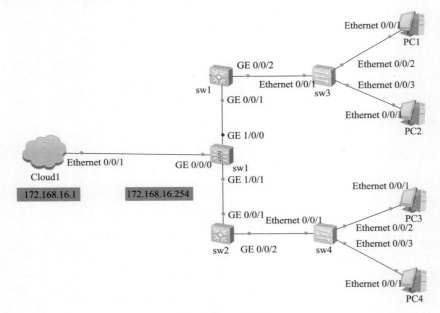

图2-26-1　拓扑图

 当点击"Cloud"进行配置时,如果绑定信息下拉条中无其他网卡信息,请卸载 WinPcap 软件,重新安装,见图 2-26-2。

图2-26-2 "cloud"配置

二、基本配置

1. 防火墙管理口配置

电脑端启用环回地址,电脑端 IP 地址:172.168.16.1。

防火墙型号采用 USG6000V,在 eNSP 模拟器中第一次启动防火墙设备的时候,需要导入设备包。见图 2-26-3。

华为防火墙默认用户名:admin,密码:Admin@123,将密码修改为:huawei@123。

图2-26-3 导入设备包

防火墙导入设备包后,双击防火墙图标,按照提示输入默认用户名以及密码。

```
Username:admin
Password:Admin@123
The password needs to be changed. Change now? [Y/N]: y
# 输入原来密码
Please enter old password: Admin@123
# 输入新密码
Please enter new password: huawei@123
# 再次输入新密码
Please confirm new password: huawei@123
Info: Your password has been changed. Save the change to survive a reboot.
…………省略…………
<USG6000V1>system-view
Enter system view, return user view with Ctrl+Z.
```

```
# 修改防火墙名字为 firewall
[USG6000V1]sysname firewall
# 进入防火墙管理接口
[firewall]interface GigabitEthernet 0/0/0
# 配置防火墙管理地址
[firewall-GigabitEthernet0/0/0]ip address 172.168.16.254 24
# 允许 ping 命令功能
[firewall-GigabitEthernet0/0/0]service-manage ping permit
```

测试：电脑（IP:172.168.16.1）-ping- 防火墙（IP：172.168.16.254）- 通，见图 2-26-4。

```
# 允许 https 访问，web 界面管理防火墙
[firewall-GigabitEthernet0/0/0]service-manage https permit
# 开启接口的访问控制管理功能
[firewall-GigabitEthernet0/0/0]service-manage enable
[firewall-GigabitEthernet0/0/0]quit
[firewall]quit
<firewall>save
```

在电脑端通过火狐浏览器登录防火墙图形化界面，浏览器输入地址：https://172.168.16.254:8443，点击允许访问，在弹出界面中。

用户名输入：admin，密码：huawei@123，见图 2-26-5。

图 2-26-4　网络测试

图 2-26-5　web界面登录

登录成功界面，见图 2-26-6。

图 2-26-6　登录界面

在 web 界面中查看刚才在命令行中配置的管理 IP 信息，见图 2-26-7，管理 IP 地址：172.168.16.254。

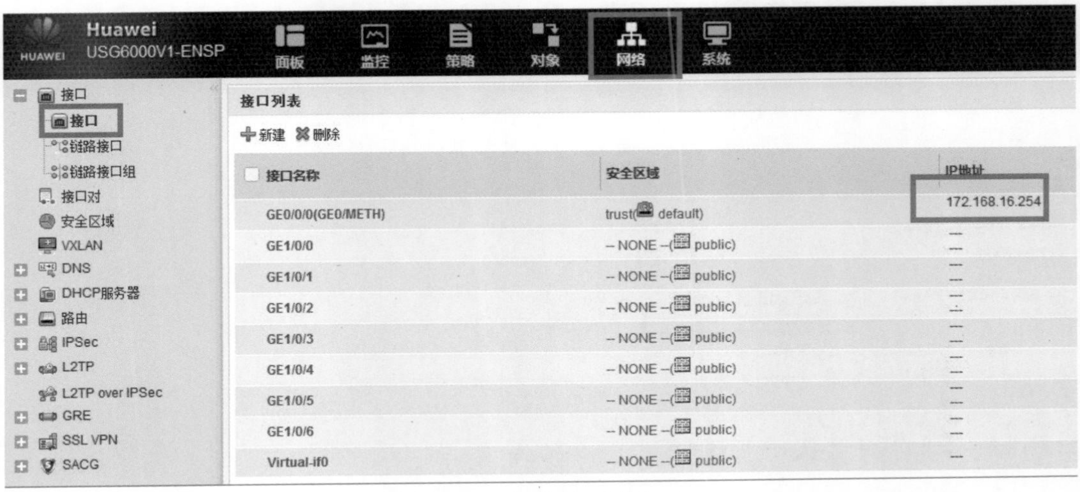

图2-26-7 管理地址

见图 2-26-8，点击"编辑"查看"启用访问管理"目前权限功能为 https 与 ping。启用 SSH/Telnet 功能，见图 2-26-9。

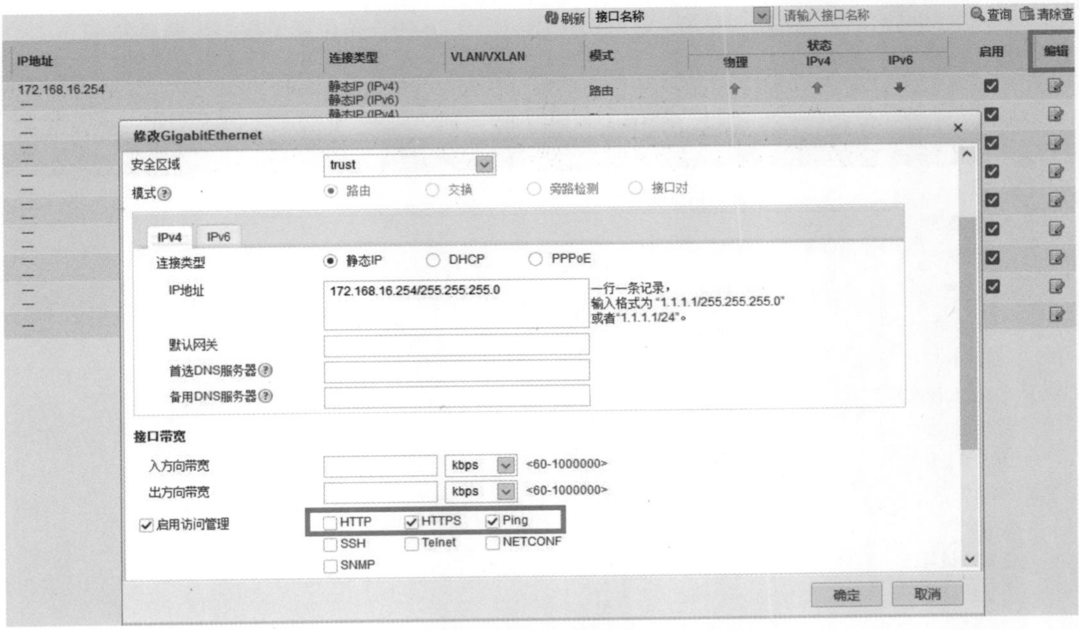

图2-26-8 启动前权限功能为https与ping

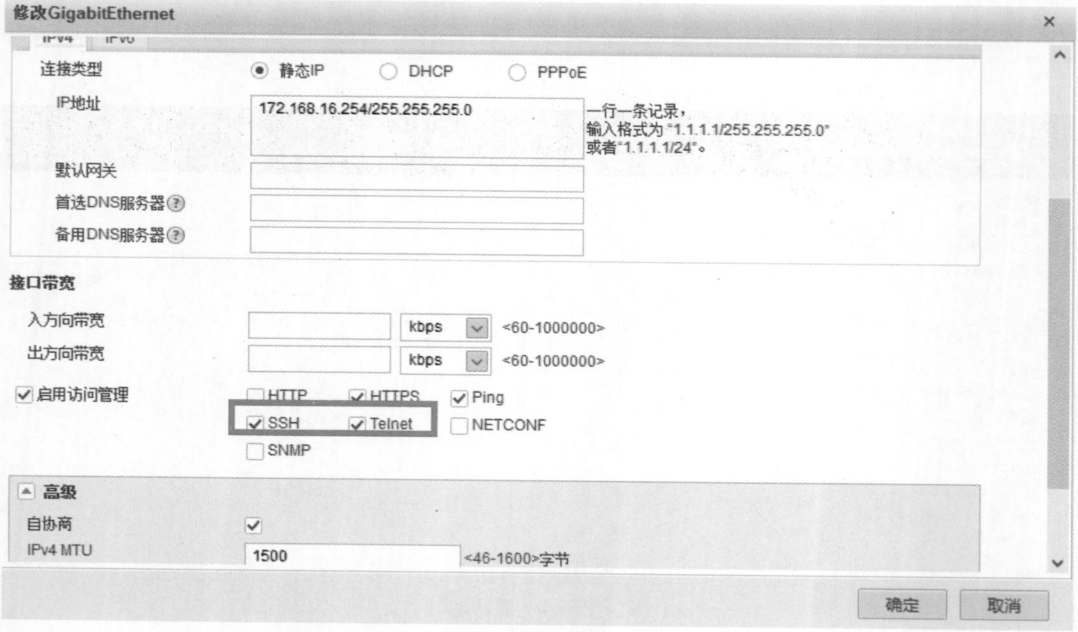

图2-26-9 启用访问管理开启SSH与Telnet

2. 创建新用户 fsm，设置访问防火墙权限

新用户 fsm 权限范围包含面板、监控、网络等菜单。

步骤1 创建角色。路径：系统 - 管理员角色见图 2-26-10。角色赋予权限见图 2-26-11。

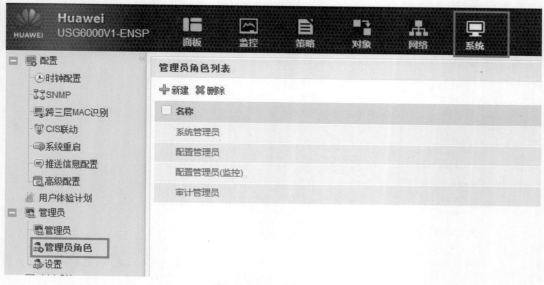

图2-26-10 创建角色

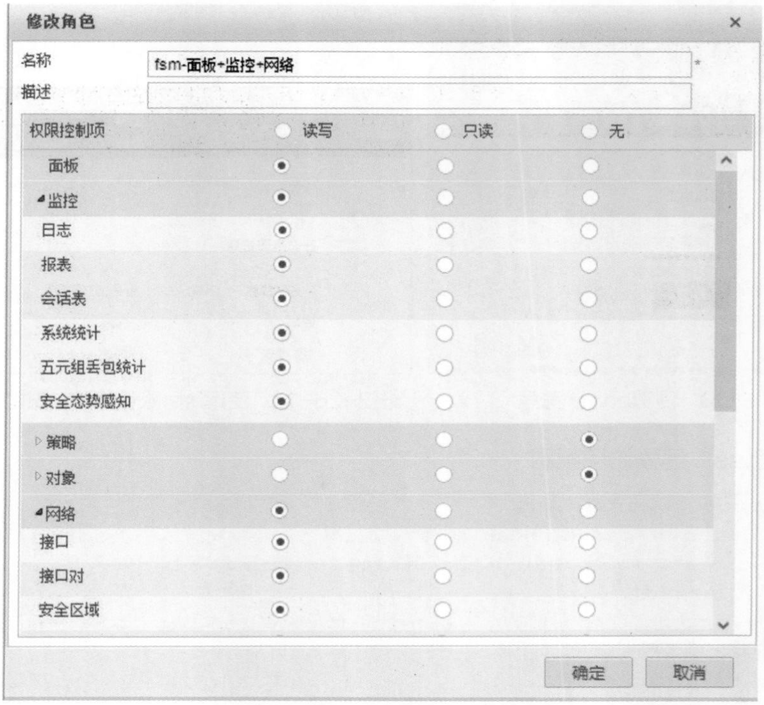

图2-26-11　角色赋予权限

步骤 2　创建用户 fsm 见图 2-26-12。

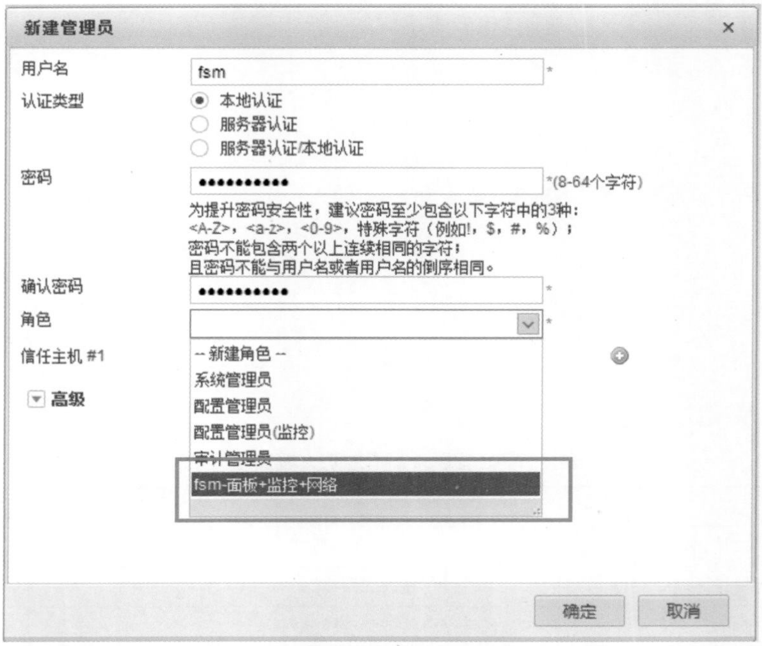

图2-26-12　创建用户

步骤 3　注销 admin 账号，见图 2-26-13。使用 fsm 账号登录，见图 2-26-14。

图2-26-13　注销admin账号

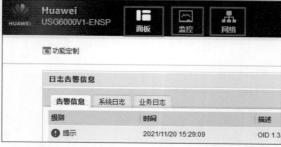

图2-26-14　使用fsm账号登录，只能看到三个菜单

3. 防火墙 ssh 远程访问配置

```
# 启用 ssh 服务（stelnet）功能
[firewall]stelnet server enable
Info: Succeeded in starting the Stelnet server.
# 管理接口配置 ssh 允许
[firewall-GigabitEthernet0/0/0]service-manage ssh permit
# 查询华为防火墙区域
[firewall]display zone
2021-11-20 15:42:58.060 +08:00
local
 priority is 100
 interface of the zone is (0):
#
trust
 priority is 85
 interface of the zone is (1):
 GigabitEthernet0/0/0
#
untrust
 priority is 5
 interface of the zone is (0):
#
dmz
 priority is 50
 interface of the zone is (0):
#
[firewall]
```

华为防火墙安全区域，如表 2-26-1。

表 2-26-1　华为防火墙安全区域

区域	含义	默认级别
Local（本地区域）	代表防火墙本身	100
Trust（信任区域）	代表内部网络	85
Untrust(非信任区域)	代表互联网或者非内部网络	5
DMZ（隔离区域）	一般是内部服务器设置区域	50

```
# 将 GigabitEthernet0/0/0 加入到 trust（以下两句）
[firewall]firewall zone trust
[firewall-zone-trust]add interface GigabitEthernet 0/0/0
# 查看防火墙 trust 区域信息
[firewall-zone-trust]display this
2021-11-20 16:36:27.080 +08:00
#
firewall zone trust
 set priority 85
 add interface GigabitEthernet0/0/0
#
return
[firewall-zone-trust]
……
# 配置能同时远程登录的终端数量
[firewall]user-interface vty 0 4
# 配置为 aaa 模式，用户名以及密码认证
[firewall-ui-vty0-4]authentication-mode aaa
# 警告内容（警告：用户界面的级别将与 aaa 用户的默认级别，应检查是否正确）
Warning: The level of the user-interface(s) will be the default level of
AAA users, please check whether it is correct.
[firewall-ui-vty0-4]
……
# 配置 ssh 远程登录用户名为 fsmfire
[firewall-aaa]manager-user fsmfire
# 允许 ssh 访问
[firewall-aaa-manager-user-fsmfire]service-type ssh
# 配置访问级别，密码为 huawei@123。
[firewall-aaa-manager-user-fsmfire]level 15
# 配置密码 huawei@123
[firewall-aaa-manager-user-fsmfire]password
Enter Password:huawei@123
Confirm Password: huawei@123
[firewall-aaa-manager-user-fsmfire]
#……生成公钥……
[firewall]rsa local-key-pair create
The key name will be: firewall_Host
The range of public key size is (2048～2048).
NOTES: If the key modulus is greater than 512,
it will take a few minutes.
# 生成秘钥的长度 2048
Input the bits in the modulus[default = 2048]:2048
Generating keys...
.+++++
........................++
....++++
...........++
[firewall]
```

使用 SecureCRT 软件，通过 ssh 协议访问防火墙，见图 2-26-15。

输入 IP：172.168.16.254，用户名：fsmfire。

输入密码：huawei@123，见图 2-26-16。
登录到防火墙命令行，见图 2-26-17。

图2-26-15　ssh协议访问防火墙　　　　　　　图2-26-16　输入密码

图2-26-17　登录到防火墙命令行

查询防火墙用户登录情况，从图 2-26-18 中可以看出，fsmfire 用户通过 ssh 访问。

图2-26-18　防火墙用户登录情况

4. 实现功能 PC1 不能 ping PC3

PC-IP 地址规划，如表 2-26-2。

表 2-26-2　PC-IP 地址规划表

PC	IP	网关	vlan
PC1	10.1.10.1/24	10.1.10.254	10
PC2	10.1.20.1/24	10.1.20.254	20
PC3	10.1.30.1/24	10.1.30.254	30
PC4	10.1.40.1/24	10.1.40.254	40

步骤1　PC1-PC4 电脑 IP 配置

PC1 配置见图 2-26-19，PC2-PC4 参照表 2-26-2 配置。

步骤2　交换机配置

```
sw3
<Huawei>system-view
```

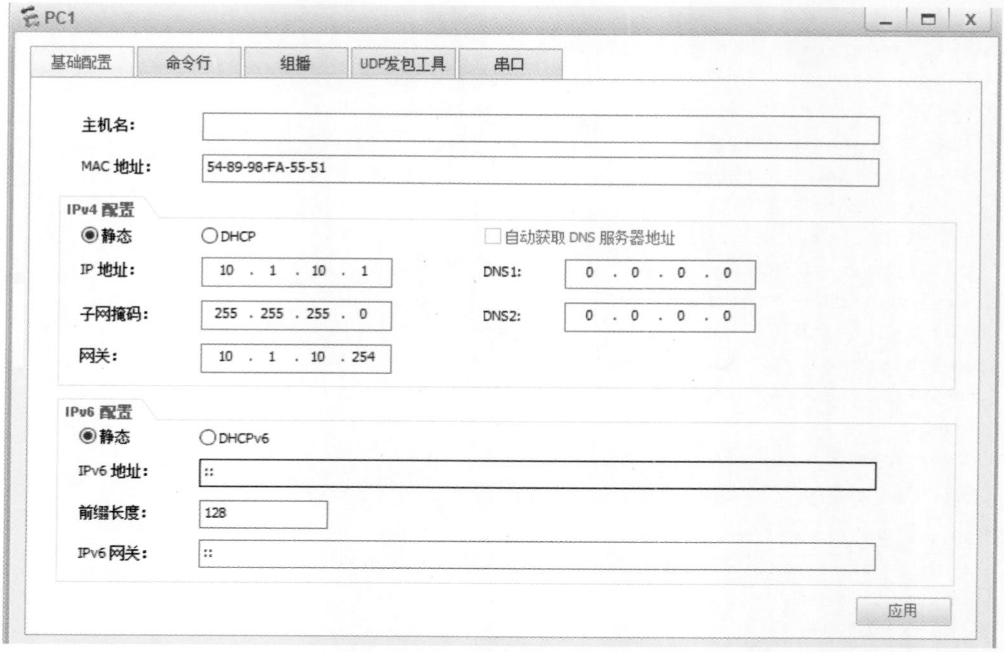

图2-26-19 PC1配置

```
[Huawei]sysname sw3
[sw3]un in en
Info: Information center is disabled.
# 创建 vlan
[sw3]vlan batch 10 20
Info: This operation may take a few seconds. Please wait for a moment...done.
[sw3]interface Ethernet 0/0/1
[sw3-Ethernet0/0/1]port link-type trunk
# 允许 vlan 10 、20 通过 Ethernet0/0/1
[sw3-Ethernet0/0/1]port trunk allow-pass vlan 10 20
[sw3-Ethernet0/0/1]quit
[sw3]interface Ethernet 0/0/2
[sw3-Ethernet0/0/2]port link-type access
# 允许 vlan 10 通过 Ethernet0/0/2
[sw3-Ethernet0/0/2]port default vlan 10
[sw3-Ethernet0/0/2]quit
[sw3]interface Ethernet 0/0/3
[sw3-Ethernet0/0/3]port link-type access
# 允许 vlan 20 通过 Ethernet0/0/3
[sw3-Ethernet0/0/3]port default vlan 20
[sw3-Ethernet0/0/3]quit
[sw3]quit
<sw3>save
sw4
<Huawei>sys
Enter system view, return user view with Ctrl+Z.
[Huawei]sysname sw4
[sw4]vlan batch 30 40
```

```
Info: This operation may take a few seconds. Please wait for a moment...done.
[sw4]interface Ethernet 0/0/1
[sw4-Ethernet0/0/1]port link-type trunk
[sw4-Ethernet0/0/1]port trunk allow-pass vlan 30 40
[sw4-Ethernet0/0/1]quit
[sw4]interface Ethernet 0/0/2
[sw4-Ethernet0/0/2]port link-type access
[sw4-Ethernet0/0/2]port default vlan 30
[sw4-Ethernet0/0/2]quit
[sw4]interface Ethernet 0/0/3
[sw4-Ethernet0/0/3]port link-type access
[sw4-Ethernet0/0/3]port default vlan 40
[sw4-Ethernet0/0/3]quit
[sw4]quit
<sw4>save
```

sw1（sw1 与防火墙之间通过 access 模式连接）

```
<Huawei>system-view
Enter system view, return user view with Ctrl+Z.
[Huawei]sysname sw1
[sw1]un in en
Info: Information center is disabled.
# 创建vlan 10、20、100
[sw1]vlan batch 10 20 100
Info: This operation may take a few seconds. Please wait for a moment...done.
[sw1]interface Vlanif 10
# 创建10.1.10.0/24 网关为10.1.10.254
[sw1-Vlanif10]ip address 10.1.10.254 24
[sw1-Vlanif10]quit
[sw1]interface Vlanif 20
[sw1-Vlanif20]ip address 10.1.20.254 24
[sw1-Vlanif20]quit
[sw1]interface Vlanif 100
[sw1-Vlanif100]ip address 10.1.100.2 30
[sw1-Vlanif100]quit
[sw1]interface GigabitEthernet 0/0/2
[sw1-GigabitEthernet0/0/2]port link-type trunk
[sw1-GigabitEthernet0/0/2]port trunk allow-pass vlan 10 20
[sw1-GigabitEthernet0/0/2]quit
[sw1]interface GigabitEthernet 0/0/1
[sw1-GigabitEthernet0/0/1]port link-type access
[sw1-GigabitEthernet0/0/1]port default vlan 100
[sw1-GigabitEthernet0/0/1]quit
[sw1]quit
<sw1>save
```

测试：PC1(IP:10.1.10.1)-ping-PC2（IP: 10.1.20.1）- 通，见图 2-26-20。

sw2（sw2 与防火墙之间通过 access 模式连接）

```
<Huawei>system-view
Enter system view, return user view with Ctrl+Z.
[Huawei]sysname sw2
```

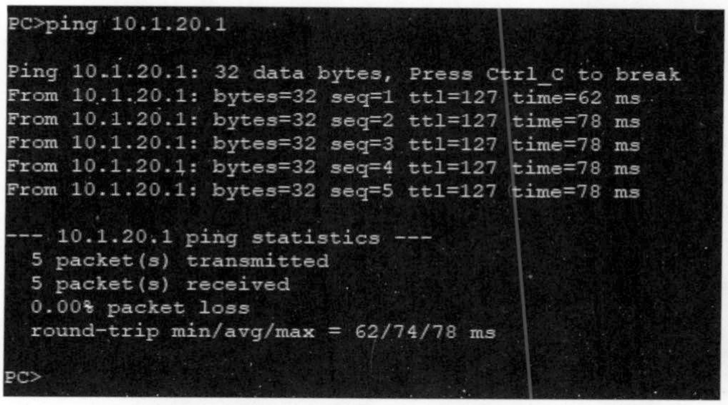

```
PC>ping 10.1.20.1
Ping 10.1.20.1: 32 data bytes, Press Ctrl_C to break
From 10.1.20.1: bytes=32 seq=1 ttl=127 time=62 ms
From 10.1.20.1: bytes=32 seq=2 ttl=127 time=78 ms
From 10.1.20.1: bytes=32 seq=3 ttl=127 time=78 ms
From 10.1.20.1: bytes=32 seq=4 ttl=127 time=78 ms
From 10.1.20.1: bytes=32 seq=5 ttl=127 time=78 ms

--- 10.1.20.1 ping statistics ---
  5 packet(s) transmitted
  5 packet(s) received
  0.00% packet loss
  round-trip min/avg/max = 62/74/78 ms
PC>
```

图2-26-20 网络测试

创建vlan 30、40、200
[sw2]vlan batch 30 40 200
Info: This operation may take a few seconds. Please wait for a moment...done.
[sw2]interface Vlanif 30
[sw2-Vlanif30]ip address 10.1.30.254 24
[sw2-Vlanif30]quit
[sw2]interface Vlanif 40
[sw2-Vlanif40]ip address 10.1.40.254 24
[sw2-Vlanif40]quit
[sw2]interface Vlanif 200
[sw2-Vlanif200]ip address 10.1.200.2 30
[sw2-Vlanif200]quit
[sw2]interface GigabitEthernet 0/0/2
[sw2-GigabitEthernet0/0/2]port link-type trunk
[sw2-GigabitEthernet0/0/2]port trunk allow-pass vlan 30 40
[sw2-GigabitEthernet0/0/2]quit
[sw2]interface GigabitEthernet 0/0/1
[sw2-GigabitEthernet0/0/1]port link-type access
[sw2-GigabitEthernet0/0/1]port default vlan 200
[sw2-GigabitEthernet0/0/1]quit
[sw2]quit
<sw2>save

测试PC3（IP:10.1.30.1）-ping-PC4（IP：10.1.40.1）- 通，见图2-26-21。

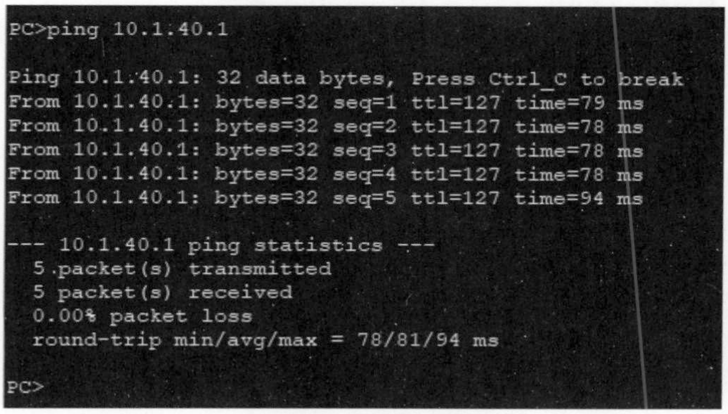

```
PC>ping 10.1.40.1
Ping 10.1.40.1: 32 data bytes, Press Ctrl_C to break
From 10.1.40.1: bytes=32 seq=1 ttl=127 time=79 ms
From 10.1.40.1: bytes=32 seq=2 ttl=127 time=78 ms
From 10.1.40.1: bytes=32 seq=3 ttl=127 time=78 ms
From 10.1.40.1: bytes=32 seq=4 ttl=127 time=78 ms
From 10.1.40.1: bytes=32 seq=5 ttl=127 time=94 ms

--- 10.1.40.1 ping statistics ---
  5 packet(s) transmitted
  5 packet(s) received
  0.00% packet loss
  round-trip min/avg/max = 78/81/94 ms
PC>
```

图2-26-21 网络测试

步骤 3　防火墙 firewall 接口区域配置

```
<firewall>system-view
Enter system view, return user view with Ctrl+Z.
[firewall]interface GigabitEthernet 1/0/0
[firewall-GigabitEthernet1/0/0]ip address 10.1.100.1 30
[firewall-GigabitEthernet1/0/0]quit
[firewall]interface GigabitEthernet 1/0/1
[firewall-GigabitEthernet1/0/1]ip address 10.1.200.1 30
[firewall-GigabitEthernet1/0/1]quit
# 将 GigabitEthernet 1/0/0 、GigabitEthernet 1/0/1 加入 trust 区域
[firewall]firewall zone trust
[firewall-zone-trust]add interface GigabitEthernet 1/0/0
[firewall-zone-trust]add interface GigabitEthernet 1/0/1
# 查询区域配置
[firewall-zone-trust]display this
2021-11-21 19:22:30.190 +08:00
#
firewall zone trust
 set priority 85
 add interface GigabitEthernet0/0/0
 add interface GigabitEthernet1/0/0
 add interface GigabitEthernet1/0/1
#
return
[firewall-zone-untrust]quit
[firewall]
```

步骤 4　配置路由

```
sw1:
[sw1]ip route-static 10.1.30.0 24 10.1.100.1
[sw1]ip route-static 10.1.40.0 24 10.1.100.1
sw2:
[sw2]ip route-static 10.1.10.0 24 10.1.200.1
[sw2]ip route-static 10.1.20.0 24 10.1.200.1
Firewall
[firewall]ip route-static 10.1.10.0 24 10.1.100.2
[firewall]ip route-static 10.1.20.0 24 10.1.100.2
[firewall]ip route-static 10.1.30.0 24 10.1.200.2
[firewall]ip route-static 10.1.40.0 24 10.1.200.2
```

以上地址可以聚合处理，也可以自行配置。

步骤 5　测试网络，见图 2-26-22。

```
PC1(10.1.10.1)-ping-PC3(10.1.30.1) - 通
PC1(10.1.10.1)-ping-PC4(10.1.40.1) - 通
```

步骤 6　通过图像化界面配置策略，实现 PC1 与 PC3 之间不能 ping 通（icmp 服务）。

路径：策略 - 安全策略 - 新建安全策略，见图 2-26-23。

新建安全策略名称：pc1-deny-pc3，源地址为 PC1，目的地址为 PC3，服务为 icmp，动作为"拒绝"。

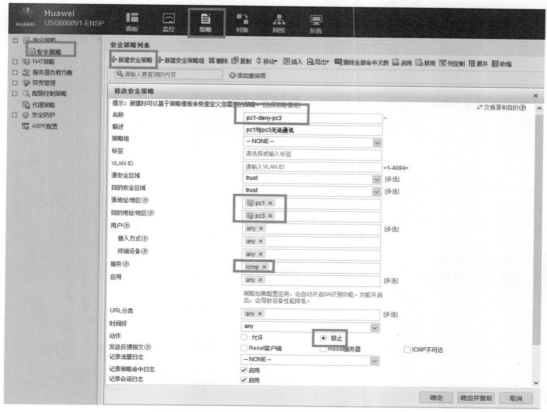

图2-26-22 网络测试

图2-26-23 新建策略（动作选择"拒绝"）

327

再次测试 PC1(10.1.10.1)-ping-PC3（10.1.30.1）- 已经不通。见图 2-26-24。

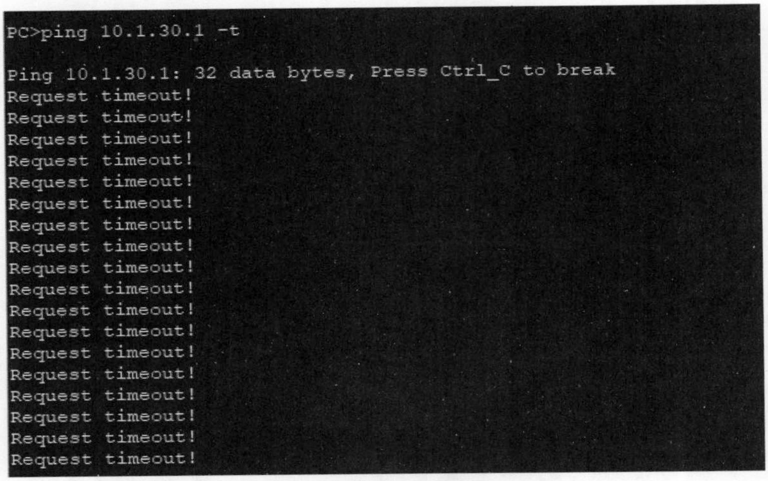

图2-26-24 网络测试

通过查看防火墙策略已经有命中次数。命中次数达到 25842，见图 2-26-25。

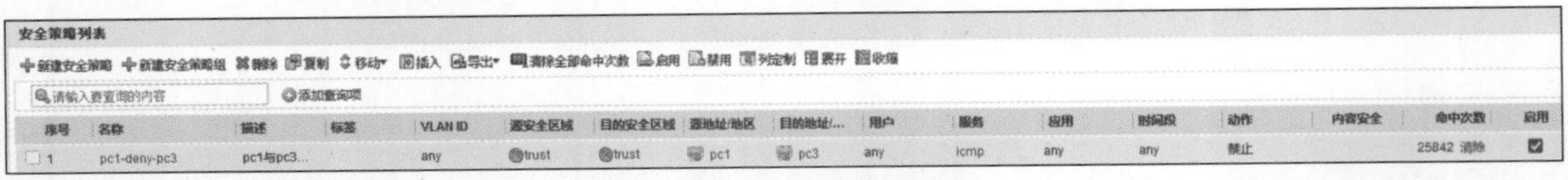

图2-26-25 命中策略

项目 27　防火墙策略路由配置

一般情况下，为了保障公司业务与因特网数据传输不中断，在企业网络出口采用多条链路（比如联通、移动等运营商）来提高链路的高可靠性。防火墙出口链路选路配置的方式有运营商路由、策略路由、静态路由等几种方式。本节重点介绍如何配置策略路由。

项目简介

防火墙出口配置策略路由主要实现出口流量负载分担，在项目中通过配置策略路由实现内网访问互联网中对方是联通（走联通链路）或者电信的 IP 地址（走电信链路），通讯流量按照策略配置要求走相关的链路。当某条链路故障后，数据通信能自动切换到另一条链路，当故障排除后，恢复策略路由转发。

一、拓扑图

拓扑图如图 2-27-1 所示，图中是企业网络两个运营商出口（比如电信与联通）通过配置策略路由实现企业内网访问互联网流量负载均衡。

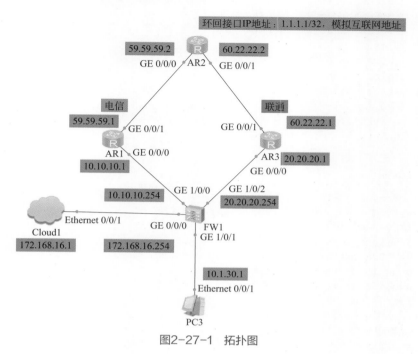

图2-27-1　拓扑图

二、基本配置

1. 防火墙 ssh 以及 GigabitEthernet1/0/0 配置。

参照防火墙基础配置项目。

增加两个接口配置

```
[firewall-GigabitEthernet1/0/0]ip address 10.10.10.254 24
[firewall-GigabitEthernet1/0/2]ip address 20.20.20.254 24
```

2. AR1 配置

```
[Huawei]sys
[Huawei]sysname dianxin
[dianxin]interface GigabitEthernet 0/0/0
[dianxin-GigabitEthernet0/0/0]ip address 10.10.10.1 24
[dianxin-GigabitEthernet0/0/0]quit
[dianxin]interface GigabitEthernet 0/0/1
# 模拟电信公网地址
[dianxin-GigabitEthernet0/0/1]ip address 59.59.59.1 24
[dianxin-GigabitEthernet0/0/1]quit
[dianxin]quit
<dianxin>save
```

3. AR2 配置

```
<Huawei>sys
```

```
Enter system view, return user view with Ctrl+Z.
[AR3]sysname AR2
[AR2]interface GigabitEthernet 0/0/0
# 模拟电信公网地址
[AR2-GigabitEthernet0/0/0]ip address 59.59.59.2 24
[AR2-GigabitEthernet0/0/0]quit
[AR2]interface GigabitEthernet 0/0/1
# 模拟联通公网地址
[AR2-GigabitEthernet0/0/1]ip address 60.22.22.2 24
[AR2-GigabitEthernet0/0/1]quit
[AR2]interface LoopBack 0
# 模拟互联网地址
[AR2-LoopBack0]ip address 1.1.1.1 32
[AR2]quit
<AR2>save
```

4. AR3 配置

```
<Huawei>system-view
Enter system view, return user view with Ctrl+Z.
[Huawei]sysname liantong
[liantong]interface GigabitEthernet 0/0/0
[liantong-GigabitEthernet0/0/0]ip address 20.20.20.1 24
[liantong-GigabitEthernet0/0/0]quit
[liantong]interface GigabitEthernet 0/0/1
# 模拟联通公网地址
[liantong-GigabitEthernet0/0/1]ip address 60.22.22.1 24
[liantong]quit
<liantong>save
```

5. 配置静态路由

```
[dianxin]ip route-static 0.0.0.0 0 59.59.59.2
[liantong]ip route-static 0.0.0.0 0 60.22.22.2
[AR2]ip route-static 10.10.10.0 24 59.59.59.1
[AR2]ip route-static 20.20.20.0 24 60.22.22.1
```

6. 防火墙区域配置

trust 区域配置：将防火墙 GigabitEthernet0/0/0、GigabitEthernet1/0/1 两个接口加入 trust 区域，见图 2-27-2。

图2-27-2　防火墙区域配置

untrust 区域配置：将防火墙 GigabitEthernet1/0/0、GigabitEthernet1/0/2 两个接口加入 untrust 区域，见图 2-27-3。

图2-27-3　防火墙区域配置

网络测试 - 通

```
[AR2]ping 10.10.10.1
  PING 10.10.10.1: 56  data bytes, press CTRL_C to break
    Reply from 10.10.10.1: bytes=56 Sequence=1 ttl=255 time=50 ms
……省略……
  --- 10.10.10.1 ping statistics ---
    5 packet(s) transmitted
    5 packet(s) received
    0.00% packet loss
    round-trip min/avg/max = 20/30/50 ms
```

网络测试 - 通

```
[AR2]ping 20.20.20.1
  PING 20.20.20.1: 56  data bytes, press CTRL_C to break
    Reply from 20.20.20.1: bytes=56 Sequence=1 ttl=255 time=70 ms
……省略……
  --- 20.20.20.1 ping statistics ---
    5 packet(s) transmitted
    5 packet(s) received
    0.00% packet loss
    round-trip min/avg/max = 10/26/70 ms
```

网络测试 - 通

```
[AR2]ping 1.1.1.1
  PING 1.1.1.1: 56  data bytes, press CTRL_C to break
    Reply from 1.1.1.1: bytes=56 Sequence=1 ttl=255 time=1 ms
……省略……
  --- 1.1.1.1 ping statistics ---
    5 packet(s) transmitted
    5 packet(s) received
    0.00% packet loss
    round-trip min/avg/max = 1/1/1 ms
```

网络测试 - 通

```
<dianxin>ping 1.1.1.1
  PING 1.1.1.1: 56  data bytes, press CTRL_C to break
    Reply from 1.1.1.1: bytes=56 Sequence=1 ttl=255 time=20 ms
……省略……
  --- 1.1.1.1 ping statistics ---
    5 packet(s) transmitted
    5 packet(s) received
    0.00% packet loss
    round-trip min/avg/max = 10/22/40 ms
```

网络测试 - 通

```
<liantong>ping 1.1.1.1
  PING 1.1.1.1: 56  data bytes, press CTRL_C to break
    Reply from 1.1.1.1: bytes=56 Sequence=1 ttl=255 time=30 ms
……省略……
  --- 1.1.1.1 ping statistics ---
    5 packet(s) transmitted
    5 packet(s) received
    0.00% packet loss
```

```
  round-trip min/avg/max = 10/20/30 ms
```

网络测试 - 通
```
[AR2]ping -a 1.1.1.1 10.10.10.1
  PING 10.10.10.1: 56   data bytes, press CTRL_C to break
    Reply from 10.10.10.1: bytes=56 Sequence=1 ttl=255 time=20 ms
……省略……
  --- 10.10.10.1 ping statistics ---
    5 packet(s) transmitted
    5 packet(s) received
    0.00% packet loss
    round-trip min/avg/max = 10/22/30 ms
```

网络测试 - 通
```
[AR2]ping -a 1.1.1.1 20.20.20.1
  PING 20.20.20.1: 56   data bytes, press CTRL_C to break
    Reply from 20.20.20.1: bytes=56 Sequence=1 ttl=255 time=20 ms
……省略……
  --- 20.20.20.1 ping statistics ---
    5 packet(s) transmitted
    5 packet(s) received
    0.00% packet loss
    round-trip min/avg/max = 20/26/30 ms
[AR2]
```

7. 防火墙配置静态路由

见图 2-27-4，红色方框内所示。

图2-27-4　防火墙配置静态路由

如下测试网络不通，原因是加入到 untrust 区域的两个接口没有放行 ping 的权限。
```
[AR2]ping 10.10.10.254
  PING 10.10.10.254: 56   data bytes, press CTRL_C to break
    Request time out
……省略……
  --- 10.10.10.254 ping statistics ---
    5 packet(s) transmitted
    0 packet(s) received
100.00% packet loss
```

接口赋予权限
```
[firewall-GigabitEthernet1/0/0]service-manage ping permit
[firewall-GigabitEthernet1/0/2]service-manage ping permit
```

再次网络测试 - 通

```
[AR2]ping 10.10.10.254
  PING 10.10.10.254: 56  data bytes, press CTRL_C to break
    Reply from 10.10.10.254: bytes=56 Sequence=1 ttl=254 time=20 ms
……省略……
  --- 10.10.10.254 ping statistics ---
    5 packet(s) transmitted
    5 packet(s) received
    0.00% packet loss
    round-trip min/avg/max = 10/22/30 ms
```

网络测试 - 通

```
[AR2]ping 20.20.20.254
  PING 20.20.20.254: 56  data bytes, press CTRL_C to break
    Reply from 20.20.20.254: bytes=56 Sequence=1 ttl=254 time=20 ms
……省略……
  --- 20.20.20.254 ping statistics ---
    5 packet(s) transmitted
    5 packet(s) received
    0.00% packet loss
    round-trip min/avg/max = 20/28/40 ms
……………………
```

8. 防火墙内部网络地址配置

```
[firewall]vlan 30
[firewall-vlan30]quit
[firewall]interface Vlanif 30
[firewall-Vlanif30]ip address 10.1.30.254 24
[firewall-Vlanif30]service-manage ping permit
[firewall-Vlanif30]quit
# 三层接口设置为二层接口
[firewall-GigabitEthernet1/0/1]portswitch
[firewall-GigabitEthernet1/0/1]port link-type access
[firewall-GigabitEthernet1/0/1]port default vlan 30
```

9. NAT 转换配置

路径：策略 -NAT 策略，见图 2-27-5。

图2-27-5　NAT策略路径

新建 NAT 策略，见图 2-27-6。

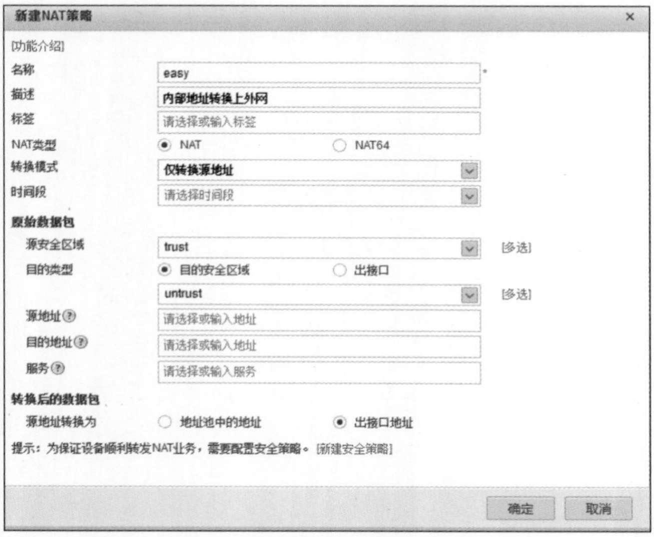

图2-27-6　新建安全策略（按照图中内容配置）

安全策略配置，放行 ping 功能，见图 2-27-7。

图2-27-7　安全策略配置（按照图中内容配置）

将 Vlanif 30 加入 trust 区域，见图 2-27-8。
网络测试 - 通
```
PC>ping 10.1.30.254
```

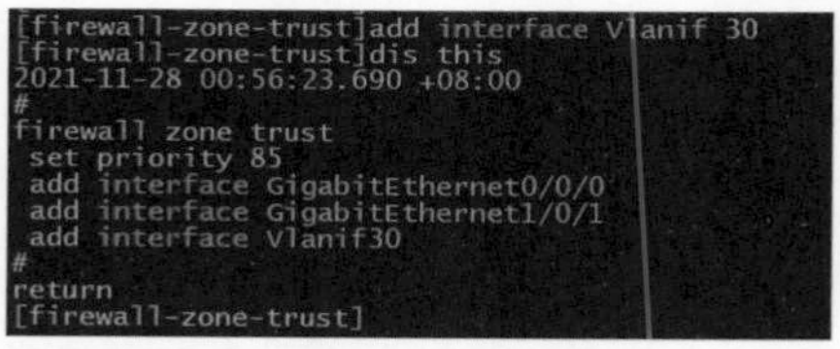

```
[firewall-zone-trust]add interface Vlanif 30
[firewall-zone-trust]dis this
2021-11-28 00:56:23.690 +08:00
#
firewall zone trust
 set priority 85
 add interface GigabitEthernet0/0/0
 add interface GigabitEthernet1/0/1
 add interface Vlanif30
#
return
[firewall-zone-trust]
```

图2-27-8　Vlanif 30加入到trust区域

```
Ping 10.1.30.254: 32 data bytes, Press Ctrl_C to break
From 10.1.30.254: bytes=32 seq=1 ttl=255 time<1 ms
……省略……
--- 10.1.30.254 ping statistics ---
  5 packet(s) transmitted
  5 packet(s) received
  0.00% packet loss
  round-trip min/avg/max = 0/0/0 ms
```

网络测试 - 通（实现内网地址与互联网通信）

```
PC>ping 1.1.1.1
Ping 1.1.1.1: 32 data bytes, Press Ctrl_C to break
From 1.1.1.1: bytes=32 seq=1 ttl=253 time=16 ms
……省略……
--- 1.1.1.1 ping statistics ---
  5 packet(s) transmitted
  5 packet(s) received
  0.00% packet loss
  round-trip min/avg/max = 16/25/31 ms
PC>
```

查看 NAT 转换会话信息，见图 2-27-9。

```
[firewall]dis firewall session table verbose source-zone trust destination-zone untrust
2021-11-28 01:05:06.190 +08:00
 Current Total Sessions : 19
 icmp  VPN: public --> public  ID: c387fd2adbd5de824161a2d5b5
 Zone: trust --> untrust  TTL: 00:00:20  Left: 00:00:07
 Recv Interface: Vlanif30
 Interface: GigabitEthernet1/0/0 NextHop: 10.10.10.1 MAC: 00e0-fcf8-740c
 <--packets: 1 bytes: 60 -->packets: 1 bytes: 60
 10.1.30.1:14437[10.10.10.254:2404] --> 1.1.1.1:2048 PolicyName: trust-ping-untrst

 icmp  VPN: public --> public  ID: c387fd2adbd5ed02ef61a2d5b4
 Zone: trust --> untrust  TTL: 00:00:20  Left: 00:00:06
 Recv Interface: Vlanif30
 Interface: GigabitEthernet1/0/0 NextHop: 10.10.10.1 MAC: 00e0-fcf8-740c
 <--packets: 1 bytes: 60 -->packets: 1 bytes: 60
 10.1.30.1:14181[10.10.10.254:2403] --> 1.1.1.1:2048 PolicyName: trust-ping-untrst

 icmp  VPN: public --> public  ID: c387fd2adbd5c1830461a2d5b2
 Zone: trust --> untrust  TTL: 00:00:20  Left: 00:00:04
 Recv Interface: Vlanif30
 Interface: GigabitEthernet1/0/0 NextHop: 10.10.10.1 MAC: 00e0-fcf8-740c
 <--packets: 1 bytes: 60 -->packets: 1 bytes: 60
 10.1.30.1:13669[10.10.10.254:2401] --> 1.1.1.1:2048 PolicyName: trust-ping-untrst

 icmp  VPN: public --> public  ID: c387fd2adbd40507c961a2d5b7
 Zone: trust --> untrust  TTL: 00:00:20  Left: 00:00:09
 Recv Interface: Vlanif30
 Interface: GigabitEthernet1/0/0 NextHop: 10.10.10.1 MAC: 00e0-fcf8-740c
 <--packets: 1 bytes: 60 -->packets: 1 bytes: 60
 10.1.30.1:14949[10.10.10.254:2406] --> 1.1.1.1:2048 PolicyName: trust-ping-untrst

 icmp  VPN: public --> public  ID: c387fd2adbd5960b2361a2d5b0
 Zone: trust --> untrust  TTL: 00:00:20  Left: 00:00:02
 Recv Interface: Vlanif30
 Interface: GigabitEthernet1/0/0 NextHop: 10.10.10.1 MAC: 00e0-fcf8-740c
 <--packets: 1 bytes: 60 -->packets: 1 bytes: 60
 10.1.30.1:1315[10.10.10.254:2399] --> 1.1.1.1:2048 PolicyName: trust-ping-untrst

 icmp  VPN: public --> public  ID: c387fd2adbd4220c5661a2d5b8
 Zone: trust --> untrust  TTL: 00:00:20  Left: 00:00:10
 Recv Interface: Vlanif30
 Interface: GigabitEthernet1/0/0 NextHop: 10.10.10.1 MAC: 00e0-fcf8-740c
 10.1.30.1:15205[10.10.10.254:2407] --> 1.1.1.1:2048 PolicyName: trust-ping-untrst
```

图2-27-9　NAT转换会话信息

10. 策略路由配置

因为是两条静态路由，优先级相同，所以也就是随机转发数据。通过配置策略路由按需转发数据。详细步骤如下：

Ip-link 配置，见图 2-27-10。

配置路径：系统 - 高可靠性 -Ip-link。建立两条用于电信与联通链路检测。

图2-27-10 Ip-link配置

电信链路检测，见图 2-27-11。

图2-27-11 电信链路检测（名称：ip-link-dianxin）

联通链路检测，见图 2-27-12。

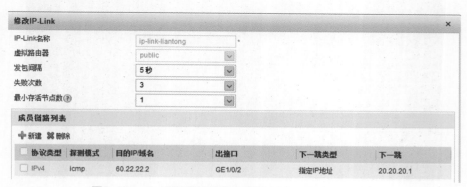

图2-27-12 联通链路检测（名称：ip-link-liantong）

查看 ip-link 状态，已经全部 up（正常）。见图 2-27-13。

图2-27-13　链路状态

11. 创建策略路由关联 ip-link

查询目前现有的默认静态路由，见图 2-27-14。

图2-27-14　静态路由

两条默认静态路由关联 ip-link，见图 2-27-15。

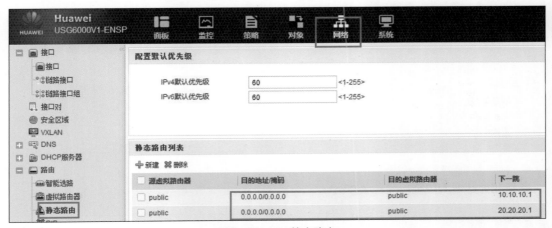

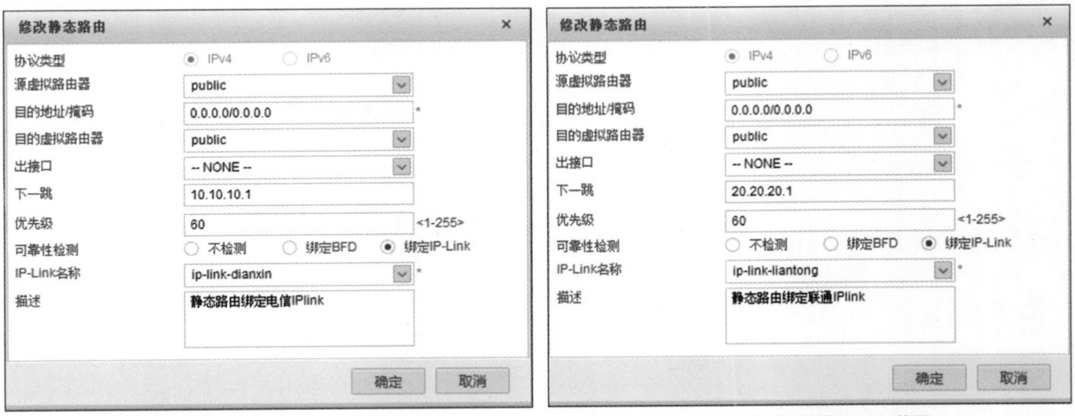

(a) 静态路由关联ip-link(电信)　　(b) 静态路由关联ip-link(联通)

图2-27-15　静态路由关联

查询静态路由关联 ip-link，见图 2-27-16。

图2-27-16　查询静态路由关联IP-link

配置策略路由如下：

路径：网络 - 路由 - 智能选路 - 策略路由，见图 2-27-17。

图2-27-17　配置策略路由

新建内部网络地址，见图 2-27-18。

图2-27-18　新建内部网络地址

创建电信地址，见图 2-27-19。

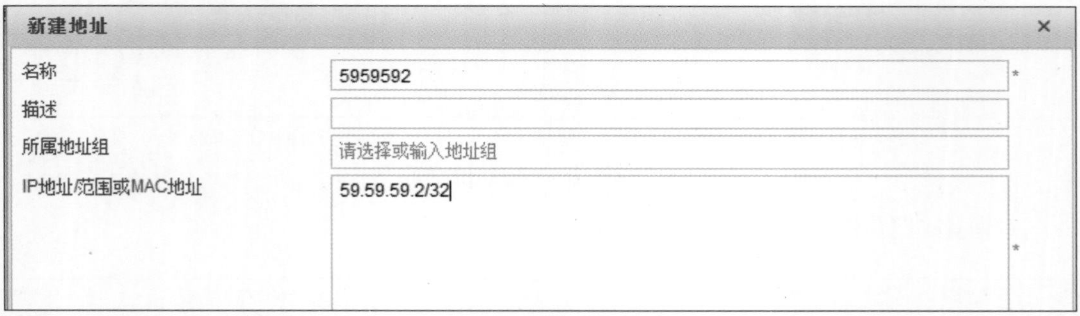

图2-27-19　新建电信地址

创建策略路由（电信），见图 2-27-20。

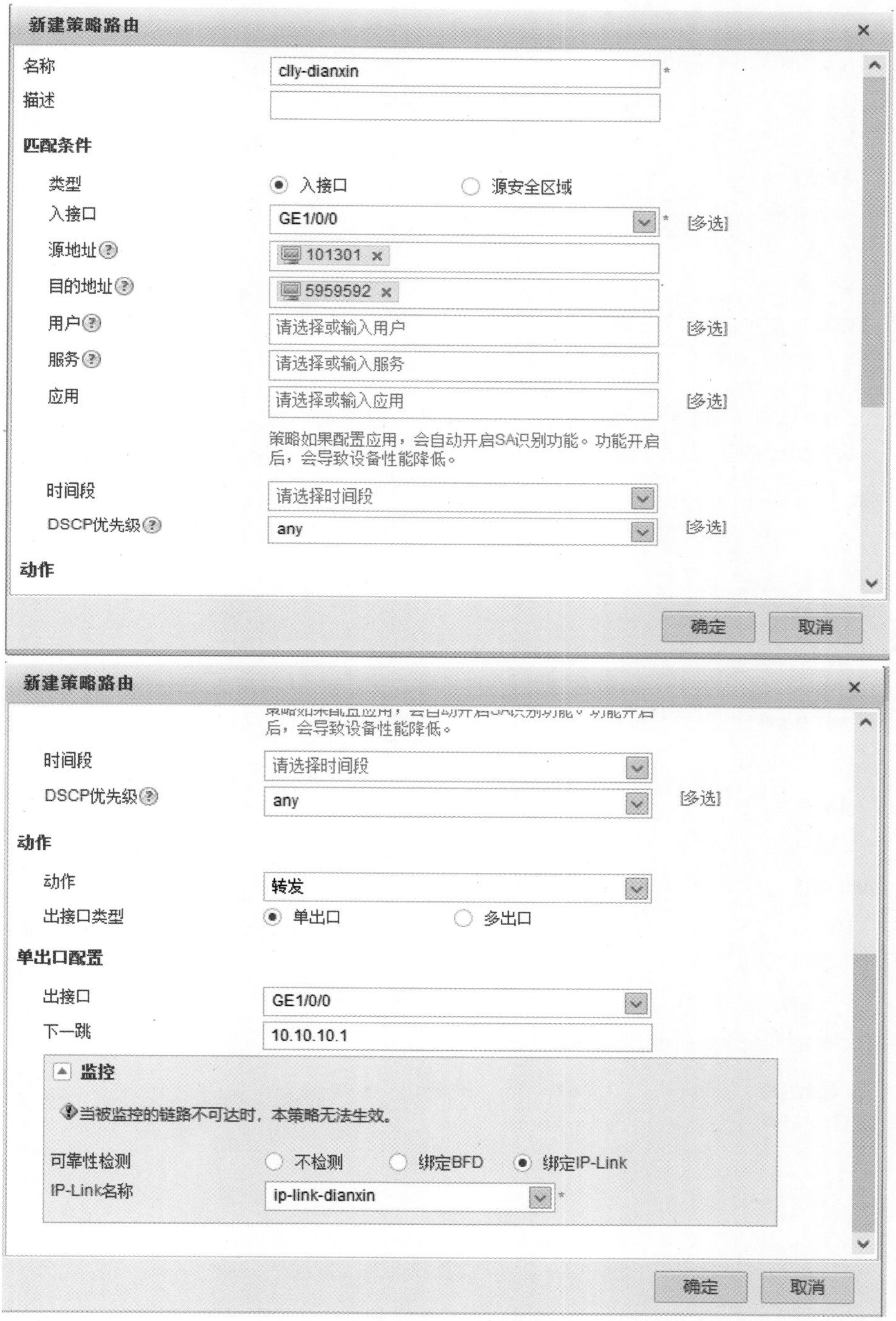

图2-27-20　创建策略路由

创建策略路由（联通），见图2-27-21。

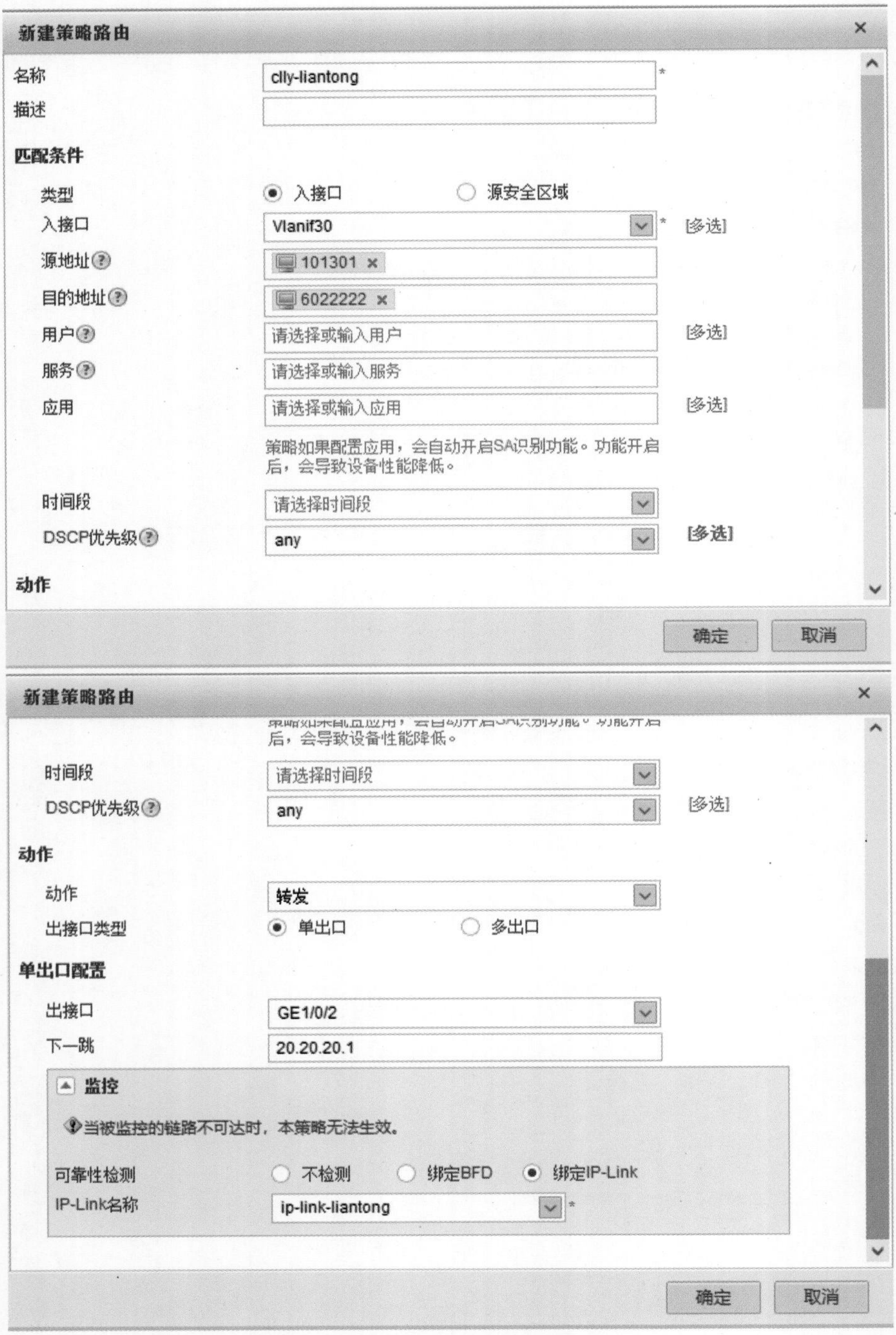

图2-27-21 创建策略路由（联通）

网络测试：
```
PC>tracert 59.59.59.2
traceroute to 59.59.59.2, 8 hops max
```

```
(ICMP), press Ctrl+C to stop
 1    *   *   *
 2  10.10.10.1    15 ms    <1 ms    16 ms
 3  59.59.59.2    16 ms    31 ms    16 ms
```

以上可以看出访问电信 IP（59.59.59.2），通信路径走电信链路。

```
PC>tracert 60.22.22.2
traceroute to 60.22.22.2, 8 hops max
(ICMP), press Ctrl+C to stop
 1    *   *   *
 2  20.20.20.1    15 ms    16 ms    16 ms
 3  60.22.22.2    15 ms    16 ms    15 ms
```

以上可以访问联通 IP（60.60.60.2），通信路径走联通链路。

测试当电信链路故障时，当访问的是电信地址自动跳转到联通链路。

```
# 将电信路由器 GigabitEthernet0/0/0 接口关闭，模拟链路故障。
[dianxin-GigabitEthernet0/0/0]shutdown
```

查看电信 ip-link，已经检测出，显示"down"状态。见图 2-27-22。

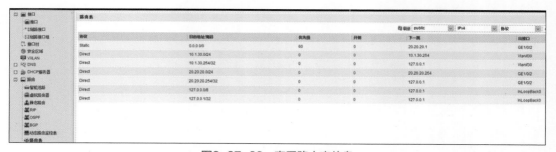

图2-27-22 查看电信ip-link

查看路由表信息，通往电信的默认静态路由已不存在。见图 2-27-23。

图2-27-23 查看路由表信息

再次网络测试

```
PC>tracert 59.59.59.2
traceroute to 59.59.59.2, 8 hops max
(ICMP), press Ctrl+C to stop
 1    *   *   *
 2  20.20.20.1    15 ms    16 ms    <1 ms
 3  59.59.59.2    16 ms    15 ms    16 ms
PC>
```

以上可以看出，访问电信地址已经自动切换到联通链路。通过静态路由与策略路由分别关联 ip-link，当检测到链路故障，路由表中删除相应的路由，策略路由功能实效。当链路故障排除后，策略路由自动恢复，路由表条目自动添加。

项目28　GRE隧道

项目29　防火墙 IPsecVPN配置

参考文献

【1】王叶，李瑞华，孟繁华.黑客攻防从入门到精通（实战篇）.2版.北京：机械工业出版社，2020.

【2】李华峰，陈虹.Wireshark网络分析从入门到实践.北京：人民邮电出版社，2019.

【3】王灵霞，刘永纯.网络管理与运维实战宝典.北京：中国铁道出版社，2016.